Crime Mapping and Spatial Data Analysis using R

Crime mapping and analysis sit at the intersection of geocomputation, data visualisation and cartography, spatial statistics, environmental criminology, and crime analysis. This book brings together relevant knowledge from these fields into a practical, hands-on guide, providing a useful introduction and reference material for topics in crime mapping, the geography of crime, environmental criminology, and crime analysis. It can be used by students, practitioners, and academics alike, whether to develop a university course, to support further training and development, or to hone skills in self-teaching R and crime mapping and spatial data analysis. It is not an advanced statistics textbook, but rather an applied guide and later useful reference book, intended to be read and for readers to practice the learnings from each chapter in sequence.

In the first part of this volume we introduce key concepts for geographic analysis and representation and provide the reader with the foundations needed to visualise spatial crime data. We then introduce a series of tools to study spatial homogeneity and dependence. A key focus in this section is how to visualise and detect local clusters of crime and repeat victimisation. The final chapters introduce the use of basic spatial models, which account for the distribution of crime across space. In terms of spatial data analysis, the focus of the book is on spatial point pattern analysis and lattice or area data analysis.

Chapman & Hall/CRC
Statistics in the Social and Behavioral Sciences Series

Series Editors
Jeff Gill, Steven Heeringa, Wim J. van der Linden, Tom Snijders

Recently Published Titles

Big Data and Social Science: Data Science Methods and Tools for Research and Practice, Second Edition
Ian Foster, Rayid Ghani, Ron S. Jarmin, Frauke Kreuter and Julia Lane

Understanding Elections through Statistics: Polling, Prediction, and Testing
Ole J. Forsberg

Analyzing Spatial Models of Choice and Judgment, Second Edition
David A. Armstrong II, Ryan Bakker, Royce Carroll, Christopher Hare, Keith T. Poole and Howard Rosenthal

Introduction to R for Social Scientists: A Tidy Programming Approach
Ryan Kennedy and Philip Waggoner

Linear Regression Models: Applications in R
John P. Hoffman

Mixed-Mode Surveys: Design and Analysis
Jan van den Brakel, Bart Buelens, Madelon Cremers, Annemieke Luiten, Vivian Meertens, Barry Schouten and Rachel Vis-Visschers

Applied Regularization Methods for the Social Sciences
Holmes Finch

An Introduction to the Rasch Model with Examples in R
Rudolf Debelak, Carolin Stobl and Matthew D. Zeigenfuse

Regression Analysis in R: A Comprehensive View for the Social Sciences
Jocelyn H. Bolin

Intensive Longitudinal Analysis of Human Processes
Kathleen M. Gates, Sy-Min Chow, and Peter C. M. Molenaar

Applied Regression Modeling: Bayesian and Frequentist Analysis of Categorical and Limited Response Variables with R and Stan
Jun Xu

The Psychometrics of Standard Setting: Connecting Policy and Test Scores
Mark Reckase

Crime Mapping and Spatial Data Analysis using R
Juan Medina Ariza and Reka Solymosi

Computational Aspects of Psychometric Methods: With R
Patricia Martinková and Adéla Hladká

For more information about this series, please visit: https://www.routledge.com/Chapman--HallCRC-Statistics-in-the-Social-and-Behavioral-Sciences/book-series/CHSTSOBESCI

Crime Mapping and Spatial Data Analysis using R

Juan Medina Ariza

Reka Solymosi

CRC Press
Taylor & Francis Group
Boca Raton London New York

CRC Press is an imprint of the
Taylor & Francis Group, **an informa** business

A CHAPMAN & HALL BOOK

Designed cover image: Juan Medina Ariza and Reka Solymosi

First edition published 2023
by CRC Press
6000 Broken Sound Parkway NW, Suite 300, Boca Raton, FL 33487-2742

and by CRC Press
4 Park Square, Milton Park, Abingdon, Oxon, OX14 4RN

CRC Press is an imprint of Taylor & Francis Group, LLC

ISBN: 978-0-367-72469-6 (hbk)
ISBN: 978-0-367-72459-7 (pbk)
ISBN: 978-1-003-15491-4 (ebk)

DOI: 10.1201/9781003154914

Typeset in Latin Modern font
by KnowledgeWorks Global Ltd.

Publisher's note: This book has been prepared from camera-ready copy provided by the authors.

To the University scheduling team,

Who booked our course into a computer cluster which had R installed and not ArcGIS, prompting these materials.

Contents

List of Figures

List of Tables

0

Preface: how to use this book

This book aims to provide the reader with an introduction to Crime Mapping and Spatial Data Analysis, using R as an engine for spatial data analysis and visualisation. Based on teaching materials developed by the authors, the book is a practical guide for those interested in crime analysis and criminology.

We imagine this to be useful for those teaching (or enrolled in) upper-level undergraduate and graduate courses in higher education, as well as analysts in professional roles embarking on further training, or looking for a guide for their work, as well as criminologists interested in crime mapping. The material has been successful in our teaching both in higher education settings, and when training crime analysts working for the police or other law enforcement agencies.

Given the source material, this book may be used as a companion text for similar course units in crime mapping, the geography of crime, environmental criminology, or crime analysis. Equally, it can be used by students, practitioners, and academics alike interested in learning more about R and its GIS and spatial analysis capacity. It is not an advanced statistics textbook, but rather an applied textbook. Someone "self-teaching" R for these purposes will find it helpful and, thus, unlike reference books, it is better to read and practice each chapter in sequence.

Crime mapping and analysis sits at the intersection of geocomputation, data visualisation and cartography, spatial statistics, environmental criminology, and crime analysis. In this book, environmental criminology, which focuses on the analysis of the spatial and geographical distribution of crime, provides the substantive background.

This text cannot make justice to each of these bodies of inquiry, professional practice, and literature. We cannot offer a comprehensive and systematic treatment of each of these areas. What we do is provide a helpful introduction to R as a way to bring together these specialties and offer adequate references to our readers so that they can deepen their understanding of each of them. For a grounding in Environmental Criminology, we recommend Bruinsma and Johnson (2018); for debates and issues in spatial criminology/crime analysis and crime and place research, we suggest Weisburd, Bernasco, and Bruinsma (2008); and for geocomputation and spatial data science, Lovelace, Nowosad, and Muenchow (2019) and M. D. Smith, Goodchild, and Longley (2007) provide great foundations. For further topics for crime analysts, Chainey and Ratcliffe (2005) and more recently Chainey (2021) are excellent as well. We also provide further resources at the end of each chapter for those who wish to delve deeper into each individual topic introduced.

Although all the examples we use concern the study of crime, there is fairly limited substantive criminology in the text. In fact, and despite the title, the volume could also be used, more generally, as a companion text for courses on social science spatial data analysis, since the techniques we cover are transferable to other substantive domains.

In the first part of this volume we introduce key concepts for geographic analysis and representation and provide the reader with the foundations needed to visualise spatial crime data. We then introduce a series of tools and techniques that are relevant to study spatial homogeneity and dependence of crime data. A key focus in this section is how to visualise and detect local clusters of crime and repeat victimisation. The final chapters introduce the use of spatial models, which account for the distribution of crime across space. In terms of spatial data analysis, the focus of the book is on spatial point pattern analysis and lattice or area data analysis. Geostatistics have fewer applications in crime analysis and research and, therefore, we do not cover this topic here.

To follow along with the exercises, we provide the data we use in a zip file. The best approach is to download this zip file into the working directory of your R project, and unzip it there. This way, you can follow our code. If you run the below code, should do this for you automatically:

```
download.file(url = "https://osf.io/5u42g/download",
              destfile = "data.zip")
unzip("data.zip", exdir = "data")
```

From then on, you can read all files from this downloaded data directory. In our code, we use the convention that all data are stored in a "data/" folder within our working directory.

We hope you enjoy this book and find it useful in your journey into crime mapping. The world of R and spatial analysis is ever-evolving, and we will try to stay updated. If you have ever suggestions, comments, concerns, or requests, do not hesitate to get in touch; or raise this as an issue in our GitHub repository: www.github.com/maczokni/crime_mapping[1].

---

[1]https://github.com/maczokni/crime_mapping

Author/editor biographies

This book is based on teaching materials developed by the authors. Professor Juan Medina Ariza is Senior Distinguished Researcher at the Department of Criminal Law and Crime Sciences at the University of Seville. Previously he was Professor of Quantitative Criminology at the University of Manchester where he taught data analysis and crime mapping for 20 years. Dr Reka Solymosi is a Senior Lecturer in Quantitative Methods at the University of Manchester where she has been teaching data analysis and crime mapping since 2016.

1

Producing your first crime map

1.1 Introduction

This chapter introduces some basic concepts and will get you started on making maps using R. We will learn how we can take crime data, and assign the appropriate geometry for our chosen unit of analysis. Spatial or environmental criminologists most often work with data that are **discrete**. This may include data represented by points (e.g., locations of a crime incident) or counts and rates of crimes within a particular geographical unit (e.g., a neighbourhood, census tract, or municipality). In other scientific disciplines the mapped data may be **continuous** across the study surface (e.g., temperature). This is less common in our case.

In these first few chapters we will focus on ways to work with discrete data. Firstly, we introduce the spatial and non-spatial R packages used frequently throughout this book, and cover some key terms around projection and coordinate reference systems which will be essential for subsequent chapters. As we will discover throughout the book, there are multiple R packages that have been developed to visualise spatial data. They all have advantages and disadvantages, but many offer similar functionality. Sometimes choosing one or the other is a matter of personal preference. We will introduce many different approaches throughout the book, to allow readers to select their own preferences. In this chapter we will focus on map-making with `ggplot2`, a general package for data visualisation (not just maps) based on the theory of the grammar of graphics (Wickham 2010). If you are not new to R, you may already be familiar with this package. Our intention in this chapter is to introduce enough background to let you to quickly produce your first map. In subsequent chapters we will further refine their look and aesthetic appeal, and discuss the many decisions that go into producing a crime map.

In this chapter we will use the following packages:

```
# Packages for reading data and data carpentry
library(readr)
library(tibble)
library(janitor)           ʼ
library(dplyr)
# Packages for handling spatial data
library(sf)
# Packages for visualisation and mapping
library(ggplot2)
library(ggspatial)
```

If you are a little rusty on R packages, what they do, how to install them, how to load them, and so on, please refer to *Appendix A: A quick intro to R and RStudio* for a quick introduction and overview of R.

1.2 Geospatial Perspective: key terms and ideas

Geospatial analysis provides a distinct perspective on the world, a unique lens through which to examine events, patterns, and processes that operate on or near the surface of our planet. Ultimately geospatial analysis concerns what happens where, and it makes use of geographic information that links features and phenomena on the Earth's surface to their locations.

We can talk about a few different concepts when it comes to spatial information. These are:

- Place
- Attributes
- Objects
- Networks

Let's discuss each in turn now.

1.2.1 Place

At the center of all spatial analysis is the concept of *place*. People identify with places of various sizes and shapes, from the parcel of land, to the neighbourhood, city, country, state or nation state. Places often have names, and people use these to talk about and distinguish them. Names can be official or unofficial. Places also change continually as people move. In geospatial analysis, the basis of rigorous and precise definition of place is a coordinate system. A **coordinate system** is a set of measurements that allows place to be specified unambiguously and in a way that is meaningful across different users, analysts, mappers, and researchers.

In Environmental Criminology, place has acquired different meanings through history. The first geographical studies of crime during the 19th century looked at variation across provinces in France (Guerry 1833), Belgium (Quetelet 1842), and England (Mayhew 1861). Later on, in the first decades of the 20th century, the Chicago School of Sociology focused on the study of neighbourhoods (Park and Burgess 1925; Shaw and McKay 1942). With the emergence of Environmental Criminology (Brantingham and Brantingham 1982; Cohen and Felson 1979; Cornish and Clarke 1986) a shift towards a greater interest in microplaces, such as street segments or particular addresses, can be seen (Eck and Weisburd 2015). Although an interest in variation across large administrative units (as "place") remains within criminology, there has been a trend towards understanding place as particular locations, and focusing on risky places for crime.

However we define "place", it is the central concept to crime mapping, and spatial data analysis. Most human activity takes place in a specific location, and accounting for this spatial information is a key task for criminologists, crime analysts, and social and data scientists more broadly.

1.2.2 Attributes

Attribute has become the preferred term for any recorded characteristic or property of a place. It is what more generally in statistics we may call a variable and in data science a feature. A place's name is an obvious example of an attribute. But there can be other pieces of information, such as number of crimes in a neighbourhood, or the GDP of a country. Within geographic information science (GIS) the term 'attributes' usually refers to records in a data table associated with individual elements in a spatial data file. These data behave exactly as data you may have encountered in past experience with non-spatial statistics. The rows represent observations, and the columns represent variables. The variables can have different levels of measurement (numeric (interval or ratio, discrete or continuous) or categorical (binary, nominal, ordinal)), and depending on what they are, you can apply different methods to making sense of them. The difference between a non-spatial data set and an attribute table associated with spatial data is that the unit of analysis will be some places or locations; each row will include elements that allow us to place it on a map.

1.2.3 Spatial objects

In spatial analysis it is customary to refer to places as objects. These objects can be a whole country, or a road. In forestry, the objects of interest might be trees, and their location will be represented as points. On the other hand, studies of social or economic patterns may need to consider the two-dimensional extent of places, which will therefore be represented as areas. These representations of the world are part of what is called the **vector data model**: a representation of the world using points, lines, and polygons. Vector models are useful for storing data that have discrete boundaries, such as country borders, land parcels, and streets. This is made up of points, lines, and areas (polygons):

- *Points*: Points are pairs of coordinates, in latitude/longitude or some other standard system. In the context of crime analysis, the typical point we work with, though not the only one, represents the specific location of a criminal event. These points form patterns we explore and analyse. Postcodes can also be represented with a single point in the middle of the postcode area, which could be mapped using latitude/longitude pair.
- *Lines*: Lines are ordered sequences of points connected by straight lines. An example might be a road, or street segment.

- *Areas* (polygons): Areas are ordered rings of points, also connected by straight lines to form polygons. It can contain holes, or be linked with separate islands. Areas can represent neighbourhoods, police districts, municipal terms, etc. You may come across the term **lattice data** to denote the type of data we work with when we observed attributes of areas and want to explore and analyse this data.

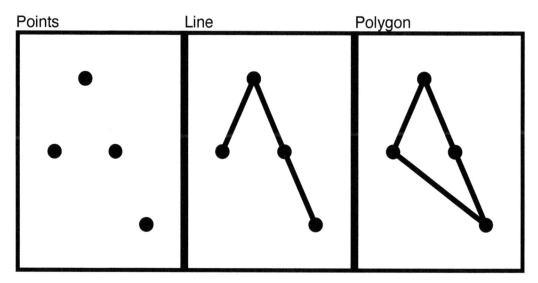

FIGURE 1.1: Points, Line, and Polygon

Spatial objects may also be stored as **raster data**. Raster data are made up of pixels (or cells), and each cell has an associated value. Simplifying slightly, a digital photograph is an example of a raster dataset where each pixel value corresponds to a particular colour. In GIS, the pixel values usually represent continuous data such as elevation above sea level, or chemical concentrations, or rainfall, etc. The key point is that all of this data is represented as a grid of (usually square or hexagonal) cells.

1.2.4   Networks

We already mentioned lines that constitute objects of spatial data, such as streets, roads, railroads, etc. Networks constitute one-dimensional structures embedded in two or three dimensions. Discrete point objects may be distributed on the network, representing phenomena such as landmarks; or observation points. Mathematically, a network forms a graph, and many techniques developed for graphs have application to networks. These include various ways of measuring a network's connectivity, or of finding the shortest path between pairs of points on a network. We elaborate on networks further in Chapter 8 of this book.

1.3 Maps and their types

Historically maps have been the primary means to store and communicate spatial data. Objects and their attributes can be readily depicted, and the human eye can quickly discern patterns and outliers in a well-designed map.

In GIS we distinguish between reference and thematic maps. A **reference map** places the emphasis on the location of spatial objects such as cities, mountains, rivers, parks, etc. You use these maps to orient yourself in space and find out the location of particular places.

Thematic maps, or statistical maps are used to represent the spatial distribution of attributes or statistics. For example, the number of crimes across different neighbourhoods. Our focus in this book is on thematic maps, but often when producing thematic maps you may use a reference map as a backdrop, as a *basemap*, to help interpretation and to provide context. In this and subsequent chapters we will introduce different types of thematic maps.

Another useful distinction is between **static** and **interactive** maps. Your traditional printed road map is an example of a static map, whereas the web application Google maps is an example of an interactive map. In an interactive map you can zoom in and out, you can select and query information about objects in an interactive fashion, etc. In this chapter we introduce the R package `ggplot2`, which excels at static maps. In other chapters we will introduce other packages, such as `leaflet`, which is particularly useful for interactive purposes, and the package `tmap`, in which we can shift our maps between static and interactive mode.

1.4 Map projections and geographic coordinate systems

Whenever we put something on a map, we need some sort of system to pinpoint the location. A coordinate system allows you to integrate any dataset with other geographical datasets within a common framework. With the help of **coordinate reference systems** (CRS) every place on Earth can be specified by a set of numbers, called coordinates. In general CRS can be divided into **projected coordinate reference systems** (also called Cartesian or rectangular coordinate reference systems) and **geographic coordinate reference systems**.

A **geographic coordinate system** is a *three-dimensional* reference system that enables you to locate any location on Earth. This coordinate system is round, and so, records locations in angular units (usually degrees). The use of geographic coordinate reference systems is very common. The most popular is called the **World Geodetic System (WGS 84)** which refers to locations using "Longitude", "Latitude", and in some cases "Elevations". For example, to communicate where is the University of Manchester, we would say it is Longitude: -2.2360724 degrees and Latitude: 53.466853 degrees.

Projected coordinate systems, or map projections, try to portray the surface of the Earth or a portion of the Earth on a *two-dimensional* flat piece of paper or computer screen. Therefore, location is recorded in linear units (usually meter). Working, for example, with data in the UK, we would use **British National Grid (BNG)**. In this case, points will be defined by "Easting" and "Northing" rather than "Longitude" and "Latitude". It basically divides the UK into a series of squares, and uses references to these to locate something. The most common usage is the six-figure grid reference, employing three digits in each coordinate to determine a 100 m square. For example, the grid reference of the 100 m square containing the summit of Ben Nevis is NN 166 712. The more digits are included, the more precisely we can refer to the location of interest.

All projections of a sphere like the Earth in a two-dimensional map involve some sort of distortion. You can't fit a three-dimensional object into two dimensions without doing so. **Projections** differ to a large extent on the kind of distortion that they introduce. This will be important later on when we are linking data from different projections, or when you look at your map and you try to figure out why it might look "squished".

The decision as to which map projection and coordinate reference system to use, depends on the regional extent of the area you want to work in, the analysis you want to do, and often the availability of data. Knowing the system you use allows you to translate your data into other systems whenever this may be necessary. Often you may have to integrate data that is provided to you in different coordinate or projected systems. As long as you know the systems, you can do this, and we will be demonstrating this throughout the book.

You will often see the notation *epsg* when referring to different coordinate systems. This refers to the EPSG registry or **EPSG Geodetic Parameter Dataset**. It is a collection of geodetic datums, spatial reference systems, Earth ellipsoids, coordinate transformations, and related units of measurement. All standard coordinate systems will have one of these codes assigned to it. So, for example, the WGS84 coordinate system corresponds the epsg key 4326, whereas the British National Grid has the key 27700. Having this common framework to identify projections makes things much easier when we want to change our data from one system to another. You can query the website https://epsg.io/ for the different keys and information associated to each of them. This EPSG dataset was originally developed by the European Petroleum Survey Group (EPSG), thus the name. You can also define the coordinate system in R using ways other than an EPSG code (for details see Lovelace, Nowosad, and Muenchow (2019)).

A word of caution about WGS84. Although it is very popular (global positioning systems rely upon it) and we will be using it a lot, it has a problem: it is a dynamic reference frame. Due to plate tectonic motion, there are changes over time and the coordinates are adjusted to account for this. The implications are that WGS84 coordinates derived by users from GPS measurements may appear to move over time. This means that we need to know the time of our spatial feature (say, the time of a criminal event) if we want the correction needed as a result of plate tectonics. The problem is not simply of data availability (say, you don't have the exact timing of the event that you want to place in a map), but of data structures. Traditionally, the file formats that we have used to store this information do not lead themselves easily to record this information so that we can then easily address these inaccuracies. Progress has been made in recent times, but you need to be aware that WGS84 is not perfect despite its popularity. Some European guidelines suggest, for example, the use of ETRS89 in Europe as a consequence of this.

1.5 Key terms summary

While the above was a whistle stop tour of some key terms, this should start you thinking about these concepts. For a deeper understanding, consult the further reading at the end of this chapter. For now, to progress, be sure that you are confident to know about:

- Spatial objects - what they are and how they are represented,
- Attributes - the bits of information that belong to your spatial objects,
- Maps and projections - especially what geographic versus projected coordinates mean, and why it's important that you know what CRS your data have.

1.6 Getting started with crime data

We live in a world awash with data and through this book we will use the different examples to show you some useful places and sources you can use to obtain your spatial data. Most geographically referenced crime data is based on crime reported to the police, although there are some notable exceptions (e.g., public health data on violence, geocoded victim survey data, etc.). Many police departments across the world make this data readily available to the public or researchers. This is more common in places like the United States, where crime mapping applications first got established, but increasingly we see publicly available crime data from other countries as well.

For our first map we will use data from the UK, which can be downloaded from the police.uk[1] website. We have downloaded some data for crime in Manchester, which is included in the data file you have downloaded to follow along (see the Preamble chapter for details). If you wanted to acquire the data yourself directly from the source, you could open the `data.police.uk/data` website and then choose the data tab, in order to manually download some data.

So whether from the website, or the provided data file, you should have some police data ready. The next step is to read this into R. How can we do this? We say, "hello R, i would like to create a new object please and I will call this new object `my_data`." We do this by typing the name we are giving the object and the assignment operator `<-`. Then on the right, hand side of the assignment operator, there is the value that we are assigning the variable. So it could be a bit of text, or it could be some function, for example when you read a csv file with the `read_csv()` function from the `readr` package.

When we read in this file, inside the function, we also need to specify *where* to read the csv from. Where should R look to find this data? This is where normally you are putting in the path to your file. Something like:

```
my_data <- read_csv("path to my file here")
```

---

[1] https://data.police.uk/data/

If you downloaded the data following the steps in the Preface chapter for this book, your data will be in your working directory, in a sub-folder called "data". It is within this sub-folder that you'll find all you need. In this case, let's look for the file `2019-06-greater-manchester-street.csv`. To read this into R we run:

```
library(readr)

crimes <- read_csv("data/2019-06-greater-manchester-street.csv")
```

If you look at the *Environment* window in the top right corner of RStudio you should see now a new object that contains a *tibble*, a particular format for dataframes, and that is called `crimes`. It will tell you how many observations (rows - and incidentally the number of recorded crimes in June 2019 within the GMP jurisdiction) and how many variables (columns) your data has. Let's have a look at the crimes dataframe with the `View()` function. This will open the data browser in RStudio.

```
View(crimes)
```

If you just want your results in the console, you can use the `glimpse()` function from the `tibble` package. This function does just that: it gives you a quick glimpse of the first few cases in the dataframe. Notice that there are two columns (Longitude and Latitude) that provide the required geographical coordinates that we need to plot this data.

```
library(tibble)

glimpse(crimes)
```

```
## Rows: 32,058
## Columns: 12
## $ `Crime ID`               <chr> NA, "aa1cc4cb0c436f46~
## $ Month                    <chr> "2019-06", "2019-06",~
## $ `Reported by`            <chr> "Greater Manchester P~
## $ `Falls within`           <chr> "Greater Manchester P~
## $ Longitude                <dbl> -2.464, -2.441, -2.44~
## $ Latitude                 <dbl> 53.61, 53.62, 53.61, ~
## $ Location                 <chr> "On or near Parking A~
## $ `LSOA code`              <chr> "E01004768", "E010047~
## $ `LSOA name`              <chr> "Bolton 001A", "Bolto~
## $ `Crime type`             <chr> "Anti-social behaviou~
## $ `Last outcome category`  <chr> NA, "Unable to prosec~
## $ Context                  <lgl> NA, NA, NA, NA, NA, N~
```

You may notice that a lot of the variable names are messy in that they have a space in them, this can cause issues, so before playing around too much with the data we want to clean this up. Luckily there is a very handy package you can use for this called `janitor` which contains the function `clean_names()`. This function will clean your variable names not only of spaces but also of special characters, and it will convert characters to lower-case, to follow tidyverse naming conventions outlined in the tidyverse style guide (Wickham 2021).

```
library(janitor)

crimes <- clean_names(crimes)
```

Now the names are much neater. You can print them all for a view using the `names()` function:

```
names(crimes)
```

```
##  [1] "crime_id"             "month"
##  [3] "reported_by"          "falls_within"
##  [5] "longitude"            "latitude"
##  [7] "location"             "lsoa_code"
##  [9] "lsoa_name"            "crime_type"
## [11] "last_outcome_category" "context"
```

1.7   From dataframes to spatial objects

Having had a chance to inspect the data set you've downloaded, let's consider what sort of spatial information we might be able to use. If you have a look at the column names, what are some of the variables which you think might have some spatial component? Have a think about each column; and how it may help to put these crimes on the map. There are a few answers here. In fact there are one each to map onto point, line, and polygon.

1.7.1   The point

First, and possibly most obvious, are the coordinates provided with each crime incident recorded. You can find this in the two columns - Longitude and Latitude. These two columns help put each crime incident on a specific point on a map. For example, let's take the very first crime incident. Here we use the `head()` function and specify that we want the first 1 rows only with n=1 parameter.

```
head(crimes, n = 1)
```

You can see that the values are -2.4644 for Longitude and 53.6125 for Latitude (*If you see fewer digits, you might have global options set to fewer significant digits than the default 7. To address this set options(digits = 7).*). These two numbers allow us to put this point on a map.

1.7.2   The line

Another column which contains information about *where* the crime happened is the aptly named *location* variable. This shows you a list of locations related to where the crimes happened. You may see a few values such as on or near XYZ street. Let's look again at the first entry.

```
head(crimes, n = 1)
```

You can see that the value is "On or near Parking Area"; this isn't great, as we might struggle to identify *which* parking area. Some other ones are more useful; let's look at the last entry for example with the `tail()` function.

```
tail(crimes, n = 1)
```

You can see that the value is "On or near Fulwood Road." This makes our crime much easier to find; we just need to locate "Fulwood Road." We might have a file of lines of all the roads of Manchester, and if we did, we can link the crime to that particular road, in order to map it.

Note: If you cannot see the whole text with printing the head or tail of the data, you can bring this up in your viewer with the View() *function.*

1.7.3   The polygon

What more? You may also have seen the column "lsoa_name" and it seems to contain what looks like names for some sort of area or place. Let's have a look at the first crime again. You see the value for LSOA name is "Bolton 001A". Bolton we know is a Borough of Greater Manchester, but what does the 001 mean?

Well, it denotes a particular geographical sub-unit within the municipality of Bolton called a **Lower Layer Super Output Area**. This is a unit of UK Census Geography[2]. The basic unit for Census Geography in the UK is an 'Output area'. This is the resolution at which we can access data from the UK Census. The Output Area (OA) is therefore the smallest unit we could use. The censuses in other countries use different names for the units for which they publish information, but the logic is similar.

There are 181,408 OAs, 34,753 LSOA and 7,201 MSOA in England and Wales. The neat thing about these census geographies is the idea that they don't change much from census to census (unlike other administrative boundaries) and in the UK case were created with statistical analysis in mind (they were designed to be as homogeneous as possible). The less neat thing is that although we use them to operationalise the concept of neighbourhood a lot, they may not bear much resemblance to what residents might think of as their neighbourhood. This is a common problem in the UK and elsewhere (that has been widely discussed in the literature, see Weisburd, Bruinsma, and Bernasco (2009) for example) when relying on census units as our proxy for community or neighbourhood; but, one that is hard to escape from, for these units are those at which key demographic variables are typically sourced and published.

---

[2]https://www.ons.gov.uk/methodology/geography/ukgeographies/censusgeography

Looking back to our crime data, we find two columns that reference LSOAs, lsoa_name and lsoa_code. We can use these to *link* our crime data to a file containing the geometries needed to put the crime data on the map. In the next section we will illustrate how.

1.7.4 Choosing the ideal unit of analysis

We see that in the situation of our crime data, we can choose between different units of analysis, whether we want our data at individual crime level (points), or aggregated to streets (line) or aggreagated to areas such as census neighbourhoods (polygons). Which should you choose? The unit of analysis at which to consider approaching our research and analytical questions will depend most largely on what is the appropriate level at which addressing the question makes sense, and at what level we can get reliable data for the variables we wish to analyse. There is no simple answer to this. Questions around levels of measurement have formed a central part of discourse in the area of crime mapping (Weisburd, Bruinsma, and Bernasco 2009). It is important that careful thought and consideration is given to this decision.

1.8 The simple features framework

We've established now that our crime data has spatial information, which we can use to put our crimes on the map. In this section, we will introduce the simple features framework as a way to do this using R. Our task is to specify a geometry for our data, which links each unit of analysis (whether that is the point, line, or polygon) to a relevant geographical representation, allowing us to put this thing on the map.

How you add geographical information will vary with the type of information we have, but in all of these, we will use the **simple features** framework. The author of the sf package, Edzer Pebesma, describes simple features as a standardized way of encoding spatial *vector data* (points, lines, polygons). The sf package is an R package for reading, writing, handling, and manipulating simple features in R, implementing the vector data handling functionality.

Traditionally spatial analysis in R were done using the sp package which creates a particular way of storing spatial objects in R. When most packages for spatial data analysis in R and for thematic cartography were first developed sp was the only way to work with spatial data in R. There are more than 450 packages that rely on sp, making it an important part of the R ecosystem. More recently sf is changing the way that R does spatial analysis. This package provides a new way of storing spatial objects in R and most recent R packages for spatial analysis and cartography are using it as the new default. It is easy to transform sf objects into sp objects and vice versa, so that those packages that still don't use this new format can be used. In this book we will emphasise the use of sf whenever possible. You can read more about the history of spatial packages and the sf package in the first two chapters of Lovelace, Nowosad, and Muenchow (2019).

Features can be thought of as "things" or objects that have a spatial location or extent; they may be physical objects like a building, or social conventions like a political state. Feature geometry refers to the spatial properties (location or extent) of a feature, and can be described by a point, point set, linestring, a set of linestrings, polygon, a set of polygons, or a combination of these. The "simple" adjective of simple features refers to the property that linestrings and polygons are built from points connected by straight line segments. Features typically also have other properties (temporal properties, colour, name, measured quantity), which are called feature attributes. For more detailed insight, we recommend Pebesma (2018).

Let's get started with making some maps using `sf`. First, make sure you install the package, and then load it with the `library()` function. We know that we have two columns, one for longitude and one for latitude, which pinpoint each crime event to a specific point, close to where it happened. Not *quite* where it happened, as the data are anonymised (more on this later), but for our purposes here, we can assume this is the location of the crime. To map these points, we can transform our ordinary dataframe into a simple features object.

To do so, we can use the `st_as_sf()` function from `sf`, into which we need to specify what we are to transform (our dataframe), where the spatial data can be found (our columns which hold the latitude and longitude information), and also what coordinate reference system the object has (see above our discussion about projections and coordinate reference systems).

Latitude and longitude coordinates specify location on the **WGS 84** CRS. We can tell R that this is our CRS of choice by including its EPSG identifier as a parameter in our function. It is handy to know the more common EPSG identifiers. For example, as mentioned above, for WGS84 the EPSG identifier is *4326*. For British National Grid, the identifier is *27700*.

Putting it all together in practice, we can create a simple features object from our dataframe using the latitude and longitude columns:

```
library(sf)
```

```
crimes_sf <- st_as_sf(crimes,     #dataframe
                     #columns with coordinates:
                     coords = c("longitude", "latitude"),
                     crs = 4326)   #crs is WGS84
```

We can see that this is now a simple features object using the `class()` function to print the result "sf":

```
class(crimes_sf)
```

```
## [1] "sf"          "tbl_df"     "tbl"         "data.frame"
```

You might also notice something else that is different between "crimes" and "crimes_sf." Have a look at the dimension (hint: look in your 'Environment' tab). In the `sf` object you will see that the information provided by the longitude and latitude variables has been "merged" into a new variable called *geometry*, that `sf` uses to store the kind of object we have (a point in this case) and where to locate it.

1.9 Plotting data with ggplot2

Now that we have this `sf` object, how can we map it? We mentioned before about the graphical package `ggplot2`. We can use this, and its syntax, in order to map spatial data using the `geom_sf()` geometry.

First, a quick refresher on `ggplot2` and the grammar of graphics. The grammar of graphics upon which this package is based defines various components of a graphic. Some of the most important are:

- **The data**: For using `ggplot2` the data has to be stored as a data-frame or tibble (`sf` objects are of class tibble).

- **The geoms**: They describe the objects that represent the data (e.g., points, lines, polygons, etc.). This is what gets drawn. And you can have various different types layered over each other in the same visualisation.

- **The aesthetics**: They describe the visual characteristics that represent data (e.g., position, size, colour, shape, transparency).

- **Facets**: They describe how data is split into subsets and displayed as multiple small graphs.

- **Stats**: They describe statistical transformations that typically summarise data.

Let's take it one step at time. Essentially, the philosophy behind this is that all graphics are made up of layers. You can build every graph from the same few components: a dataset, a set of geoms visual marks that represent data points, and a coordinate system. Take this example from our crimes dataframe. You have a table such as:

TABLE 1.1: Example data table

crime_type	n
Robbery	463
Shoplifting	1479
Theft from the person	718

You then want to plot this. To do so, you want to create a plot that combines the following layers:

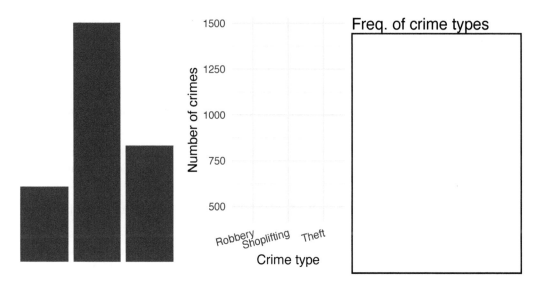

FIGURE 1.2: Layers of ggplot

This will result in a final plot:

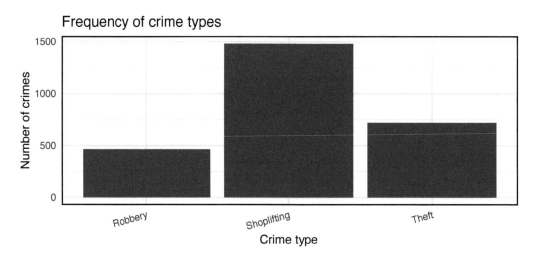

FIGURE 1.3: Final plot

Taking our crime data as an example, we would build up our plot as follows. First let's create a small illustrative dataset of only the crimes of "Robbery", "Shoplifting", and "Theft from the person".

```
library(dplyr)

df <- crimes %>%
  filter(crime_type %in% c("Robbery","Shoplifting", "Theft from the person")) %>%
  group_by(crime_type) %>%
```

```
 count()

df
```

```
## # A tibble: 3 x 2
## # Groups:   crime_type [3]
##   crime_type                n
##   <chr>                 <int>
## 1 Robbery                 463
## 2 Shoplifting            1479
## 3 Theft from the person   718
```

Now let us add the layers to our ggplot object. First, the data layer:

```
library(ggplot2)

ggplot(df, aes(x = crime_type, y = n))
```

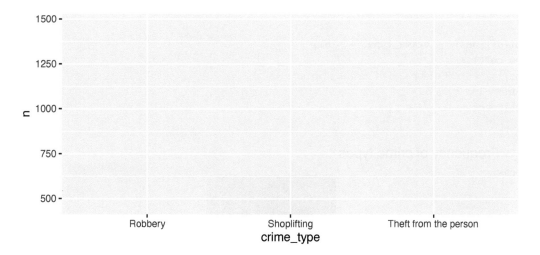

FIGURE 1.4: Data layer

Then add the geometry (in this case, `geom_col()`):

```
ggplot(df, aes(x = crime_type, y = n)) +
  geom_col()
```

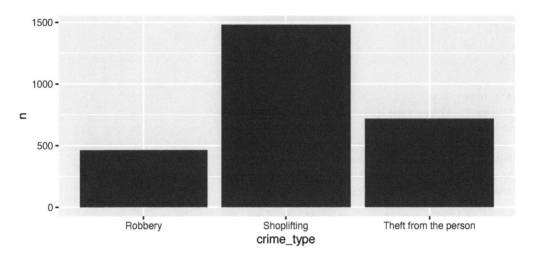

FIGURE 1.5: Geometry layer

Then our annotations:

```
ggplot(df, aes(x = crime_type, y = n)) +
  geom_col() +
  labs(title = "Frequency of crime types") +
  xlab("Crime type") +
  ylab("Number of crimes")
```

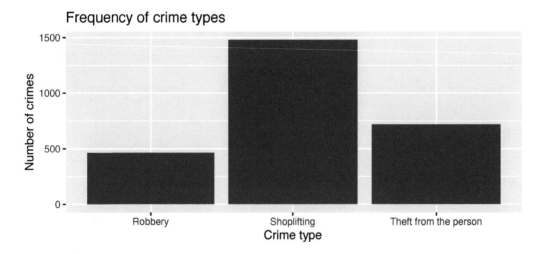

FIGURE 1.6: Annotations layer

We can also specify our themes, by using a custom theme such as `theme_minimal()`, and by adding our own specifications within an additional `theme()` function:

```
ggplot(df, aes(x = crime_type, y = n)) +
  geom_col() +
  labs(title = "Frequency of crime types") +
  xlab("Crime type") +
  ylab("Number of crimes") +
  theme_minimal() +
  theme(axis.text.x = element_text(angle = 15, hjust = 1),
        panel.border = element_rect(colour = "black", fill=NA, size=1))
```

FIGURE 1.7: Customise themes

We can add further specifications within the aes() (aesthetics) function, where we add additional layers from our data, to represent even more information in our charts, for example the outcomes for each crime.

```
# create new dataframe (df) including last_outcome_category variable

df <- crimes %>%
  filter(crime_type %in% c("Robbery", "Shoplifting", "Theft from the person")
        ) %>%
  group_by(crime_type, last_outcome_category) %>%
  count()

# plot including new variable with fill= parameter in aes() function

ggplot(df, aes(x = crime_type, y = n, fill = last_outcome_category)) +
  geom_col() +
  labs(title = "Frequency of crime types") +
  xlab("Crime type") +
  ylab("Number of crimes") +
  theme_minimal() +
```

```
theme(axis.text.x = element_text(angle = 15, hjust = 1),
      panel.border = element_rect(colour = "black", fill=NA,  size=1))
```

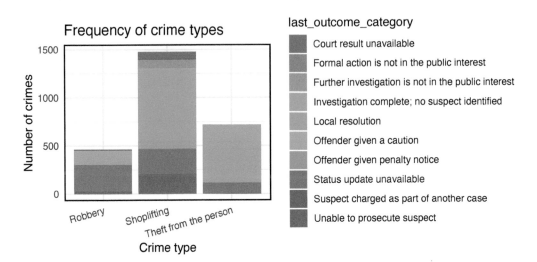

FIGURE 1.8: Additional layers

We can be explicit about the colours we want to use with the function `scale_fill_brewer()`, which we can also use to rename our legend. Don't worry too much at this point about where the palette comes from; in the chapter 5 on cartography we will discuss colour in more detail.

```
ggplot(df, aes(x = crime_type,  y = n, fill = last_outcome_category)) +
  geom_col() +
  labs(title = "Frequency of crime types") +
  xlab("Crime type") +
  ylab("Number of crimes") +
  theme_minimal() +
  theme(axis.text.x = element_text(angle = 15, hjust = 1),
        panel.border = element_rect(colour = "black", fill=NA, size=1)) +
  scale_fill_brewer(type = "qual", palette = 3, name = "Outcome")
```

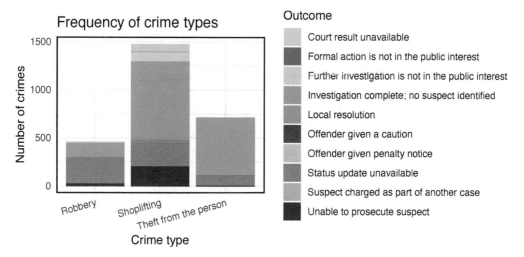

FIGURE 1.9: Customise colour

There are many more options. While this here is not the greatest graph you'll ever see, it illustrates the process of building up your graphics the ggplot way. Do read up on `ggplot2` for example in Wickham and Grolemund (2017). In later chapters, we will talk more about visualisation, colour choice, and more!

1.10 Mapping crime data as points

So how can we use this for spatial data? We can use the `geom_sf()` function to do so. Using `geom_sf` is slightly different to other geometries, for example how we used `geom_col()` above. First we initiate the plot with the `ggplot()` function but don't include the data in there. Instead, it is in the geometry where we add the data. And second we don't need to specify the mapping of x and y, since this is in the geometry column of our spatial object. Like so:

```
ggplot() +
  geom_sf(data = crimes_sf)
```

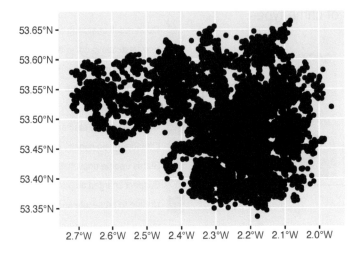

FIGURE 1.10: Using geom sf for plotting

And here we have a map of each point in our dataset, each recorded crime in June 2019 in Greater Manchester. Would you call this a map though? While it is presenting spatial data, there is not a lot of meaning being communicated. Point maps generally can be messy, and their uses are specific to certain situations and cases, usually when you have fewer points. But here, these points are especially devoid of any meaning, as they are floating in a graph grid. So let's give it a **basemap**.

We can do this by adding a layer to our graph object. Specifically, we will use the `annotation_map_tile()` from the `ggspatial` package. This provides us with a static Open Street Map layer behind our data, giving it (some) more context. Remember to load the package (and install if you haven't already). And then use the `annotation_map_tile()` function, making sure to place it before the `geom_sf` points layer, so the background map is placed first, and the points on top of that:

```
library(ggspatial)

ggplot() +
 annotation_map_tile() +
  geom_sf(data = crimes_sf)
```

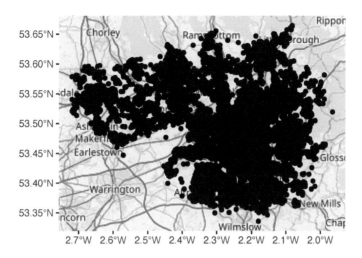

FIGURE 1.11: Adding a basemap

So what you see above behind the points is what we call a **basemap**. The term basemap is seen often in GIS and refers to a collection of GIS data and/or orthorectified imagery that form the background setting for a map. The function of the basemap is to provide background detail necessary to orient the location of the map. Basemaps also add to the aesthetic appeal of a map. **Basemaps** are essentially reference maps that may give us context and help with the interpretation. You can see above the *Open Street Map* Basemap. This is one option, but there are others.

Let's leave the points for now and move on to how we might map our lines and polygons.

1.11 Mapping crime data as polygons

What about our other two columns, location, and LSOAs? Well, to put these on the map, we need a geometry representation of them. We need boundary data representing the areas we want to map. We will learn in this section where you may find, download and turn them into sf objects, and how to link our dataframe as attribute data in order to be able to map them.

1.11.1 Finding boundary data

In this section you are going to learn how you take one of the most popular data formats for spatial objects, the **shapefile**, and read it into R. The shapefile was introduced by ESRI, the developers and vendors of ArcGIS. And although many other formats have developed since and ESRI no longer holds the same market position it once occupied (though they're still the player to beat), shapefiles continue to be one of the most popular formats you will encounter in your work.

We are going to obtain shapefiles for British census geographies. For this activity, we will focus on the polygon (LSOA) rather than the lines of the streets, but the logic is more or less the same.

Census boundary data are a digitised representation of the underlying geography of the census. Census Geography is often used in research and spatial analysis because it is divided into units based on population counts, created to form comparable units, rather than other administrative boundaries such as wards or police force areas. However, depending on your research question and the context for your analysis, you might be using different units.

The hierarchy of the census geographies in the UK goes from Country to Local Authority to Middle Layer Super Output Area (MSOA) to Lower Layer Super Output Area (LSOA) to Output Area (in other countries you have similar levels in the census hierarchies):

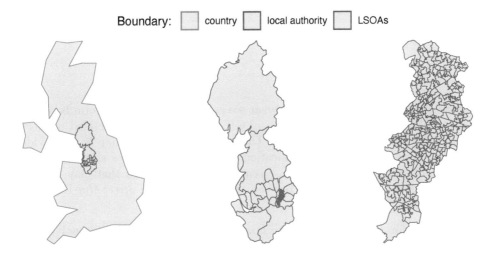

FIGURE 1.12: Hierarchy of Census Geographies in the UK

Here we will get some boundaries for Manchester. We know that our crime data has a column for LSOA, so we can proceed with this as our appropriate unit of analysis.

To get some boundary data, you will need to be able to source **geographic data** for your study area. In this case, since we are working with data in the UK, we can use the UK Data Service website to find this. There is a simple Boundary Data Selector[3] tool which you could use. Other countries also have such repositories, and in some cases where they do not, other resources, such as Open Street Map, or natural earth (and associated R package rnaturalearth) can be handy resources. In *Appendix C: Sourcing geographical data for crime analysis*, we elaborate more on these, with some examples.

For now, you can turn to the data folder which you have downloaded from this book's repository. (If unsure, refer to the *Preface*). Specifically, we're looking for a folder called BoundaryData within the "data" folder. As mentioned, we will be working with a **shapefile**, but in the case of this format we're actually working with *many files*. When we download a shapefile, we usually get many files, all bundled inside a folder which contains them all. It is important for these files to be kept together, in the same folder. If you have a look

---

[3]https://borders.ukdataservice.ac.uk/bds.html

inside this BoundaryData folder, you will notice that there are 4 files with the same name "england_lsoa_2011", but with different extensions.

NOTE: It is important that you keep all these files in the same location as each other! They all contain different bits of information about your shapefile (and they are all needed). Below is a list of the files you might see in this collection:

- .shp — shape format; the feature geometry itself, this is what you see on the map
- .shx — shape index format; a positional index of the feature geometry to allow seeking forwards and backwards quickly
- .dbf — attribute format; columnar attributes for each shape, in dBase IV format.
- .prj — projection format; the coordinate system and projection information, a plain text file describing the projection using well-known text format

Sometimes there might be more files associated with your shapefile as well, but we will not cover them here. So unlike when you work with spreadsheets and data in tabular form, which typically is just all included in one file, when you work with shapefiles, you have to live with the required information living in separate files that need to be stored together. So, being tidy and organised is even more important when you carry out projects that involve spatial data.

1.11.2 Reading shapefiles into R

To read in your data into R, you will need to know the path to where you have saved it. Ideally this will be in your data folder in your project directory.

Let's create an object and assign it our shapefile's name:

```
# Remember to use the appropriate pathfile in your case
shp_name <- "data/BoundaryData/england_lsoa_2011.shp"
```

Make sure that this is saved in your working directory, and you have set your working directory. Now use the `st_read()` function from the `sf` package to read in the shapefile (the one with the `.shp` extension):

```
manchester_lsoa <- st_read(shp_name)
```

Notice that you don't have to read in any of the other files (e.g., `.proj` or `.dbf`); it is enough that they are there silently in the folder from which you read in the `.shp` extension file. Now you have your spatial data file. Notice how running the function sends to the console some metadata about your data. You have a polygon, with 282 rows, and the CRS is the projected British National Grid. You can have a look at what sort of data it contains, the same way you would view a dataframe, with the `View()` function:

```
View(manchester_lsoa)
```

And of course, since it's spatial data, you can map it using the `geom_sf()` function, as we did with our points:

```
ggplot() +
  geom_sf(data = manchester_lsoa)
```

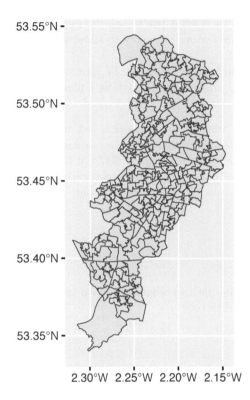

FIGURE 1.13: Plot of Manchester's LSOAs

Great, we now have an outline of the LSOAs in Manchester. Notice how the shape is different to that of the points in our crime data, since here we only obtained the data for the city of Manchester (Manchester Local Authority) rather than for the whole metropolitan area, *Greater* Manchester which includes all the other local authorities aside from Manchester city as well.

1.11.3 Data wrangling with dplyr

In order to map crimes to LSOAs, we might want to take a step back and think about the unit of analysis at which our data are collected. In our original dataframe of crimes, we saw that each crime incident is one row. So the unit of analysis is each crime. Since we were looking to map each crime at the location it happened, we used the latitude and longitude supplied for each incident, providing a geometry for mapping each crime incident. However, when we are looking to map our data to LSOA level, we need to match the crime data to the geometry we wish to display.

Have a look at the "manchester_lsoa" object we mapped above. How many rows (observations) does it have? You can check this by looking in the Environment pane, or by using the `nrow()` function.

```
nrow(manchester_lsoa)
```

You can see this has 282 rows. This means we have geometries for 282 LSOAs. On the other hand, our crimes dataframe has 32058 rows, one for each crime (observation). So how can we match these up? The answer lies in thinking about what it is that our map using LSOAs as our unit of analysis will be able to tell us. Think of other maps of areas. What are they usually telling you? Usually we expect to see crimes per neighbourhood, something like this. So our unit of analysis needs to be LSOA, and for each one we need to know how many crimes occurred in that area.

To achieve this, we will wrangle our data using functions from the `dplyr` package. This is a package for conducting all sorts of operations with dataframes. We are not going to cover the full functionality of `dplyr` (which you can consult in the `dplyr` vignette (Wickham et al. 2021)), but we are going to cover three different very useful elements of dplyr: the `select` function, the `group_by` function, and the piping operator.

The `select()` function provides you with a simple way of subsetting columns from a dataframe. So, say, we just want to use one variable, "lsoa_code", from the "crimes" dataframe and store it in a new object we could write the following code. This variable tells us the LSOA in which the crime took place and it is essential if we want to group our crimes by LSOA (by using this standard method of merging information from datasets):

```
new_object <- select(crimes, lsoa_code)
```

We can also use the `group_by()` function for performing group operations. Essentially this function asks R to group cases within categories and then do something with those grouped cases. So, say, we want to count the number of cases within each LSOA, we could use the following code:

```
#First we group the cases by LSOA code and
# store this organised data into a new object
grouped_crimes <- group_by(new_object, lsoa_code)

#Then we count the number of cases within each category
# using the summarise function to print the results
summarise(grouped_crimes, count = n())

#We create a new dataframe with these results
crime_per_LSOA <- summarise(grouped_crimes, count = n())
```

As you can see, we can do what we wanted, create a new dataframe with the required info. But if we do this, we are creating many objects that we don't need, one at each step. Instead there is a more efficient way of doing this, without so many intermediate steps clogging up our environment with unnecessary objects. That's where the piping operator comes handy. The piping operator is written like `%>%` and it can be read as "and then". Look at the code below:

```
#First we say create a new object called crime_per_lsoa,
# and then select only the LSOA.code column to exist in this object,
# and then group this object by the LSOA.code,
# and then count the number of cases within each category.
# This is what I want in the new object.

crimes_per_lsoa <- crimes %>%
  group_by(lsoa_code) %>%
  summarise(count=n())
```

Essentially we obtain the same results but with more streamlined and elegant code, and not needing additional objects in our environment. As of version 4.1.0 of R, there is also a new piping operator denoted by the |> symbol that can also be used and that does not depend on the `magrittr` package (as `%>%` does).

And now we have a new object, "crimes_per_lsoa". If we have a look at this one, we can now see that each row represents one LSOA, and next to it we have a variable for the number of crimes from each area. We created a new dataframe from a frequency table, and as each row of the crimes data was one crime, the frequency table tells us the number of crimes which occurred in each LSOA.

Those of you playing close attention might note that there are still more observations in this dataframe (1671) than in the "manchester_lsoas" one (282). Again, this is because "crimes_per_lsoa" also includes data from census areas in municipalities of the metropolitan area of *Greater* of Manchester other than Manchester city local authority.

1.11.4   Join data to sf object

Our next task is to link our crimes data to our `sf` spatial object to help us map this. Notice anything similar between the data from the shapefile and the frequency table data we just created? Do they share a column? Indeed, you might notice that the "lsoa_code" field in the crimes data matches the values in the "code" field in the spatial data. In theory we could join these two data tables.

So how do we do this? Well what you can do is to link one data set with another. Data linking is used to bring together information from different sources in order to create a new, richer dataset. This involves identifying and combining information from corresponding records on each of the different source datasets. The records in the resulting linked dataset contain some data from each of the source datasets. Most linking techniques combine records from different datasets if they refer to the same entity (an entity may be a person, organisation, household or even a geographic region.)

You can merge (combine) rows from one table into another just by pasting them in the first empty cells below the target table—the table grows in size to include the new rows. And if the rows in both tables match up, you can merge columns from one table with another by pasting them in the first empty cells to the right of the table—again, the table grows, this time to include the new columns.

Merging rows is pretty straightforward, but merging columns can be tricky if the rows of one table don't always line up with the rows in the other table. By using `left_join()` from the dplyr package, you can avoid some of the alignment problems. The `left_join()` function will return all rows from x, and all columns from x and y. Rows in x with no match in y will have NA values in the new columns. If there are multiple matches between x and y, all combinations of the matches are returned.

So we've already identified that both our crimes data, and the spatial data contain a column with matching values, the codes for the LSOA that each row represents. **You need a unique identifier to be present** for each row in all the data sets that you wish to join. This is how R knows what values belong to what row. What you are doing is matching each value from one table to the next, using this unique identified column, that exists in both tables. For example, let's say we have two data sets from some people in Hawkins, Indiana. In one data set we collected information about their age. In another one, we collected information about their hair colour. If we collected some information that is unique to each observation, and this is the *same* in both sets of data, for example their names, then we can link them up, based on this information. Something like this:

	name	age
1	Barb	16
2	Steve	16
3	Mike	13

	name	hair_colour
1	Barb	Red
2	Steve	Brown
3	Mike	Black

FIGURE 1.14: Two dataframes with common unique identifier

And by doing so, we produce a final table that contains all values, lined up *correctly* for each individual observation, like this:

	name	age	hair_colour
1	Barb	16	Red
2	Steve	16	Brown
3	Mike	13	Black

FIGURE 1.15: Joined dataframe

This is all we are doing, when merging tables: joining the two data sets using a common column. Why are we using *left* join though? There is a whole family of join functions as part of dplyr which join datasets. There is also a `right_join()`, and an `inner_join()` and an `outer_join()` and a `full_join()`. But here we use `left_join()`, because that way we keep all the rows in x (the left-hand side dataframe), and join to it all the matched columns in y (the right-hand side dataframe).

So let's join the crimes data to the spatial data, using `left_join()`. We have to tell the function what are the dataframes we want to join, as well as the names of the columns that contain the matching values in each one. This is "code" in the "manchester_lsoa" dataframe and "lsoa_code" in the "crimes_per_lsoa" dataframe. Like so:

```
manchester_lsoa <- left_join(manchester_lsoa, crimes_per_lsoa,
                             by = c("code"="lsoa_code"))
```

Now if you have a look at the data again, you will see that the column of number of crimes (count) has been added on. You may not want to have to go through this process all the time that you want to work with this data. One thing you could do is to save the "manchester_lsoa" object as a physical file in your machine. You can use the st_write() function from the sf package to do this. If we want to write into a shapefile format, we would do as shown below. Make sure you save this file, for we will come back to it in subsequent chapters.

```
st_write(manchester_lsoa, "data/BoundaryData/manchester_crime_lsoa.shp")
```

Note: if you get an error which says the layer already exists, this means there is already a file with this name in that location. You can add the parameter append = FALSE *to overwrite this file, or give it a slightly different name.*

1.11.5   Putting polygon data on the map

Now that we have joined the crimes data to the geometry, you can use this to make our map. Remember our original empty map? Well now, since we have the column (variable) for number of crimes here, we can use that to shade the polygons based on how many crimes there are in each LSOA. We can do this by specifying the fill= parameter of the geom_sf function.

```
ggplot() +
geom_sf(data = manchester_lsoa, aes(fill = count))
```

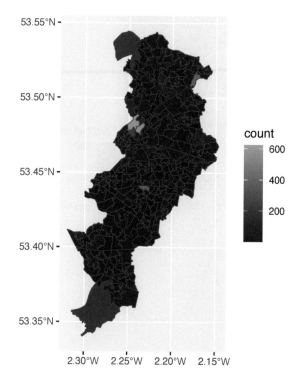

FIGURE 1.16: Number of crimes in Manchester by LSOA

Just like we did with the ggplot bar charts we created earlier, we can adjust the opacity of our thematic map, with the `alpha =` parameter, add a basemap, with the function `annota-tion_map_tile()` and adjust the colour scheme, with the `scale_fill_gradient2()` function.

```
ggplot() +
  annotation_map_tile() +  # add basemap
geom_sf(data = manchester_lsoa,
        aes(fill = count),
        alpha = 0.7) + # alpha sets the opacity
  #use scale_fill_gradient2() for a different palette
  # and name the variable on the legend
  scale_fill_gradient2(name ="Number of crimes")
```

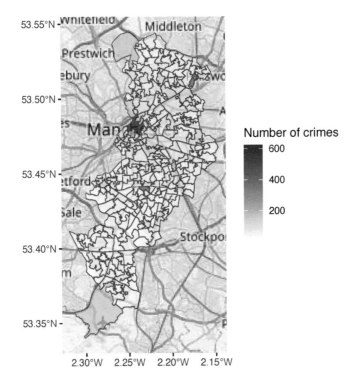

FIGURE 1.17: Number of crimes by LSOA with adjusted parameters

In subsequent chapters we will play around with other packages in R that you can use to produce this kind of maps. Gradually we will discuss the kind of choices you can make in order to select the adequate type of representation for the type of spatial data and question you have, and the aesthethic choices that are adequate depending on the purposes and medium in which you will publish your map. But for now, we can rejoice: here is our very first crime map of the book!

1.12   Summary and further reading

In this chapter we had a play around with some regular old crime data and discovered how we can use the `sf` package in R to assign it a geometry (both at point and polygon level), and how that can help us visualise our results. If you want to explore other spatial data, you can explore Moraga et al. (2021) for short tutorials in how to get and visualise different kinds of spatial data.

We covered some important concepts such as projections and coordinate reference systems, and we had a go at acquiring shapefiles which can help us visualise our data. We had a think about units of analysis, and how that will affect how we visualise our data. In the next chapter we will spend a bit of more time discussing how to make good choices when producing maps.

There are a number of introductory texts to geographic information systems and analysis —
eg., O'Sullivan and Unwin (2010), Bolstad (2019) — that provide adequate background
to some of the key concepts we introduce in this chapter . The report by Harries (1999)
produced for the National Institute of Justice still provides a good general introduction for
basic ideas around crime mapping and can be accessed for free online. Chapter 3 of Haining
and Li (2020) offers invaluable observations on the nature of spatial and spatial-temporal
attribute data, often using examples from crime research. The chapter by Radil (2016) on
the spatial analysis of crime offers a very succinct review of what will be the focus of most
of this book. From a more domain-knowledge point of view, the early chapters of Boba
(2013) and Chainey and Ratcliffe (2005) set the stage for the use of GIS as part of the
crime analysis process, whereas Bruinsma and Johnson (2018), edited handbook provides
an excellent introduction to environmental criminology (that provides the theoretical and
empirical backbone to spatial analysis of crime). Chainey (2021) is a more recent addition,
but focuses on ArcGIS. Finally, Healy (2019) provides a general introduction to data visualisation with R using the `ggplot2` package. Although Healy's text is not just about mapping,
it offers a very practical and helpful introduction to using `ggplot2` where you can learn how
to further customise your maps and other charts. For specifics of `ggplot2` refer to Wickham
(2010).

This is a book about maps and spatial analysis, but clearly one of the first questions you
need to ask yourself is whether producing a map is the right answer to your question. Just
because your data is spatial doesn't mean you need a map for every question you pose to
this data. There may be other forms of data visualisation that are more appropriate for
exploring and summarising the story you want to tell with your data. If you are uncertain
about whether you need a map or other kind of plot, books such as Cairo (2016), Camoes
(2016), or Schwabish (2021) provide useful guidance.

2

Basic geospatial operations in R

2.1 Introduction

In this chapter we get our hands dirty with **spatial manipulation of data**. Thus far, our data manipulation exercises (using `dplyr`) were such that you might be familiar with, from any earlier exposures to data analysis. For example, linking datasets using a common column is a task which you can perform on spatial or non-spatial data. These are referred to as **attribute operations**. However, in this chapter we will explore some exercises in data manipulation which are specific to *spatial* data analysis. We will be learning some key spatial operations: a set of functions that allow you to create new and manipulate spatial data.

The main objectives for this chapter are that by the end you will have:

- met a new format for accessing boundary data, called **geoJSON**.
- carried out **spatial operations** such as:
    - **subset** points that are within a certain area,
    - created new polygons by generating **buffers** around points,
    - counted the number of points that fall within a polygon (known as **points in polygon**),
    - finding the **nearest feature** in one data set to observations in another data set, and
    - **measured distance** between points in a map.
- made interactive point map with leaflet.
- used **geocoding** methods to translate text fields such as addresses into geographic coordinates.

These are all very useful tools for the spatial crime analyst, and we will hope to demonstrate this by working through an example project. The packages we will use in this chapter are:

```
# Packages for reading data and data carpentry
library(readr)
library(dplyr)
library(janitor)
library(units)
library(purrr)

# Packages for handling spatial data and for geospatial carpentry
library(sf)
library(tidygeocoder)
library(crsuggest)
```

```
# Packages for mapping and visualisation
library(leaflet)
library(RColorBrewer)

# Packages providing accesss to spatial data
library(osmdata)
```

2.2   Exploring the relationship between alcohol outlets and crime

The main example we will work through most of the chapter considers the assumption that licenced premises which serve alcohol are associated with increased crimes. We might have some hypotheses about why this may be.

One theory might be that some of these serve as *crime attractors*.

> Crime attractors are particular places, areas, neighbourhoods, districts which create well-known criminal opportunities to which strongly motivated, intending criminal offenders are attracted because of the known opportunities for particular types of crime. Examples might include bar districts; prostitution areas; drug markets; large shopping malls, particularly those near major public transit exchanges; large, insecure parking lots in business or commercial areas. The intending offender goes to rough bars looking for fights or other kinds of 'action' (Brantingham and Brantingham 1995, 7).

On the other hand, it is possible that these areas are *crime generators*.

> Crime generators are particular areas to which large numbers of people are attracted for reasons unrelated to any particular level of criminal motivation they might have or to any particular crime they might end up committing. Typical examples might include shopping precincts; entertainment districts; office concentrations; or sports stadiums (Brantingham and Brantingham 1995, 8).

It is possible that some licensed premises attract crimes, due to their reputation. However, it is also possible that some of them are simply located in areas that are busy, attract lots of people for lots of reasons, and crimes occur as a result of an abundance of opportunities instead.

Whatever the mechanism, the first step to identifying crime places is to examine whether certain outlets have more crimes near them than others. We can do this using open data, some R code, and the spatial operations discussed above. We will return to data from Manchester, UK for this example; however, as we will be using Open Street Map, you can easily replicate this for any other location where you have point-level crime data.

2.3 Acquiring relevant data

We will be using three different sources of data in this chapter. First, we will acquire our crime data, which is what we used in the previous chapter, so this should be familiar. Then we will meet the new format for boundary data, geoJSON. And finally, we will look at Open Street Map for data on our points of interest.

2.3.1 Reading in crime data

The point-level crime data for Greater Manchester can be found in the file 2019-06-greater-manchester-street.csv we used in Chapter 1. We can import it with the read_csv() function from the readr package.

```
crimes <- read_csv("data/2019-06-greater-manchester-street.csv")
```

If you replicate the exercises for getting to know the dataset from the previous chapter, you might notice that in this case the column names are slightly different. For example, Latitude and Longitude are spelled with uppercase "L". You should always familiarise yourself with your dataset to make sure you are using the relevant column names. You can see just the column names using the names() function like so:

```
names(crimes)
```

```
##  [1] "Crime ID"              "Month"
##  [3] "Reported by"           "Falls within"
##  [5] "Longitude"             "Latitude"
##  [7] "Location"              "LSOA code"
##  [9] "LSOA name"             "Crime type"
## [11] "Last outcome category" "Context"
```

This is because last time, we cleaned these names using the clean_names() function from the janitor package. Let's do this again, as these variable names are not ideal, with their capital letters and spacing...so messy!

```
# clean up the variable names
crimes <- crimes %>% clean_names()

#print them again to see
names(crimes)
```

```
##  [1] "crime_id"              "month"
##  [3] "reported_by"           "falls_within"
##  [5] "longitude"             "latitude"
##  [7] "location"              "lsoa_code"
##  [9] "lsoa_name"             "crime_type"
## [11] "last_outcome_category" "context"
```

Much better! Now let's get some boundary data for Manchester.

2.3.2   Meet a new format: geoJSON

GeoJSON is an open standard format designed for representing simple geographical features, along with their non-spatial attributes. It is based on JSON, the JavaScript Object Notation. It is a format for encoding a variety of geographic data structures and is the most common format for geographical representation in the web. Unlike ESRI shapefiles, with GeoJSON data everything is stored in a single file.

Geometries are shapes. All simple geometries in GeoJSON consist of a type and a collection of coordinates. The features include points (therefore addresses and locations), line strings (therefore streets, highways and boundaries), polygons (countries, provinces, tracts of land), and multi-part collections of these types. GeoJSON features need not represent entities of the physical world only; mobile routing and navigation apps, for example, might describe their service coverage using GeoJSON. To tinker with GeoJSON and see how it relates to geographical features, try geojson.io, a tool that shows code and visual representation in two panes.

Let's read in a geoJSON spatial file, again from the web. This particular geoJSON represents the wards of Greater Manchester.

```
manchester_ward <- st_read("data/wards.geojson")
```

Let's now select only the city centre ward, using the `filter()` function from the `dplyr` package. Notice how we're using the values in the data (or attribute) table; therefore, this is an *attribure operation*.

```
city_centre <- manchester_ward %>%
  filter(wd16nm == "City Centre")
```

Let's see how this looks. In Chapter 1 we learned about plotting maps using the `ggplot()` function from the `ggplot2` package. Here, let's look at a different way, which can be useful for quick mapping — usually used when just checking in with our data to make sure it looks like what we were expecting.

With the `st_geometry()` function in the `sf` package, we can extract only the geometry of our object, stripping away the attributes. If we include this inside the `plot()` function, we get a quick, minimalist plot of the geometry of our object.

```
#extract geometry only
city_centre_geometry <- st_geometry(city_centre)

#plot geometry object
plot(city_centre_geometry)
```

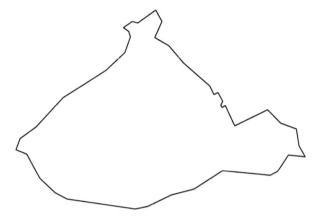

FIGURE 2.1: Plot City Centre ward geometry

Now we could use this geometry to make sure that our points are in fact only licensed premises in the city centre. This will be your first spatial operation. Excited? Let's do this!

2.3.3   Open Street Map and points of interest

To map licenced premises, we will be accessing data from Open Street Map, a database of geospatial information built by a community of mappers, enthusiasts and members of the public, who contribute and maintain data about all sorts of environmental features, such as roads, green spaces, restaurants and railway stations. You can see all about open street map on their online mapping platform[1]. One feature of Open Street Map, unlike Google Map, is that the underlying data is openly available for download *for free*. In R, we can take advantage of a package written specifically for querying Open Street Map's API, called osmdata. For more detail on how to use this, see Langton and Solymosi (2018), and *Appendix C: sourcing geographical data for crime analysis* of this book.

If we load the package osmdata we can use its functions to query the Open Street Map API. To find out more about the capabilities of this package, see Padgham et al. (2017). While this is outside the scope of our chapter here, you may want to explore osmdata more, as it is an international database, and has lots of data that may come in handy for research and analysis.

Here we focus specifically on Manchester. To retrieve data for a specific area, we must create a **bounding box**. You can think of the bounding box as a box drawn around the area that we are interested in (in this case, Manchester, UK) which tells the Open Street Map API that we want everything *inside* the box, but nothing *outside* the box.

So, how can we name a bounding box specification to define the study region? One way to do this is through a search term. Here, we want to select Greater Manchester, so we can use the search term "greater manchester united kingdom" within the getbb() function (stands for *get bounding box*). Using this function, we can also specify what format we want the data to be in. In this case, we want a spatial object, specifically an sf polygon object (from the sf package), which we name bb_sf.

---

[1]https://www.openstreetmap.org/

```
bb_sf <- getbb(place_name = "greater manchester united kingdom",
               format_out = "sf_polygon")
```

We can see what this bounding box looks like by plotting it (notice this time we combine the plot() and st_geometry() functions to save needing to create a new object):

```
plot(st_geometry(bb_sf))
```

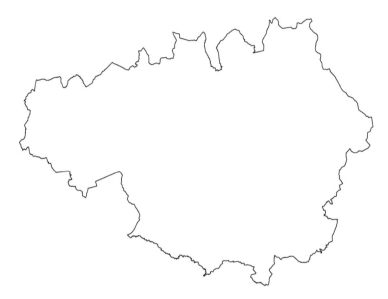

FIGURE 2.2: Outline of Greater Manchester boundary

We can see the bounding box takes the form of Greater Manchester. We can now use this to query data from the Open Street Map API using the opq() function. The function name is short for 'Overpass query', which is how users can query the Open Street Map API using search criteria.

Besides specifying what area we want to query with our bounding box object(s) in the opq() function, we must also define the feature which we want returned. Features in Open Street Map are defined through 'keys' and 'values'. Keys are used to describe a broad category of features (e.g., highway, amenity), and values are more specific descriptions (e.g., cycleway, bar). These are tags which contributors to Open Street Map have defined. A useful way to explore these is by using the comprehensive Open Street Map Wiki page on map features[2] (see also *Appendix C: sourcing geographical data for crime analysis* of this book).

We can select what features we want using the add_osm_feature() function, specifying our key as 'amenity' and our value as 'bar'. We also want to specify what sort of object (what class) to get our data into, and as we are still working with spatial data, we stick to the sf format, for which the function is osmdata_sf(). Here, we specify our bounding box as the bb_sf object we created above. If you use the bounding box obtained through getbb() one can subsequently trim down the outputs from add_osm_feature() using trim_osmdata(). For instance, we could add trim_osmdata(bb_poly = bb_sf) to our initial query.

---

[2]https://wiki.openstreetmap.org/wiki/Map_Features

```
osm_bar_sf <- opq(bbox = bb_sf) %>%      # select bounding box
  add_osm_feature(key = 'amenity', value = 'bar') %>% # select features
  osmdata_sf()            # specify class
```

The resulting object `osm_bar_sf` contains lots of information. We can view the contents of the object by simply executing the object name into the **Console**.

```
osm_bar_sf
```

This confirms details like the bounding box coordinates. It also provides information on the features collected from the query. As one might expect, most information relating to bar locations has been recorded using points (i.e. two-dimensional vertices, coordinates) of which we have 791 at the time of writing. We also have around fifty polygons. For now, let's extract the point information.

```
osm_bar_sf <- osm_bar_sf$osm_points
```

We now have an `sf` object with all the bars in our study region mapped by Open Street Map volunteers, along with ~90 variables of auxiliary data, such as characteristics of the bar (e.g., `brewery`, or `cocktail`) as well as address and contact information, amongst many others. Of course, it is up to the volunteers whether they collect all these data, and in many cases, they have not added information (you may see lots of missing values if you look at the data). Given the work relies on volunteers, there are unavoidably some accuracy issues (including how up-to-date the information may be). Nevertheless, when the details are recorded, they provide rich insight and local knowledge that we may otherwise be unable to obtain.

One column we should consider is the `name` which tells us the name of the bar. There are missing values here as well, and for this example, we will choose to exclude those lines where there is no name included, as we would like at least a little bit of context around our bars. To do this, perform another *attribute operation* using the `filter()` function:

```
osm_bar_sf <- osm_bar_sf %>% filter(!is.na(name))
```

We are still left with 283 bars in our data set.

2.4 Attribute operations

We mentioned above that we were using **attribute operations**. These are changes to the data which we make based on manipulation of elements in the attribute table. For example, the use of `filter()` is an attribute operation, because we rely on the data in the attribute table in order to accomplish this task.

For example, let's say we want to focus only on violent crime. To do this, we use the information in the attribute table, namely the values for the `crime_type` variable for each observation (crime) in our data set.

```
crimes <- crimes %>%
  filter(crime_type == "Violence and sexual offences")
```

With the above, we select only those crimes (rows of the attribute table) where the crime_type variable meets a certain criteria (takes the value of "Violence and sexual offences"). **Spatial operations** on the other hand manipulate the *geometry* part of our data. We rely on the spatial information to accomplish the tasks of interest. In the next section, we will work through some examples of these.

2.5   Spatial operations

Spatial operations are a vital part of geocomputation. Spatial objects can be modified in a multitude of ways based on their location and shape. For a comprehensive overview of spatial operations in R, we recommend chapter 4 of Lovelace, Nowosad, and Muenchow (2019).

> Spatial operations differ from non-spatial operations in some ways. To illustrate the point, imagine you are researching road safety. Spatial joins can be used to find road speed limits related with administrative zones, even when no zone ID is provided. But this raises the question: should the road completely fall inside a zone for its values to be joined? Or is simply crossing or being within a certain distance sufficient? When posing such questions, it becomes apparent that spatial operations differ substantially from attribute operations on dataframes: the type of spatial relationship between objects must be considered. (Lovelace, Nowosad, and Muenchow (2019))

So you can see, we can do exciting spatial operations with our spatial data, which we cannot with the non-spatial stuff.

2.5.1   Reprojecting coordinates

It is important to recall here some of the learning from the previous chapter on map projections and coordinate reference systems. We learned about ways of flattening out the earth, and ways of making sense of what that means for how to be able to point to specific locations in our maps. **Coordinate Reference System** (CRS) is this method of how to refer to locations with our data. You might use a *Geographic Coordinate System*, which tells you where your data are located on the surface of the Earth. The most commonly used one is the **WGS 84**, where we define our locations with latitude and longitude points. The other type is a *Projected Coordinate System* which tells the data how to draw on a flat, two-dimensional surface (such as a computer screen). In our case here, we will often encounter the **British National Grid** when working with British data. Here our locations are defined with Eastings and Northings.

So why are we talking about this?

It is important to note that spatial operations that use two spatial objects rely on both objects having the same coordinate reference system.

If we are looking to carry out operations that involve two different spatial objects, they need to have the same CRS! Funky weird things happen when this condition is not met, so beware! So how do we know what CRS our spatial objects are? Well the sf package contains a handy function called st_crs() which let us check. All you need to pass into the brackets of this function is the name of the object you want to know the CRS of.

Let's check what is the CRS of our crimes:

```
st_crs(crimes)
```

```
## Coordinate Reference System: NA
```

You can see that we get the CRS returned as NA. Can you think why? Have we made this into a spatial object? Or is this merely a dataframe with a latitude and longitude column? The answer is really in the question here. So we need to convert this to a sf object, or a spatial object, and make sure that R knows that the latitude and the longitude columns are, in fact, coordinates.

```
crimes <- st_as_sf(crimes, coords = c("longitude", "latitude"),
                   crs = 4326, agr = "constant", na.fail = FALSE)
```

In the st_as_sf() function, we specify what we are transforming (the name of our dataframe), the column names that have the coordinates in them (longitude and latitude), the CRS we are using (4326 is the code for WGS 84, which is the CRS that uses latitude and longitude coordinates (remember BNG uses Easting and Northing)), and finally *agr*, the attribute-geometry-relationship, specifies for each non-geometry attribute column how it relates to the geometry, and can have one of following values: "constant", "aggregate", "identity". The option "constant" is used for attributes that are constant throughout the geometry (e.g., land use), "aggregate" where the attribute is an aggregate value over the geometry (e.g., population density or population count), "identity" when the attributes uniquely identifies the geometry of particular "thing", such as a building ID or a city name. The default value, NA_agr_, implies we don't know. Finally we can specify the parameter na.fail which decides what to do when one or more of our observations have missing (NA) values in their coordinates. The default setting is for the entire process to fail in this case. However, if we set this to FALSE it simply means we will lose those observations. Essentially, we are getting rid of those violent crimes which have not been geocoding, excluding them from our analysis.

Now let's check the CRS of this spatial version of our licensed premises:

```
st_crs(crimes)
```

We can now see that we have this coordinate system as WGS84. We need to then make sure that any other spatial object with which we want to perform spatial operations is also in the same CRS. Let's look at our city centre ward boundary file:

```
st_crs(city_centre)
```

We see that this is in fact in a projected coordinate system, namely the British National Grid we mentioned. To make them align, we can **re-project** this object into the *WGS84* geographic coordinate system. To do this, we can use the `st_transform()` function.

```
city_centre <- st_transform(city_centre, crs = 4326)
```

Now we can check the projection again:

```
st_crs(city_centre)
```

And we can also check whether the CRS of the two objects match:

```
st_crs(crimes) == st_crs(city_centre)
```

```
## [1] TRUE
```

It is true! Finally, to check our bar data from Open Street Map:

```
st_crs(osm_bar_sf)
```

You will see this is also in WGS84. Since all our data are in this CRS, we can now move on to carry out some spatial operations.

But before doing so, we need to raise an important point. R is still undergoing the adaptation to new standards in the way CRS data is stored and handled. Before 2020 you could specify the CRS data with a EPSG code or with a `proj4string` reference, which look more like gibberish to the non-initiated but provided more flexibility (see Lovelace, Nowosad, and Muenchow (2019) for details). But there were problems with the old PROJ4. As we noted in the first chapter, it only allows for static reference frames (when on Earth everything is moving all the time). So PROJ the main library for performing conversions between systems has changed. It now discourages the old `proj4string` format and has changed to the OGC WKT2. When you consult the https://epsg.io/ site, you can see whenever you select a projection it translates to different standards (including OGC WKT). This is one of those changes that long-term are really good. But because the cross-dependencies between spatial R packages were being updated at different speed, this had an impact on usability. Although we will stick to the more familiar EPSG notation, we strongly recommend reading the following technical notes by Pebesma and Bivand (Pebesma and Bivand (2020), Bivand (2019), Bivand (2020)), and to be particularly alert to this issue.

2.5.2   Subsetting points

Recall above that we wanted to focus our efforts on the City Centre ward of Manchester; however, for our bounding box to download OSM data we used Greater Manchester. If we were to plot our bars, we would see that we have many which fall outside of the City Centre ward:

```
plot(st_geometry(osm_bar_sf), col = 'red')
plot(st_geometry(city_centre),  add = TRUE)
```

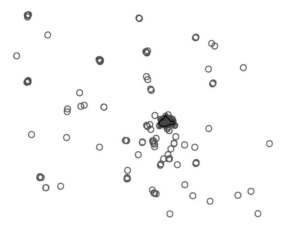

FIGURE 2.3: Many bars (in red) fall outside City Centre ward

This is also the case for our crimes data:

```
plot(st_geometry(osm_bar_sf), col = 'red')
plot(st_geometry(crimes), col = 'blue', add = TRUE)
plot(st_geometry(city_centre), add = TRUE)
```

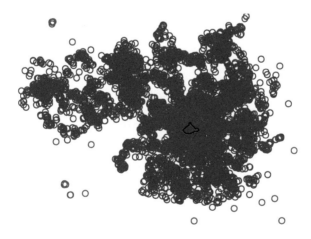

FIGURE 2.4: Most crimes (in blue) fall outside City Centre ward

So if we really want to focus on City Centre, we should create spatial objects for the crimes and the bars which include only those which fall within the City Centre ward boundary.

First, we check whether they have the same CRS.

```
st_crs(city_centre) == st_crs(crimes)
```

```
## [1] TRUE
```

We do indeed, as we made sure in the previous section. Now we can move on to our spatial operation, where we select only those points within the City Centre polygon. To do this, we first make a list of intersecting points to the polygon, using the `st_intersects()` function. This function takes two arguments: first, the polygon which we want to subset our points within, and second, the points which we want to subset. We then use the resulting "cc_crimes" object to subset the crimes object to include only those which intersect (return TRUE for intersects):

```
# intersection
cc_crimes <- st_intersects(city_centre, crimes)
# subsetting
cc_crimes <- crimes[unlist(cc_crimes),]
```

Have a look at this new "cc_crimes" object in your environment. How many observations does it have? Is this now fewer than the previous "crimes" object? Why do you think this is?

(Hint: you're removing everything that is outside the City Centre polygon)

We can plot this again to have a look:

```
plot(st_geometry(city_centre))
plot(st_geometry(cc_crimes), col = 'blue',  add = TRUE)
```

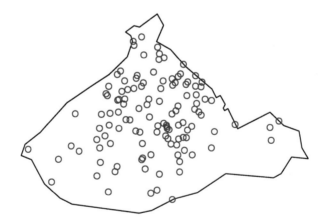

FIGURE 2.5: Crimes inside the City Centre ward

We have successfully performed our first spatial operation. We managed to subset our points dataset of crimes to include only those crimes which are located inside the polygon for City Centre.

We can do the same for the bars:

```
# intersection
cc_bars <- st_intersects(city_centre, osm_bar_sf)
# subsetting
cc_bars <- osm_bar_sf[unlist(cc_bars),]
```

We can see that of the previous 283 bars, 114 are within the City Centre ward. We can plot our data now:

```
plot(st_geometry(city_centre))
plot(st_geometry(cc_bars), col = 'red', add = TRUE)
plot(st_geometry(cc_crimes), col = 'blue', add = TRUE)
```

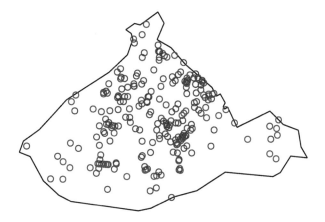

FIGURE 2.6: Bars and crimes within City Centre

2.5.3 Building buffers

So we now have our bars and our violent crimes in Manchester City Centre. Let's go back to our original question. We want to know about crime in and around our areas of interest, in this case our bars. But how can we count this? We have our points that are crimes, right? Well... How do we connect them to our points that are licensed premises?

One approach is to build a buffer around our bars, and say that we will count all the crimes which fall within a specific radius of this bar. What should this radius be? Well, this is where your domain knowledge as criminologist or crime analyst comes in. How far away would you consider a crime to still be related to this pub? 400 metres? 500 metres? 900 metres? 1 km? What do you think? This is one of those *it depends* questions, where there is no universal right answer; instead it will depend on the environment, question, and contextual factors. Theory will have an important role to play, for example whether the processes are conceptualised as micro-level, or meso-level. Whatever buffer you choose, you should justify and make sure that you can defend when someone might ask about it, as the further you reach obviously the more crimes you will include, and these might alter your results.

So, let's say we are interested in all crimes that occur within 400 metres of each licensed premise, within the study area. We chose 400 metres here as this is often the recommended distance for accessible bus stop guidance, so basically as far as people should walk to get to a bus stop. So in this case, we want to take our points, which represent the licensed premises, and build buffers of 400 metres around them.

You can do with the `st_buffer()` function. We pass two arguments to our function: the item which we want to buffer (the points in our 'cc_bars' object) and the size of this buffer. Let´s quickly illustrate:

```
prem_buffer <- st_buffer(cc_bars, 1)
```

You might get a warning here, saying *"st_buffer does not correctly buffer longitude/latitude datadist is assumed to be in decimal degrees (arc_degrees)."*. This message indicates that sf assumes a distance value (our size of the buffer, specified as '1' above) is given in degrees. This is because we have our data in a Geographic Coordinate System (lat/long data in WSG 48).

If we want to calculate the size of our buffer in a meaningful distance on our 2D surfaces, we can transform to a Projected Coordinate System, such as British National Grid. Let's do this now:

```
#The code for BNG is 27700
bars_bng <- st_transform(cc_bars, 27700)
```

Now we can try again, with metres, specifying our indicated 400 metres radius:

```
bars_buffer <- st_buffer(bars_bng, 400)
```

Let´s see how that looks:

```
plot(st_geometry(bars_buffer))
plot(st_geometry(bars_bng), add = T)
```

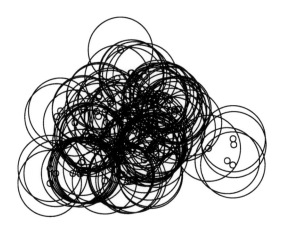

FIGURE 2.7: Bars with 400 metres buffers

That should look nice and squiggly. We can see it looks like there is *quite* a lot of overlap here. Should we maybe consider smaller buffers? Let's look at 100 metre buffers:

```
bar_buffer_100 <- st_buffer(bars_bng, 100) # create 100m buffer

# plot new buffers
plot(st_geometry(bar_buffer_100))
plot(st_geometry(bars_bng), add = T)
```

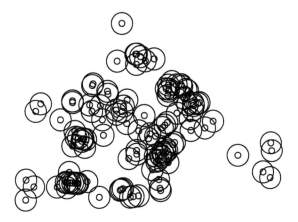

FIGURE 2.8: Bars with 100 metres buffers

Quite a bit of overlap, but this is possibly because the licensed premises are close together in the City Centre. We will discuss how to deal with this later on. For now, let's go with these 100 metre buffers, and where a crime falls into an area of overlap, we will count it towards both premises. Where a crime falls within multiple buffers, it will count towards all the bars associated with those buffers.

The next step will be to count the number of crimes which fall into each buffer. Before we move on though, remember the CRS for our crimes is WGS 48 here, so we will need to convert our buffer layer back to this:

```
buffer_WGS84 <- st_transform(bar_buffer_100, 4326)
```

Now let's just have a look:

```
plot(st_geometry(buffer_WGS84))
plot(st_geometry(cc_crimes), col = 'blue', add = T)
```

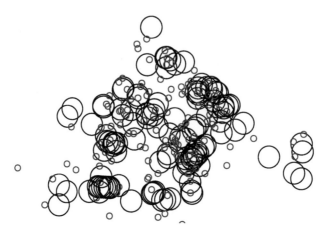

FIGURE 2.9: Crimes around the 100 metre buffer polygons

Some crimes fall inside some buffers, others not so much. Well, let's get to our next spatial operation to be able to determine how many crimes happened in the 100 metre radius of each bar in Manchester City Centre.

2.5.4 Counting points within a polygon

When you have a polygon layer and a point layer and want to know how many or which of the points fall within the bounds of each polygon, you can use this method of analysis. In computational geometry, the **point-in-polygon** (PIP) problem asks whether a given point in the plane lies inside, outside, or on the boundary of a polygon. As you can see, this is quite relevant to our problem, wanting to count how many crimes (points) fall within 100 metres of our licensed premises (our buffer polygons). We can achieve this with the st_join() function, which spatially joins the bar name to each crime, and then we could find the frequency of each name (hence the count(name)). This returns a frequency table of the number of crimes within the buffer of each bar, saved in the crimes_per_prem object.

```
crimes_per_prem <- cc_crimes %>%
  st_join(buffer_WGS84, ., left = FALSE) %>%
  count(name)
```

You now have a new dataframe, crimes_per_prem, which has a column for the name of the bars, a column for the number of violent crimes that fall within the buffer, and a column for the geometry. Take a moment to look at this table. Use the View() function. Which premises have the most violent crimes? Let's see the bar with the most crimes:

```
crimes_per_prem %>%
  filter(n == max(n)) %>%
  select(name, n)
```

The bar with the highest number of crimes is Crafty Pig with 59 crimes. Keep this in mind for the next section.

So in this case, we used the PIP approach, and counted the number of points which fell into each polygon. We saw earlier, with the buffers, that they often overlapped with one another. This means that a crime may have been counted multiple times. This resulting data therefore tells us: *How many crimes happened within 100 metres of each bar.* This is one way to approach the problem, but not the only way. In our next spatial operation, we will calculate distances in order to explore another way.

2.5.5 Distances: Finding the nearest point

Another way to solve this problem is to assign each crime (point) to the closest possible bar (other point). That is, look at the distances between crimes' locations and the locations of all the bars in Manchester, and then, from those, choose the bar which is the closest. Then, we can assign this bar as the location for that crime.

We can achieve this using the `st_nearest_feature()` function. This function takes our two sf objects, and for each row of the first one (x = cc_crimes), simply returns the index of the nearest features from the second one (y = cc_bars). We combine with the `mutate()` function in order to create a new variable which contains this index for each crime. Let's illustrate:

```
crime_w_bars <- cc_crimes %>%
  mutate(nearest_bar = st_nearest_feature(cc_crimes, cc_bars))
```

If we now have a look at this new object "crime_w_bars", we can see it is our crimes data. But we have a new column, which contains the index of the closest bar in the cc_bars dataframe, right at the end. So for example, the first point there, the nearest bar is that in location 84 (vectors in R are 1-indexed, not 0-indexed like many other languages). If we wanted to look at what is on the 84th row, we may call:

```
cc_bars[84]
```

Note: you may get a different number, if since time of writing new bars have been added, or taken away in the OSM database. You can continue with a different number to follow along!

However, this returns all the 90+ variables for this row. If we want only the name, we can query for the 84th row and the 2nd column (which is name):

```
cc_bars[84, 2]
```

You can see the name is "Dive". For that first crime, in our dataset, the nearest bar is "Dive" bar. Now, instead of going through this process manually for each point, we can use the index to subset within our `mutate()` function:

```
crime_w_bars  <- cc_crimes %>%
  mutate(nearest_bar = cc_bars[st_nearest_feature(cc_crimes, cc_bars),2])
```

Now we have new information in this `nearest_bar` column, the name of the nearest bar, and the geometry. We actually don't need the geometry for now, as we will simply be counting the frequency of each bar, which we can join back to our `cc_bars` object, which has a geometry, so we can extract the `$name` element only, and remove the geometry. Like so:

```
crimes_per_prem_2 <- crime_w_bars %>%   # create new crimes_per_prem_2 object
  st_drop_geometry() %>%                # drop (remove) the geometry
  group_by(nearest_bar$name) %>%        # group by to find frequency of each bar
  summarise(num_crimes = n()) %>%       # count number of crimes
  rename(name = `nearest_bar$name`)     # rename variable to 'name'
```

To tie this back to our spatial object "cc_bars", we can use the `left_join()` function:

```
crimes_per_prem_2 <- left_join(cc_bars, crimes_per_prem_2,
                      by = c("name" = "name"))
```

Let's see the bar with the most crimes with this approach:

```
crimes_per_prem_2 %>%
  filter(num_crimes == max(num_crimes, na.rm = TRUE)) %>%
  select(name, num_crimes)
```

```
## Simple feature collection with 1 feature and 2 fields
## Geometry type: POINT
## Dimension:     XY
## Bounding box:  xmin: -2.236 ymin: 53.48 xmax: -2.236 ymax: 53.48
## Geodetic CRS:  WGS 84
##           name num_crimes                 geometry
## 1 Crafty Pig          50 POINT (-2.236 53.48)
```

The bar with the highest number of crimes is still Crafty Pig, but now with 50 crimes. This means there is a difference in the number of crimes attributed to this bar between the two approaches. Clearly there are 9 crimes which fell within the buffer in the first approach, but were closer to another bar in the dataset, and were instead attributed to that one using this approach.

So which is better? This is once again up to you as the researcher and analyst to decide. They do slightly different things, and so will answer slightly different questions. With the nearest feature approach, instead of talking about the number of crimes within some distance to the bar, we are instead talking about, for each crime, the closest venue. This might mean that we could be attributing crimes that happen quite far from the venue to it, just because it's the closest within our dataset. However, we are counting each crime only once. Pros and cons need to be weighed to make decisions like these.

2.5.6 Measuring distances

Let's have a look at this bar called "Crafty Pig". We can select this from the `cc_bars`, buffers, and crimes

```
cp <- cc_bars %>% filter(name == "Crafty Pig")

cp_buffer <- bar_buffer_100 %>% filter(name == "Crafty Pig") %>%
  st_transform(4326) # transform CRS

cp_crimes <- crime_w_bars %>% filter(nearest_bar$name  == "Crafty Pig")
```

We can use `mapply()` and the `st_union()` function to draw a line between each crime and the closest bar (Crafty Pig in this case):

```
dist_lines <- st_sfc( # Create simple feature geometry list column
  mapply(   # apply function to multiple objects
    function(a,b){                    # specify function parametres
      st_cast(st_union(a,b),"LINESTRING") # specify function
      },
    cp_crimes$geometry,  # input a for function
    cp_crimes$nearest_bar$geometry,  # input b for function
    SIMPLIFY=FALSE)) # don't attempt to reduce the result
```

We can then plot these to get an idea of what we're looking at:

```
plot(st_geometry(cp_buffer))
plot(st_geometry(cp), col = "black", add = TRUE)
plot(st_geometry(cp_crimes), col = "blue", add = TRUE)
plot(st_geometry(dist_lines), add = TRUE)
```

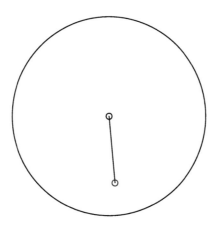

FIGURE 2.10: Crimes near the Noho bar in Manchester

Note: make sure to run all 4 lines above in one batch.

So we can see that all these crimes happened only *one* location, which is within the 100 metre buffer. But how far exactly are they?

You can use the `st_distance()` function to answer this question. We wrap this in the `mutate()` function in order to create a new column called *distance* which will contain for each row (each crime) the distance between that and its nearest bar (in this case Crafty Pig).

```
cp_crimes <- cp_crimes %>%
  mutate(distance = st_distance(geometry, nearest_bar$geometry))
```

Having a look at our newly created variable, we can see that the crime locations are 67.094514757815 away from the Crafty Pig bar.

One thing you might find strange about the data is: why are all these crimes geocoded on top of one another? This is how the open data are released, using geo-masking by snapping crime locations to a geo-mask (a set of points). This is done to ensure anonymity in the data (see Tompson et al. (2015) for more detail on this). In non-anonimysed data, you might expect to see a little less overlap in your crime locations. Then with variation in distances between crimes and their nearest bars, we could use these distances to inform a buffer width for example. Anyway, we will return to distances a little later with a better dataset. But now, let's move on to putting these outcomes on a map, which will help us further investigate the case of the Crafty Pig!

2.6   Plotting interactive maps with leaflet

In Chapter 1, we introduced the `ggplot2` package for making maps in R. In this chapter, we are going to introduce `leaflet` as one way to easily make some neat maps. It is the leading open-source JavaScript library for mobile-friendly interactive maps. It is very popular, used by websites ranging from *The New York Times* and *The Washington Post* to GitHub and Flickr, as well as GIS specialists like OpenStreetMap, Mapbox, and CartoDB. We will also make use of the `RColorBrewer` package.

To make a map, load the `leaflet` and `RColotBrewer` libraries.

```
library(leaflet) #for mapping
library(RColorBrewer) #for getting nice colours for your maps
```

Then create a map with this simple bit of code:

```
m <- leaflet() %>% addTiles()
```

And just print it:

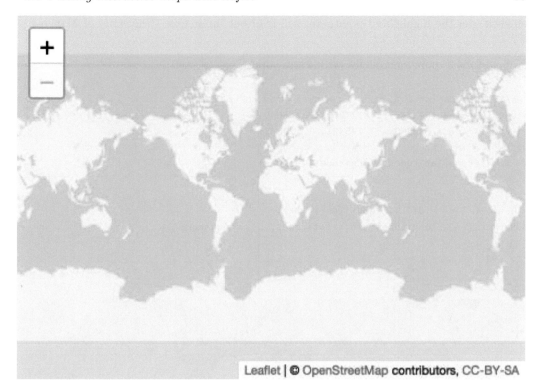

FIGURE 2.11: A first interactive map

It is not a super useful map, but it was really easy to make! You might of course want to add some content to your map.

You can add a point manually:

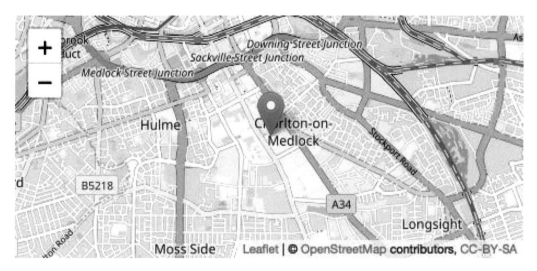

FIGURE 2.12: Mapping the Univeristy of Manchester

If you click over the highlighted point, you will read our input text "University of Manchester".

You can add many points manually, with some popup text as well:

```
# create dataframe of latitude, longitude, and popups
latitudes <- c(53.464987, 53.472726, 53.466649)
longitudes <- c(-2.230899, -2.245481, -2.243421)
popups <- c("You are here", "Here is another point", "Here is another point")
df <- data.frame(latitudes, longitudes, popups)

# create leaflet map
m <- leaflet(data = df) %>% addTiles()  %>%
  addMarkers(lng=~longitudes, lat=~latitudes, popup=~popups)

#print leaflet map
m
```

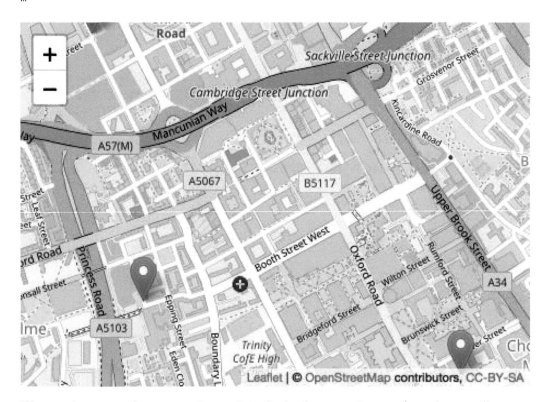

We can also map polygons, not just points. Let's plot our crimes on/near bars to illustrate. To do this, we can return to our buffers where we counted the number of crimes within 100 metres of each bar/ licensed premise (the "crimes_per_prem" object).

First, let's pick a colour palette. We do this with the colorBin() function. We will discuss colour choices in maps in Chapter 5; for now, let's just pick the *"RdPu"* palette. We should also specify the domain = parameter (what value to use for shading, in this case n, bins =, the number of crimes), the number of bins (in this case 5, we will discuss this in detail in the coming chapters as well), and pretty = to use pretty breaks (this may actually mess with the number of bins specified in the bins parameter; but again, for now this is OK).

Let's create this palette and save in an object called `pal` for palette:

```
pal <- colorBin("RdPu", domain = crimes_per_prem$n, bins = 5, pretty = TRUE)
```

Now we can make a leaflet map, where we add these polygons (buffers) with the `addPoly-gons()` function, and call our palette, specifying again the variable to use for shading, as well as some other parametres. One to highlight specifically is the `label` parameter. This allows us to use a variable as a label for when a user clicks on our polygon (buffer). Here we specify the name of the bar with `label = ~as.character(name)`. This way we not only shade each buffer with the number of crimes which fall inside it, but also include a little popup label with the name of the establishment:

```
leaflet(crimes_per_prem) %>%
  addTiles() %>%
  addPolygons(fillColor = ~pal(n), fillOpacity = 0.8,
              weight = 1, opacity = 1, color = "black",
              label = ~as.character(name)) %>%
  addLegend(pal = pal, values = ~n, opacity = 0.7,
            title = 'Violent crimes', position = "bottomleft")
```

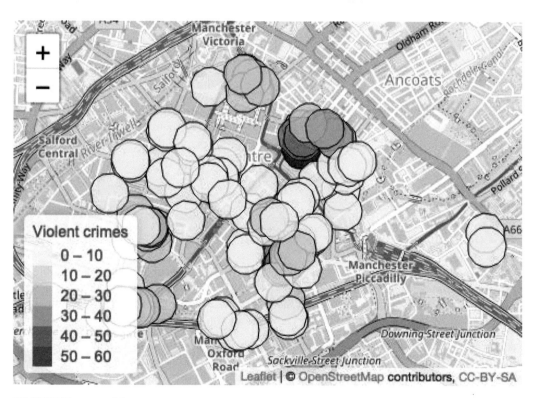

FIGURE 2.13: Mapping crimes around bars in Manchester with leaflet

It's not the neatest of maps, with all these overlaps, but we will talk about prettifying maps further down the line. You can, however, pan and zoom, and investigate to find our most high-crime venue, the Crafty Pig. And here, with this background information, we can solve the puzzle. You see, the Crafty Pig appears to be the nearest venue to an area in Manchester City Centre called *Piccadilly Gardens*[3] which is an area known for high levels of crime and anti social-behaviour. Therefore, it is likely that we are erroneously attributing many of these crimes to the Crafty Pig venue, as they may be taking place in Piccadilly Gardens instead. It is important, therefore, to think about any unintended consequences of the spatial operations we carry out, and how these might affect the conclusions which we draw from our crime mapping exercises.

Finally, let's say we want to save our interactive map, while keeping it interactive. You can do this by clicking on the export button at the top of the plot viewer, and choose the *Save as Webpage* option saving this as a `.html` file:

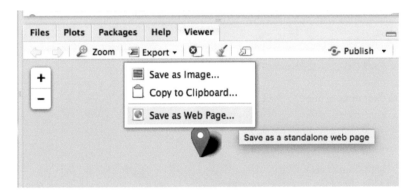

FIGURE 2.14: Save leaflet map as interactive html document

Then you can open this file with any type of web browser (safari, firefox, chrome) and share your map that way. You can send this to your friends, and make them jealous of your fancy map-making skills.

2.7 Geocoding

We were making use of point of interest data from Open Street Map above, but it is possible that we have a dataset of bars from official, administrative data sets, but that are not geocoded. In this case, we may have a list of bars with an associated address, which is clearly *some* sort of spatial information, but how would you put this on a map?

One solution to this problem is to geocode these addresses. We can use the package `tidy-geocoder` to achieve this. This package takes an address given as character values, for example "221B Baker Street, Marylebone, London NW1 6XE" and returns coordinates, geocoding this address. Let's say we have a dataframe of addresses (in this case, only one observation):

```
addresses <- data.frame(name = "Sherlock Holmes",
                        address = "221B Baker Street, London, UK")
```

---

[3]https://en.wikipedia.org/wiki/Piccadilly_Gardens

We can then use the `geocode()` function from the `tidygeocoder` package to get coordinates for this address. We have to specify the column which has the address (in this case *address*), and the method to use for geocoding. See the help file for the function for the many options. For example, if you are in the USA, you may use "census". Since we are global, we will use "osm", which uses nominatim (OSM) to provide worldwide coverage. So, given the above example:

```
addresses %>% geocode(address, method = 'osm')
```

```
## Passing 1 address to the Nominatim single address geocoder

## Query completed in: 1 seconds

## # A tibble: 1 x 4
##   name            address                  lat   long
##   <chr>           <chr>                  <dbl>  <dbl>
## 1 Sherlock Holmes 221B Baker Street, Lond~  51.5 -0.158
```

To illustrate on scale, let's have a look at another source of data of bars. Manchester City Council have an Open Data Catalogue[4] on their website, which you can use to browse through what sort of data they release to the public. Like in many city open data portals, there are some more and some less interesting datasets made available here. It's not quite as impressive as the open data from some of the cities in the US such as New York[5] or Dallas[6], but we'll take it.

One interesting dataset, especially for our questions about the different alcohol outlets, is the Licensed Premises dataset[7]. This details all the currently active licenced premises in Manchester.

There are a few ways we can download this data set. On the manual side of things, we can simply right-click on the download link from the website, save it to your computer, and read it in from there, by specifying the file path. Remember, if you save it in your R *working directory*, then you just need to specify the file name, as the working directory folder is where R will first look for this file.

So do this now. Save the file to your working directory. For me, this will be in a sub-folder called 'data':

```
lic_prem <- read_csv('data/LicensedPremises.csv') %>%
  clean_names()
```

You will likely get some warnings when reading this data, but you can safely ignore those. You can always check if this worked by looking to your global environment on the right-hand side and seeing if this *lic_prem* object has appeared. If it has, you should see it has 65535 observations (rows) and 36 variables (columns).

Let's have a look at what this dataset looks like. You can use the `View()` function for this:

---

[4] http://open.manchester.gov.uk/open/homepage/3/manchester_open_data_catalogue
[5] https://opendata.cityofnewyork.us/
[6] https://www.dallasopendata.com/
[7] http://www.manchester.gov.uk/open/downloads/file/169/licensed_premises

```
View(lic_prem)
```

We see that there is a field for "premisesname" which is the name of the premise, and two fields, "locationtext" and "postcode" which refer to address information. To geocode these, let's create a new column which combines the address and post code, and then use the `geocode()` function introduced above. This will take a while for the whole 65535 addresses dataset; so just for illustration purposes, we take the first 30.

```
lic_prem <- lic_prem %>%
  slice(1:30) %>%    # Select first 30 venues
  # Create new complete_address column from locationtext and postcode using paste()
  mutate(complete_address = paste(locationtext, postcode, sep=", ")) %>%
  geocode(complete_address, method = 'osm')  # geocode with osm method
```

```
## Passing 30 addresses to the Nominatim single address geocoder
```

```
## Query completed in: 30.3 seconds
```

Now we have these licenced premises geocoded, with brand new latitude and longitude information! We can use this to make a leaflet map of our venues!

```
# make sure coordinates are numeric values
lic_prem$latitude <- as.numeric(lic_prem$lat)
lic_prem$longitude <- as.numeric(lic_prem$long)
```

```
# create map
leaflet(data = lic_prem) %>%
  addTiles() %>%
  addMarkers(lng=~longitude, lat=~latitude, popup=~as.character(premisesname),
             label = ~as.character(premisesname))
```

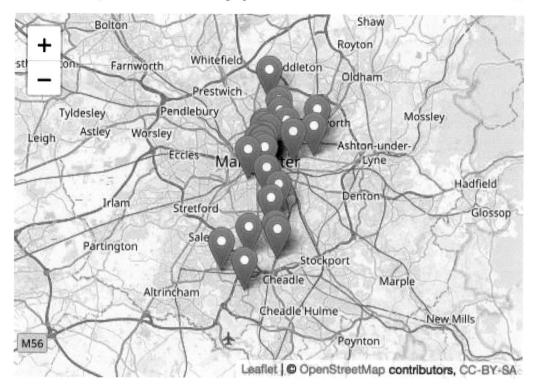

FIGURE 2.15: Map of (some) geocoded licenced premises in Manchester

Geocoding may come in handy when we have address data, or something similar but no geometry to use to map it.

2.8 Measuring distance more thoroughly

Before we end the chapter, we want to return to the spatial operation of measuring the distance between points. This may be important, for example to those researchers focusing on studying the journey to crime by offenders. In this area, a common parameter studied is the average distance to crime from their home locations. In order to estimate these parametres, we first need to have a way to generate the distances. In this section, we will use another data set (this time from Madrid, Spain) to show a simpler example to look at the issue of geographical distance.

2.8.1 How far are police stations in Madrid?

To illustrate how to measure distance, we will download data from the city of Madrid in Spain. Specifically we will obtain a csv file with the latitude and longitude of the police stations and a geoJSON file with the administrative boundary for the city of Madrid. Both are available from the data provided with this book (see *Preamble* section). We will also turn the .csv into a sf object with the appropriate coordinate reference system for this data, following the steps we've outlined in Chapter 1 and earlier in this chapter.

```
#read csv data
comisarias <- read_csv("data/nationalpolice.csv")
#set crs, read into sf object, and assign crs
polCRS <- st_crs(4326)
comisarias_sf <- st_as_sf(comisarias, coords = c("X", "Y"), crs = polCRS)

#create unique id for each row
comisarias_sf$id <- as.numeric(rownames(comisarias_sf))

#Read as sf boundary data for Madrid city
madrid <- st_read("data/madrid.geojson")
```

We went through a lot of steps there, so it's worth to check in and plot our data to make sure that everything looks the way we expect. To practice with leaflet map some more, let's plot using `leaflet()` function:

```
leaflet(comisarias_sf) %>% addTiles() %>%
  addMarkers(data = comisarias_sf) %>%
  addPolygons(data = madrid)
```

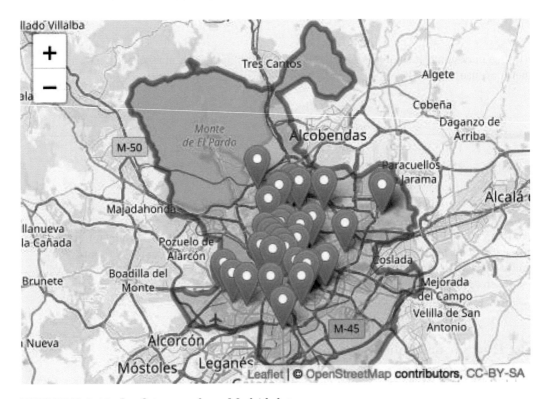

FIGURE 2.16: Leaflet map of our Madrid data

We can clearly see here that there are areas of the municipality that are far away from any national police station, the North West part of the city which you can see is a green area but noticeably also the South East, which is mostly urban and in fact is the location of a known shanty town and open drug market ("Cañada Real", you can read about it in the award-winning book by Briggs and Monge-Gamero (2017)).

2.8.2 Distance in geographical space

There are many definitions of distance in data science and spatial data science. A common definition of distance is the **Euclidean distance**, which simply is the length of a segment connecting two points in a two-dimensional place. Because of the distortions caused by projections on a flat surface, a straight line on a map is not necessarily the shortest distance. Thus, another common definition used in geography is the **great circle distance**, which corresponds to an arc linking two points on a sphere and takes into account the spherical shape of the world. The great circle distance is useful, for example, to evaluate the shortest path when intercontinental distances are concerned. For applications which require a network distance measure, Euclidean distance is generally inadequate. Instead, **Manhattan distance** (also called Taxicab, Rectilinear, or City block metric) is a commonly used metric. We return to dealing with spatial data on networks in Chapter 8.

We can compute both with the `st_distance` function of the `sf` package. This function can be used to measure the distance between two points, between one point and others or between all points. In the latter case we obtain a symmetric matrix of distances ($N \times N$), taken pairwise between the points in our dataset. In the diagonal we find the combinations between the same points giving all null values.

Say, we want to measure the distance between the main police headquarters ("Jefatura Superior de Policia", row 34) and three other stations (say row 1, row 10, and row 25 in our dataset). We could use the following code for that:

```
# calculate distance
dist_headquarters <- st_distance(slice(comisarias_sf, 34),
                        slice(comisarias_sf, c(1, 10, 25)))

dist_headquarters # distance in metres


## Units: [m]
##        [,1] [,2] [,3]
## [1,] 8124 5086 8555
```

The result is a matrix with a single row or column (depending on the order of the spatial objects) with a class of *units*. Often we may want to re-express these distances in a different unit. For this purpose the `units` package offers useful functionality, through the `set_units()` function.

```
set_units(dist_headquarters, "km")


## Units: [km]
##         [,1]  [,2]  [,3]
## [1,] 8.124 5.086 8.555
```

We can compute the distance between all police stations as well.

```
# calculate distance
m_distance <- st_distance(comisarias_sf)

# matrix dimensions
dim(m_distance)
```

```
## [1] 34 34
```

If you want to preview the top of the matrix, you can use:

```
head(m_distance)
```

2.8.3   A practical example to evaluate distance

For this practical example, we will look at the Madrid data. Earlier, we read in the relevant geoJSON file and stored in the object madrid. The CRS is EPSG 4326 and therefore it is geographic projection, with distance expressed in degrees.

```
st_crs(madrid)
```

For this we will transform and reproject to EPSG 2062, which is one of the appropriate projected coordinate system for Madrid[8]. How do you figure out this if you don't want to check the EPSG site? There is a convenient package crsuggest developed by Kyle Walker (2021) that does this job for you. The function suggest_crs() from this package will attempt to list of candidate projections for your sf object. And we will see that top of that list is the 2062 EPSG.

```
suggested_crs <- suggest_crs(madrid)
head(suggested_crs, n = 1)
```

```
## # A tibble: 1 x 6
##   crs_code crs_name        crs_type crs_gcs crs_units
##   <chr>    <chr>           <chr>    <dbl>   <chr>
## 1 2062     Madrid 1870 (Mad~ project~  4903   m
## # ... with 1 more variable: crs_proj4 <chr>
```

Now we can transform our madrid object appropriately:

```
madrid_metres <- st_transform(madrid, crs = 2062)
```

Before we saw that, some areas of Madrid are nowhere near a police station. Let's say we want to get precise about this and we want to know how far different parts of the city of Madrid are from a police station, and we want to be able to show this in a map. Solving this means we have to define "parts of the city". What we will do is to divide the city of

---

[8]https://epsg.io/?q=Spain%20kind%3APROJCRS

Madrid into different cells of 250 metres within a grid using the `st_make_grid` function of
the `sf` package.

```
madrid_grid <- st_make_grid(madrid_metres,  cellsize = 250)

#only extract the points in the limits of Madrid
madrid_grid <- st_intersection(madrid_grid, madrid_metres)
```

We can plot the results, to see that everything has gone according to plan:

```
plot(madrid_grid)
```

FIGURE 2.17: Madrid divided into 250 metres grid cells

With so many "small" cells, the paper printed version of this map may just look completely
black. But it is simply composed of many small cells. So how do we look at distance from
police stations here? We can measure the distance between each grid cell to all 39 police
stations. To estimate the distance to the nearest police station, we will find the minimum
distance value for each grid, i.e., the distance to the nearest station.

```
comisarias_sf_metres <- st_transform(comisarias_sf, crs = 2062)

distances <- st_distance(comisarias_sf_metres,
                         st_centroid(madrid_grid)) %>%
  as_tibble()
```

If you view the new object "distances", you will see there is a row for each police station
and a column representing each of the 10082 cells in our grid. For using these distances in a
leaflet map, we will project back into 4326. And then we will compute the shortest distance
for each cell.

```
# Compute distances
police_distances <- data.frame(
```

```
# We want grids in a WGS 84 CRS:
us = st_transform(madrid_grid, crs = 4326),
# Extract minimum distance for each grid
distance_km = map_dbl(distances, min)/1000,
# Extract the value's index for joining with the location info
location_id = map_dbl(distances, function(x) match(min(x), x))) %>%
# Join with the police station table
left_join(comisarias_sf, by = c("location_id" = "id"))
```

We now have a dataframe of distances, for each grid, to the nearest police station. We can have a look at the distribution of these distances through plotting them on a histogram

```
# Plot and examine distances

hist(police_distances$distance_km, main = 'Distance to nearest police station')
```

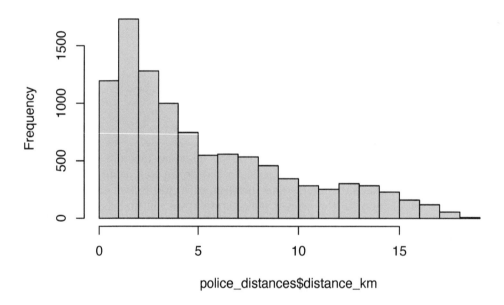

FIGURE 2.18: Histogram of distance to nearest police station

Now we are ready to use this data to plot a map. We can get creative, and use the `makeIcon()` function from leaflet to assign an image as our icon, rather than these regular pointer icons we've been seeing. To do this, first we will adjust some aesthetics.

We get the URL for the icon (we break into part 1 (_pt1) and part 2 (_pt2) so it can be seen in the textbook), and pass this to the `makeIcon()` function.

```
# Create more appropriate icon, taking it from Wikipedia commons
icon_url_pt1 <- "https://upload.wikimedia.org/wikipedia/commons/"
icon_url_pt2 <- "a/ad/189-woman-police-officer-1.svg"

# create icon, adjusting size
police_icon <- makeIcon(paste0(icon_url_pt1, icon_url_pt2),
                        iconWidth = 12, iconHeight = 20)
```

Now we also want to create a colour scale. We can, for example, group our grids into quantiles. To do so, we can use the `quantile()` function:

```
# Bin ranges for a nicer colour scale
bins <- quantile(police_distances$distance_km)
# Create a binned colour palette
pal <- colorBin(c("#0868AC", "#43A2CA", "#7BCCC4", "#BAE4BC", "#F0F9E8"),
                domain = police_distances$distance_km,
                bins = bins, reverse = TRUE)
```

Now let's create the map.

```
full_map <- leaflet() %>%
  addTiles() %>%
  addMarkers(data = comisarias_sf, icon = ~police_icon,
             group = "Police stations") %>%
  addPolygons(data = police_distances[[1]],
              fillColor = pal(police_distances$distance_km),
              fillOpacity = 0.8, weight =0, opacity = 1, color = "transparent",
              group = "Distances",
              highlight = highlightOptions(weight = 2.5, color = "#666",
                                           bringToFront = TRUE, opacity= 1),
              popupOptions = popupOptions(autoPan = FALSE, closeOnClick = TRUE,
                                          textOnly = T)) %>%
  addLegend(pal = pal, values = (police_distances$distance_km),
            opacity = 0.8, title = "Distance (Km)", position= "bottomright")

# print the map
full_map
```

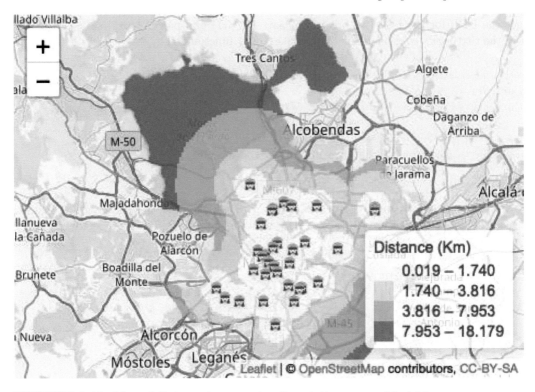

FIGURE 2.19: Map of distance to nearest police station across Madrid

And there you go. Just remember something. It is easy to misinterpret data and maps. You always need to care a great deal about measurement, quality of your data, and other potential issues affecting interpretation. When it comes to distance, and the movements of people and law enforcement personnel, for example, physical distance is not trivial, but time to arrival is also important and this is determined by factors other than Euclidean distance (e.g., availability and speed of transport, physical barriers, etc.). Our representation is always as good as the data we have. In Spain there are two other police forces (Guardia Civil, that patrols rural areas, and municipal civil, with jurisdiction for local administrative enforcement) that we are not representing here (that is, our data is incomplete). And we are not plotting the police stations in the nearby municipalities that are part of Madrid metropolitan area, around the edges.

2.9   Summary and further reading

In this chapter we explored the differences between attribute and spatial operations, and we made great use of the latter in practicing **spatial manipulation of data**. These spatial operations allow us to manipulate our data in a way that lets us study what is going on with crime at micro-places. At the start of the chapter, we introduced the idea of different crime places. To read further about crime attractors vs. crime generators, turn to the recommended readings by Brantingham and Brantingham (1995) and Newton (2018). There have since been more developments, for example about crime radiators and absorbers as well (see Bowers (2021) to learn more).

We covered how to subset points within an area, build buffers around geometries, count the number of points in a point layer that fall within each polygon in a polygon layer, find the nearest feature to a set of features, and turn non-spatial information such as an address into something we can map using geocoding. These spatial operations are just some of many, but we chose to cover as they have the most frequent application in crime analysis.

For those interested to learn more, spatial operations are typically discussed in standard GIS textbooks, such as those we have recommended in previous chapters. You could see, for example, chapter 9 of Bolstad (2019). But probably the best follow-up to what we discuss here is chapter 4 and 5 of Lovelace, Nowosad, and Muenchow (2019), for it provides a systematic introduction to how to perform these spatial operations with R. There is an online book in development by two giants of the R spatial community, Edzer Pebesma and Roger Bivand, which at the time of writing is best cited as Pebesma and Bivand (2021). The first chapters of the book provide a strong backbone to understand sf objects in greater detail, coordinate systems, and key concepts for spatial data science.

3

Mapping ratcs and counts

3.1 Introduction

In Chapter 1 we showed you fairly quickly how to create maps by understanding how data may have spatial elements, and how that can be linked to geometries. In this chapter instead we will get to know how to think about thematic maps, and how to apply your learning to creating your own maps of this variety. In the process we will discuss various types of thematic maps and the issues they raise.

Thematic maps focus on representing the spatial pattern of a variable of interest (e.g., crimes, trust in the police, etc.) and they can be used for exploration and analysis or for presentation and communication to others. There are different types of thematic maps depending on how the variable of interest is represented. In this and the next chapters, we will introduce some of these types of particular interest for crime analysis, the different challenges they pose, and some ideas that may help you to choose the best representation for your data. Critically, we need to think about the quality of the data we work with, for adequate measurement is the basis of any data analysis.

We will introduce in particular two common types of thematic maps used for mapping quantitative variables: **choropleth (thematic) maps** and **proportional symbol maps**. In previous chapters we introduced two separate packages for creating maps in R (`ggplot2` and `leaflet`). Here we will introduce a third package for creating maps: `tmap`. The libraries we use in this chapter are as follow:

```
# Packages for reading data and data carpentry
library(readr)
library(dplyr)
library(janitor)

# Packages for data exploration
library(skimr)

# Packages for handling spatial data and for geospatial carpentry
library(sf)

# Packages for mapping and visualisation
library(tmap)
library(tmaptools)

# Packages for spatial analysis
```

```
library(DCluster)

# Packages with spatial datasets
library(geodaData)
```

3.2 Thematic maps: key terms and ideas

3.2.1 Choropleth maps

Choropleth maps display variation in areas (postal codes, police districts, census adminis-trative units, municipal boundaries, regional areas, etc.) through the use of the colour that fills each of these areas in the map. A simple case can be to use a light-to-dark colour to represent less to more of the quantitative variable. They are appropriate to compare values across areas.

Most thematic maps you encounter are classified choropleth maps. They group the values of the quantitative variable into a number of classes, typically between 5 and 7. We will return to this later in the chapter. It is one of the most common forms of statistical maps, and like pie charts, they are subject to ongoing criticism. Tukey (1979) concluded they simply do not work. And there are indeed a number of known problems with choropleth maps:

1. They may not display variation within the geographical units being employed. You are imposing some degree of distortion by assuming all parts of an area display the same value. Our crime map may show a neighbourhood as secure, when there is a part of this neighbourhood that has a high level of crime, and vice versa.

2. Boundaries of geographical areas are to a large extent arbitrary (and unlikely to be associated with major discontinuities in your variable of interest). In crime analysis we very often use census administrative units, but these rarely represent natural neighbourhoods.

3. They work better if areas are of similar size. Areas of greater size may be more heterogeneous internally than those of smaller size; that is, they potentially have the largest error of representation. Also, visual attention may be drawn by areas that are large (if size is not the variable used to create a ratio).

3.2.2 What do we use choropleth maps for?

3.2.2.1 Crime rates

In this section we discuss crime rates, for they are often the key quantity represented in this kind of maps. In a choropleth map, you "can" show raw totals (absolute values), for example, number of crimes, or derived values (ratios), such as crime per 100,000 inhabitants. But as a general rule, you should restrict choropleth maps to show derived variables such as rates or percentages (Field 2018). This is because the areas we often use are of different size and this may introduce an element of confusion in the interpretation of choropleth maps. The size of, say, a province or a county, "has a big effect on the amount of colour shown on the map, but unit area may have little relationship, or even an inverse relationship, to base populations and related counts." (Brewer 2006: S30).

Mapping counts, that is, mapping where a lot of the crime incidents concentrate is helpful, but we may want to understand as well if this is simply a function of the spatial distribution of the population at risk. As Ratcliffe (2010) suggests, "practitioners often recognize that a substantial density of crime in a location is sufficient information to initiate a more detailed analysis of the problem", but equally we may want to know if "this clustering of crime is meaningfully non-random, and if the patterns observed are still present once the analysis has controlled for the population at risk" (pp. 11-12).

A map of rates essentially aims to provide information into geographic variation of crime risk, understood as the probability that a crime may occur. Maps of rates are ultimately about communicating the risk of crime, with greater rates suggesting a higher probability of becoming a victim of crime.

In social science, the denominator on mapped ratios is typically some form of population size, which typically we will want as current as possible. In other fields, some measure of area size may be a preferred choice for the denominator. However, a great deal of discussion in crime analysis has focused on the choice of the right denominator. Authors relate to this as the **denominator dilemma**: "the problem associated with identifying an appropriate target availability control that can overcome issues of spatial inequality in the areal units used to study crime" (Ratcliffe 2010). The best measure for your denominator is one which captures opportunities. If for example you are interested in residential burglary, it makes sense to use number of inhabited households as your denominator (rather than population size). Whatever denominator you choose, you will usually want to make a case as to why that is the best representation of the opportunities for the crime type you're interested in.

As noted, population is a common choice, but it is not always the one that best captures crime opportunities. Population is also highly mobile during the day. People do not stay in the areas where they live: they go to work, they go to school, they travel for tourism purposes, and in doing so they alter the population structure of any given area. As Ratcliffe (2010) highlights with an example "the residential population (as is usually available from the census) tells the researcher little about the real number of people outside nightclubs at 2 am". Geographers and criminologists, thus, distinguish between the standard measures of population that relate to people that live in an area provided by the census and government statistical authorities, and the so-called **ambient population** that relates to people that occupy an area at a given time and typically are a bit more difficult to source (see for example: Andresen (2011)).

As we will see later, one of the key problems with mapping rates is that the estimated rates can be problematic if the enumeration areas we are studying have different population counts (or whatever it is we are counting in the denominator). When this happens, and those population counts produce small samples, we may have rates for some locations (those with more population) that are better estimated than others and, therefore, are less subject to noise.

Aside from problems with the denominator, we may also have problems with the numerator. In crime analysis, the key variable we map is geocoded crime, that is crime for which we know its exact location. The source for this variable tends to be crime reported to and recorded by the police. Yet, we know this is an imperfect source of data. A very large proportion of crime is not reported to the police. In England and Wales, for example, it is estimated that around 60% of crime is unknown to the police. And this is an average! The unknown figure of crime is larger for certain types of crime (e.g., interpersonal violence, sexual abuse, fraud, etc.). What is more, we know that there are community-level attributes that are associated with the level of crime *reporting* to the police (Goudriaan, Witterbrood, and Nieuwbeerta 2006).

Equally, not all police forces or units are adept at properly *recording* all crimes reported to the police. An in-depth study conducted in England and Wales concluded that over 800,000 crimes reported to the police were not properly recorded (an under-recording of 19%, which is higher for violence and sexual abuse, 33% and 26% respectively) (Her Majesty Inspectorate of Constabulary 2014). There are, indeed, institutional, economic, cultural and political factors that may shape the quality of crime recording across different parts of the area we want to map out. Although all this has been known for a while, criminologists and crime analysts are only now beginning to appreciate how this can affect the quality of crime mapping and the decision making based on this mapping (Buil-Gil, Medina, and Shlomo 2021; Hart and Zandbergen 2012).

Finally, the quality of the geocoding process, which varies across crime type, also comes into the equation. Sometimes, there are issues with positional accuracy or the inability to geocode an address. Some authors suggest we need to be able to geocode at least 85% of the crime incidents to get accurate maps (Ratcliffe 2004); otherwise, geocoded crime records may be spatially biased. More recent studies offer a less conservative estimate depending on the level of analysis and number of incidents (Andresen et al. 2020); although some authors, on the contrary, argue for a higher hit rate and consider that hot spot detection techniques are very sensitive to the presence of non-geocoded data (Briz-Redón, Martínez-Ruiz, and Montes 2020b).

Although most of the time crime analysts are primarily concerned with mapping crime incidence and those factors associated with it, increasingly we see interest in the spatial representation of other variables of interest to criminologists such as fear of crime, trust in the police, and more. In these cases, the data may come from surveys and the problem that may arise is whether we have a sample size large enough to derive estimates at the geographical level that we may want to work with. When this is not the case, methods for small area estimation are required (for details and criminological examples see Buil-Gil et al. (2019)). There has also been recent work mapping perception of place and fear of crime at micro-levels (see Solymosi, Bowers, and Fujiyama (2015) and Solymosi et al. (2020)).

3.2.2.2   Prevalence vs. incidence

There are many concepts in science that acquire multiple and confusing meanings. Two you surely will come across when thinking about rates are: incidence and prevalence. These are well defined in epidemiology. Criminology and epidemiology often use similar tools and concepts, but not in this case. In criminological applications, these terms often are understood differently and in, at least, two possible ways. Confused? You should be!

In public health and epidemiology, prevalence refers to proportion of persons who have a condition at or during a particular time period; whereas incidence refers to the proportion or rate of persons who develop a condition during a particular time period. The numerator for prevalence is *all* cases during a given time period that have the condition; whereas the numerator for incidence is all *new* cases. What changes is the numerator and the key dimension in which it changes is time.

In criminology, on the other hand, you will find at least two ways of defining these terms. Those that focus on studying developmental criminology define prevalence as the percentage of a population that engages in crime during a specified period (number of offenders per population in a given time); while offending incidence refers to the frequency of offending among those criminally active during that period (number of offences per active offenders in a given time). For these criminologists the (total) crime rate in a population is the product of the prevalence and the incidence (Blumstein et al. 1986).

Confusingly, though, you will find authors and practitioners that consider incidence as equivalent to the (total) crime rate, as the number of crimes per population, and prevalence, as the number of victims per population. To make things more confusing, sometimes you see criminologists defining incidence as the number of crimes during a time period (say, a year) and prevalence as the number of victims during the lifetime. To avoid this confusion when producing maps, *is probably best to avoid these terms and simply refer to crime rate or victimisation rate and be very clear in your legends about the time period covered for both.*

3.2.3 Proportional and graduate symbol maps: mapping crime counts

Proportional symbol maps, on the other hand, are used to represent quantitative variables for either areas or point locations. Each area gets a symbol and the size represents the intensity of the variable. It is expected with this type of map that the reader could estimate the different quantities mapped out and to be able to detect patterns across the map (Field 2018). The symbols typically will be squares or circles that are scaled "in proportion to the square root of each data value so that symbol areas visually represent the data values" (Brewer 2006: S29). Circles are generally used since they are considered to perform better in facilitating visual interpretation.

We often use proportional symbol maps to represent count data (e.g., number of crimes reported in a given area). A common problem with proportional symbol maps is symbol congestion/overlap, especially if there are large variations in the size of symbols or if numerous data locations are close together.

A similar type of map uses graduated symbol maps. As with choropleth classing, the symbol size may represent data ranges. They are sometimes used when data ranges are too great to practically represent the full range on a small map.

3.3 Creating proportional symbol maps

In this chapter we are going to introduce the `tmap` package. This package was developed to easily produce thematic maps. It is inspired by the `ggplot2` package and the layered grammar of graphics. It was written by Martjin Tennekes, a Dutch data scientist. It is fairly user friendly and intuitive. To read more about `tmap`, see Tennekes (2018).

In tmap each map can be plotted as a static map (*plot mode*) or shown interactively (*view mode*). We will start by focusing on static maps.

For the purpose of demonstrating some of the functionality of this package, we will use the Manchester crime data generated in the first chapter. We have now added a few variables obtained from the Census that will be handy later on. You can obtain the data in a GeoJSON format from the companion data to this book (see *Preamble* if you still need to download this data!). Now let's import the data, saved in the file manchester.geojson using the st_read() function from the sf package.

```
manchester <- st_read("data/manchester.geojson", quiet=TRUE)
```

Every time you use the tmap package you will need a line of code that specifies the spatial object you will be using. Although originally developed to handle sp objects only, it now also has support for sf objects. For specifying the spatial object, we use the tm_shape() function and inside we specify the name of the spatial object we are using. Its key function is simply to do just that, to identify our spatial data. On its own, this will do nothing apparent. No map will be created. We need to add functions to specify what we are doing with that spatial object, how we want to represent it. If you try to run this line on its own, you'll get an error that you must "Specify at least one layer after each tm_shape" (or it might say "Error: no layer elements defined after tm_shape" depending on your R version).

```
tm_shape(manchester)
```

The main plotting method consists of elements that we can add. The first element is the tm_shape() function specifying the spatial object, and then we can add a series of elements specifying layers in the visualisation. They can include polygons, symbols, polylines, raster, and text labels as base layers.

In tmap you can use tm_symbols for this. As noted, with tmap you can produce both static and interactive maps. The interactive maps rely on leaflet. You can control whether the map is static or interactive with the tmap_mode() function. If you want a static map, you pass plot as an argument; if you want an interactive map, you pass view as an argument.

Let's create a static map first:

```
tmap_mode("plot") # specify we want static map

tm_shape(manchester) +
  tm_bubbles("count")    # add a 'bubbles' geometry layer
```

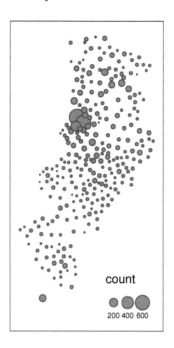

FIGURE 3.1: Bubble map of crime in each LSOA in Manchester

In the map produced, each bubble represents the number of crimes in each LSOA within the city of Manchester. We can add the borders of the census areas for better representation. The `border.lwd` argument set to NA in the `tm_bubbles()` is asking R not to draw a border to the circles. Whereas `tm_borders()` brings back a layer with the borders of the polygons representing the different LSOAs in Manchester city. Notice how we are modifying the transparency of the borders with the alpha parameter. In addition, we are adding a `tm_layout()` function that makes explicit where and how we want the legend.

```
#use tm_shape function to specify spatial object
tm_shape(manchester) +
  #use tm_bubbles to add the bubble visualisation,
  # but set the 'border.lwd' parameter to NA,
  # meaning no symbol borders are drawn
  tm_bubbles("count", border.lwd=NA) +
  #add the LSOA border outlines using tm_borders,
  # but set their transparency using the alpha parameter
  # (0 is totally transparent, 1 is not at all)
  tm_borders(alpha=0.1) +
  #use tm_layout to make the legend look nice
  tm_layout(legend.position = c("right", "bottom"),
            legend.title.size = 0.8, legend.text.size = 0.5)
```

```
## The legend is too narrow to place all symbol sizes.
```

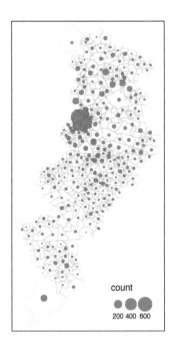

FIGURE 3.2: Bubble map with aesthetic settings

There are several arguments that you can pass within the `tm_bubble()` function that control the appearance of the symbols. The `scale` argument controls the symbol size multiplier number. Larger values will represent the largest bubble as larger. You can experiment with the value you find more appropriate. Another helpful parameter in `tm_bubble()` is `alpha`, which you can use to make the symbols more or less transparent as a possible way to deal with situations where you may have a significant degree of overlapping between the symbols. You can play around with this and modify the code we provide above with different values for these two arguments.

By default, the symbol area sizes are scaled proportionally to the data variable you are using. As noted above, this is done by taking the square root of the normalized data variable. This is called mathematical scaling. However, "it is well known that the perceived area of proportional symbols does not match their mathematical area; rather, we are inclined to underestimate the area of larger symbols. As a solution to this problem, it is reasonable to modify the area of larger circles in order to match it with the perceived area" (Tanimura, Kuroiwa, and Mizota 2006). The `perceptual = TRUE` option allows you to use a method that aims to compensate for how the default approach may underestimate the area of the larger symbols. Notice the difference between the map we produced and this new one.

```
tm_shape(manchester) +
  tm_bubbles("count", border.lwd=NA, perceptual = TRUE) + # add perceptual = TRUE
  tm_borders(alpha=0.1) +
  tm_layout(legend.position = c("right", "bottom"),
            legend.title.size = 0.8, legend.text.size = 0.5)
```

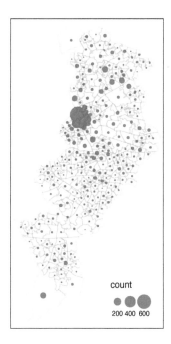

FIGURE 3.3: Bubble map with perceptual parameter set to TRUE

We can see that we get much larger bubbles in the City Centre region of Manchester local authority. There is a chance that this is due to a greater number of people present in this area. Unless we control for this, by mapping rates, we may not be able to meaningfully tell the difference.

3.4 Mapping rates rather than counts

3.4.1 Generating the rates

We have now seen the importance of mapping rates rather than counts of things, and that is for the simple reason that population is not equally distributed in space. That means that if we do not account for how many people are somewhere, we end up mapping population size rather than our topic of interest.

The manchester object we are working with has a column named respop that includes the residential population in each of the LSOA areas. These areas for the whole of the UK have an average population of around 1500 people. We can see a similar picture for Manchester city, with an average closer to 1700 here.

```
hist(manchester$respop, main = "Distribution of residential population across LSOAs")
```

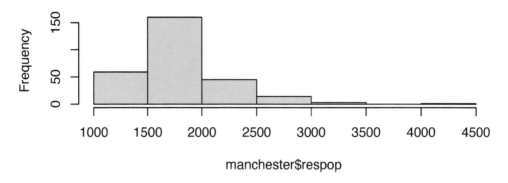

FIGURE 3.4: Histogram of respop variable

We also have a variable `wkdpop` that represents the workday population. This variable re-distributes the usually resident population to their places of work, while those not at work are recorded at their usual residence. The picture it offers is much more diverse than the previous variable and in some ways is closer to the notion of an "ambient population" (with some caveats: for example, it does not capture tourists, people visiting a place to shop or to entertain themselves in the night time economy, etc.).

```
hist(manchester$wkdpop, main = "Distribution of workday population across LSOAs")
```

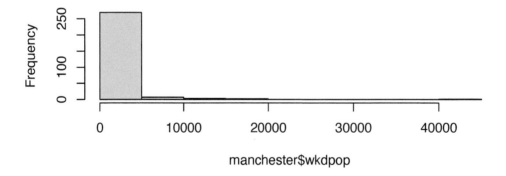

FIGURE 3.5: Histogram of wkdpop variable

The distribution of the workday population is much more skewed than the residential population, with most LSOAs having only 5,000 or fewer workday population, but the busiest LSOAs attracting over 40,000 people during the work day. This gives an idea of the relevance of the "denominator dilemma" we mentioned earlier. In this section we will create rates of crime using both variables in the denominator to observe how the emerging picture varies.

To create new variables, we can use the `mutate()` function from the `dplyr` package. This is a very helpful function to create new variables in a dataframe based on transformations or mathematical operations performed on other variables within the dataframe. In this function, the first argument is the name of the data frame, and then we can pass as arguments all new variables we want to create as well as the instructions as to how we are creating those variables.

First, we want to create a rate using the usual residents, since crime rates are often expressed by 100,000 inhabitants we will multiply the division of the number of crimes by the number of usual residents by 100,000. We will then create another variable, `crimr2`, using the workday population as the denominator. We will store this new variables in our existing `manchester` dataset. You can see that below we then specify the name of a new variable `crimr1` and then tell the function we want that variable to equal (for each case) the ratio of the values in the variable `count` (number of crimes) by the variable `respop` (number of people residing in the area). We then multiply the result of this division by 100,000 to obtain a rate expressed in those terms. We can do likewise for the alternative measure of crime.

```
manchester <- mutate(manchester,
                # crime per residential population
                crimr1 = (count/respop)*100000,
                # crime per workday population
                crimr2 = (count/wkdpop)*100000)
```

And now we have two new variables: one for crime rate with residential population as a denominator, and another with workplace population as a denominator.

3.4.2   Creating a choropleth map with tmap

The structure of the grammar for producing a choropleth map is similar to what we use for proportional symbols. First, we identify the object with `tm_shape()` and then we use a geometry to be represented. We will be using the `tm_polygons()` passing as an argument the name of the variable with our crime rate.

```
tm_shape(manchester) +
  tm_polygons("crimr1")
```

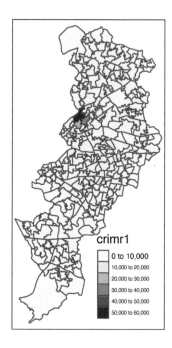

FIGURE 3.6: Crime rate per residential population

We have used `tm_polygons()` but we can also add the elements of a polygon map using different functions that break down what we represent here. In the map above, you see the polygons have a dual representation, the borders are represented by lines and the colour is mapped to the intensity of the quantitative variable we are displaying. With darker colours representing more of the variable, the areas with more crimes.

Instead of using `tm_polygon()`, we can use the related functions `tm_fill()`, for the colour inside the polygons, and `tm_borders()`, for the aesthetics representing the border of the polygons. Say, we find the borders distracting and we want to set them to be transparent. In that case we could just use `tm_fill()`.

```
tm_shape(manchester) +
  tm_fill("crimr1")
```

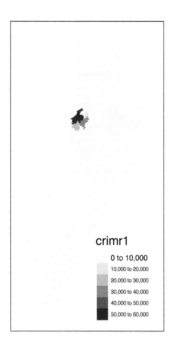

FIGURE 3.7: Crime rate per residential population mapped with tm_fill() function

As you can see here, the look is a bit cleaner. However, we don't need to get rid of the borders completely. Perhaps we want to make them a bit more translucent. We could do that by adding the border element but making the drawing of the borders less pronounced.

```
tm_shape(manchester) +
  tm_fill("crimr1") +
  tm_borders(alpha = 0.1)
```

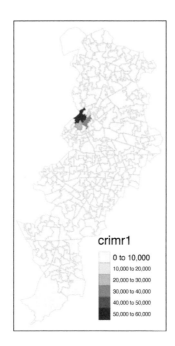

FIGURE 3.8: Crime rate per residential population mapped with tm_fill() and borders

The alpha parameter that we are inserting within `tm_borders()` controls the transparency of the borders, we can go from 0 (totally transparent) to 1 (not transparent). You can play around with this value and see the results.

Notice as well that the legend in this map is not very informative and could be improved in terms of aesthetics. We can add a title within the `tm_fill` to clarify what "crimr1" means, and we can use the `tm_layout()` function to control the appearance of the legend. This latter function, `tm_layout()`, allows you to think about many of the more general cosmetics of the map.

```
tm_shape(manchester) +
  tm_fill("crimr1", title = "Crime per 100,000 residents") +
  tm_borders(alpha = 0.1) +
  tm_layout(main.title = "Crime in Manchester City, Nov/2017",
            main.title.size = 0.7 ,
            legend.outside = TRUE,  # Place legend outside of map
            legend.title.size = 0.8)
```

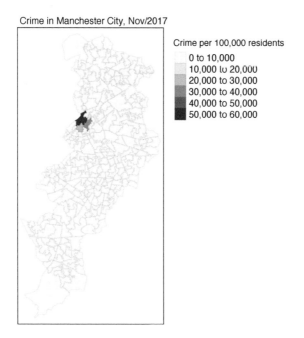

FIGURE 3.9: Crime rate per residential population mapped with aesthetics considered

We can also change the style of the maps we produce, for example by making them more friendly to colour-blind people. We can use the `tmap_style()` function to do so. Once you set the style, the subsequent maps will stick to this style. If you want to reverse to a different style when using `tmap`, you need to do so explicitly.

```
current_style <- tmap_style("col_blind")
```

See how the map changes.

```
tm_shape(manchester) +
  tm_fill("crimr1", title = "Crime per 100,000 residents") +
  tm_borders(alpha = 0.1) +
  tm_layout(main.title = "Crime in Manchester City, Nov/2017",
            main.title.size = 0.7 ,
            legend.outside = TRUE,  # Takes the legend outside the main map
            legend.title.size = 0.8,
            )
```

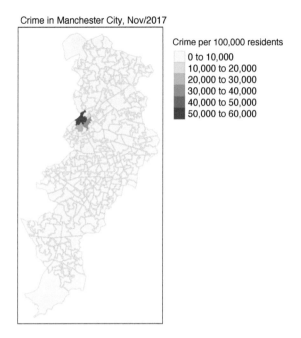

Crime in Manchester City, Nov/2017

FIGURE 3.10: Crime rate per residential population with col_blind style

We will discuss more about map aesthetics in Chapter 5. For now, let's talk about another important decision when it comes to thematic maps, the choice of classification system.

3.5   Classification systems for thematic maps

In thematic mapping, you have to make some key decisions, the most important one being how to display your data. When mapping a quantitative variable, we often have to "bin" this variable into groups. For example in the map we made below, the default binning applied was to display LSOAs grouped into those with a number of 0 to 10,000 crimes per 100,000 residents, then from 10,000 to 20,000, and so on. But why these? How were these groupings decided upon?

The quantitative information is usually classified before its symbolization in a thematic map. Theoretically, accurate classes that best reflect the distributional character of the data set can be calculated. There are different ways of breaking down your data into classes. We will discuss each in turn here.

The equal interval (or equal step) classification method divides the range of attribute values into equally sized classes. What this means is that the values are divided into equal groups. Equal interval data classification subtracts the maximum value from minimum value in your plotted variable and then divides this by the number of classes to get the size of the intervals. This approach is best for continuous data. The range of the classes is structured so that it covers the same number of values in the plotted variable. With highly skewed variables, this plotting method will focus our attention in areas with extreme values. When mapping crime, it will create the impression that everywhere is safe, save a few locations. The graphic below shows five classes and the function shows the locations (represented by points) that fall within each class.

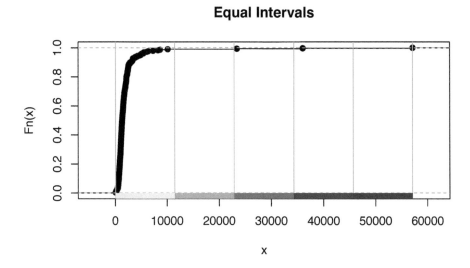

FIGURE 3.11: Results for equal intervals classification

The quantile map bins the same count of features (areas) into each of its classes. This classification method places equal numbers of observations into each class. So, if you have five classes you will have twenty percent of the areas within each class. This method is best for data that is evenly distributed across its range. Here, given that our variable is highly skewed, notice how the top class includes areas with fairly different levels of crime, even if it does a better job at separating areas at the lower end of the crime rate continuum. Several authors consider that quantile maps are inappropriate to represent the skewed distributions of crime (Harries 1999).

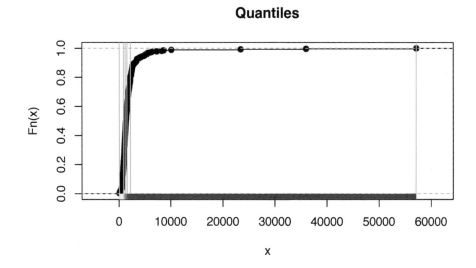

FIGURE 3.12: Results for quantiles classification

The natural breaks (or Jenks) classification method utilizes an algorithm to group values in classes that are separated by distinct break points. It is an optimisation method which takes an iterative approach to its groupings to achieve least variation within each class. Cartographers recommend to use this method with data that is unevenly distributed but not skewed toward either end of the distribution.

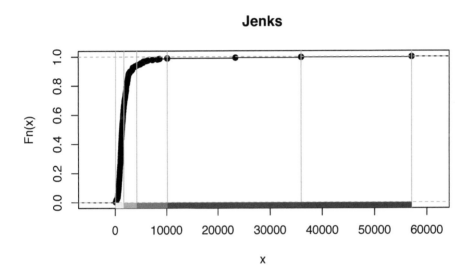

FIGURE 3.13: Results for Jenks classification

The standard deviation map uses the standard deviation (standardised measure of observations' deviation from the mean) to bin the observations into classes. This classification method forms each class by adding and subtracting the standard deviation from the mean of the dataset. It is best suited to be used with data that conforms to a normal distribution.

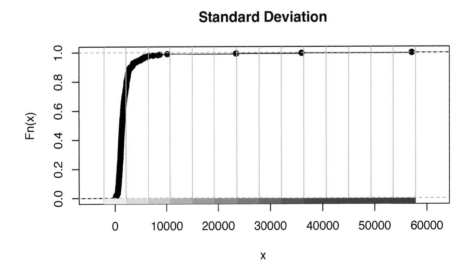

FIGURE 3.14: Results for standard deviation classification

The headtails method This uses a method proposed by Jiang (2013) as a solution for variables with a heavy tail distribution, like we often have with crime. "This new classification scheme partitions all of the data values around the mean into two parts and continues the process iteratively for the values (above the mean) in the head until the head part values are no longer heavy-tailed distributed. Thus, the number of classes and the class intervals are both naturally determined" (Jiang 2013, 482).

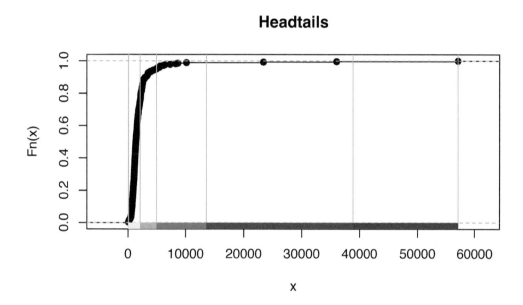

FIGURE 3.15: Results for headtails classification

Not only do you need to choose the classification method, you also need to decide on the number of classes. This is critical. The convention in cartography is to choose between 5 to 7 classes, although some authors would say 5 to 6 (Field 2018). Less than five and you loose detail, more than 6 or 7 and the audience looking at your map starts to have problems to perceive the differences in the symbology and to understand the spatial pattern displayed.

If you want to lie with a map, it would be very easy by using a data classification scheme that conveys the message that you want to get across. It is, thus, important that your decisions are based on good practice and are impartial.

For comparing the effects of using different methods, we can use **small multiples**. Small multiples is simply a way of reproducing side by sides similar maps for comparative purposes. To be more precise, small multiples are *sets of charts of the same type, with the same scale, presented together at a small size and with minimal detail, usually in a grid of some kind*. The term was popularized by Edward Tufte, appearing first in his *Visual Display of Quantitative Information* in 1983 (Tufte 2001).

There are different ways of creating small multiples with tmap as you could see in the vignettes for the package, some of which are quicker but a bit more restricted. Here we are going to use tmap_arrange(). With tmap_arrange() first we need to create the maps we want and then we arrange them together.

Let's make five maps, each one using a different classification method: Equal Interval, Quantile, Natural Breaks (Jenks), Standard Deviation, and Headtails. For each map, instead of visualising them one-by-one, just assign them to a new object. Let's call them `map1`, `map2`, `map3`, `map4`, `map5`. So let's make `map1`. This will create a thematic map using equal intervals.

```
map1 <- tm_shape(manchester) +
  #Use tm_fill to specify variable, classification method,
  # and give the map a title
  tm_fill("crimr1", style="equal", title = "Equal") +
  tm_layout(legend.position = c("left", "top"),
            legend.title.size = 0.7, legend.text.size = 0.5)
```

Now create `map2`, with the jenks method often preferred by geographers.

```
map2 <- tm_shape(manchester) +
  tm_fill("crimr1", style="jenks", title = "Jenks") +
  tm_layout(legend.position = c("left", "top"),
            legend.title.size = 0.7, legend.text.size = 0.5)
```

Now create `map3`, with the quantile method often preferred by epidemiologists.

```
map3 <- tm_shape(manchester) +
  tm_fill("crimr1", style="quantile", title = "Quantile") +
  tm_layout(legend.position = c("left", "top"),
            legend.title.size = 0.7, legend.text.size = 0.5)
```

Let's make `map4`, standard deviation map, which maps the values of our variable to distance to the mean value.

```
map4 <- tm_shape(manchester) +
  tm_fill("crimr1", style="sd", title = "Standard Deviation") +
  tm_layout(legend.position = c("left", "top"),
            legend.title.size = 0.7, legend.text.size = 0.5)
```

And finally let's make `map5`, which is handy with skewed distributions.

```
map5 <- tm_shape(manchester) +
  tm_fill("crimr1", style="headtails", title = "Headtails") +
  tm_layout(legend.position = c("left", "top"),
            legend.title.size = 0.7, legend.text.size = 0.5)
```

Notice that we are not plotting the maps, we are storing them into R objects (map1 to map5). This way they are saved, and you can call them later, which is what we need in order to plot them together using the `tmap_arrange()` function. So if you wanted to map just `map3` for example, all you need to do, is call the `map3` object. Like so:

```
map3
```

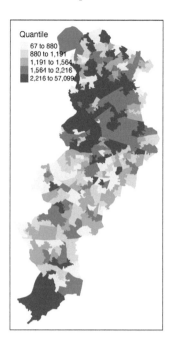

FIGURE 3.16: Quantile classification of crime in Manchester

But now we will plot all maps together, arranged using the `tmap_arrange()` function. Like so:

```
# deploy tmap_arrange to plot these maps together
tmap_arrange(map1, map2, map3, map4, map5)
```

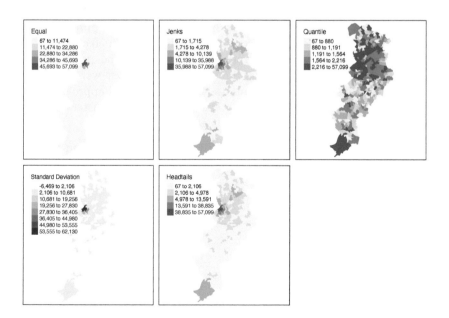

FIGURE 3.17: Compare maps using the different classifications of crime in Manchester

As we can see, using data-driven, natural breaks (as the Jenks or Headtails method do, in different ways) to classify data is useful when mapping data values that are not evenly distributed, since it places value clusters in the same class. The disadvantage of using this approach is that it is often difficult to make comparisons between maps (for example, of different crimes or for different time periods) since the classification scheme used is unique to each data set.

There are some other classification methods built into `tmap` which you can experiment with if you'd like. Your discrete gradient options are "cat", "fixed", "sd", "equal", "pretty", "quantile", "kmeans", "hclust", "bclust", "fisher", "jenks", "dpih", "headtails", and "log10_pretty". A numeric variable is processed as a categorical variable when using "cat", i.e., each unique value will correspond to a distinct category.

Taken from the help file, we can find more information about these, for example the "kmeans" style uses kmeans clustering technique (a form of unsupervised statistical learning) to generate the breaks. The "hclust" style uses hclust to generate the breaks using hierarchical clustering and the "bclust" style uses bclust to generate the breaks using bagged clustering. These approaches are outside the scope of what we cover, but just keep in mind that there are many different ways to classify your data, and you must think carefully about the choice you make, as it may affect your readers' conclusions from your map.

Imagine you were a consultant working for one of the political parties in the city of Manchester. Which map would you choose to represent to the electorate the situation of crime in the city if you were the party in control of the local government? and which map would you choose if you were working for the opposition? As noted above, it is very easy to mislead with maps and, thus, this means the professional map maker has to abide by strict deontological criteria and take well-justified impartial decisions when visualising data.

Cameron (2005) suggests the following:

> To know which classification scheme to use, an analyst needs to know how the data are distributed and the mapping objective. If the data are unevenly distributed, with large jumps in values or extreme outliers, and the analyst wants to emphasize clusters of observations that house similar values, use the natural breaks classification approach. If the data are evenly distributed and the analyst wants to emphasize the percentage of observations in a given classification category or group of categories, use the quantile classification approach. If the data are normally distributed and the analyst wants to represent the density of observations around the mean, use the equal interval approach. If the data are skewed and the analyst wants to identify extreme outliers or clusters of very high or low values, use the standard deviation classification approach.

No matter the choice you make, be sure to be explicit about why you made it, and the possible implications it has for your maps.

3.6   Interactive mapping with tmap

So far we have been producing static maps with `tmap`. But this package also allows for interactive mapping by linking with leaflet. To change whether the plotted maps are static or interactive, we need to use the `tmap_mode()` function. The default is `tmap_mode("plot")`, which corresponds to static maps. If we want to change to interactive display, we need to change the argument we pass to `tmap_mode("view")`.

```
tmap_mode("view")
```

```
## tmap mode set to interactive viewing
```

When you use `tmap`, R will remember the mode you want to use. So once you specify `tmap_mode("view")`, all the subsequent maps will be interactive. It is only when you want to change this behaviour that you would need another `tmap_mode()` call. When using the interactive view, we can also add a basemap with the `tm_basemap()` function and passing as an argument a particular source for the basemap. Here we specify OpenStreetMap, but there are many other choices[1].

Let's explore the distribution of the two alternative definitions of crime rates in an interactive way.

```
tm_shape(manchester) +
  tm_fill("crimr1", style="jenks", palette= "Reds",
          title = "Crime per residential pop", alpha = 0.6) +
  tm_basemap(leaflet::providers$OpenStreetMap)
```

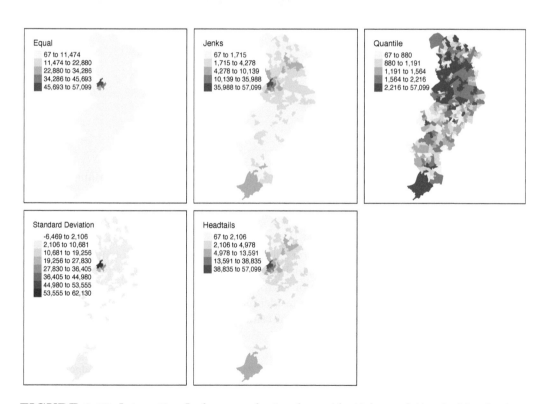

FIGURE 3.18: Interactive Jenks map of crime by residential population in Manchester

If you are following along, you can now scroll down to see that the crime rate is highest in the city centre of Manchester, but there are also pockets of high level of the crime rate in

---

[1]http://leaflet-extras.github.io/leaflet-providers/preview

the North East of the city, Harpurhey and Moston (areas of high levels of deprivation) and in the LSOA farthest to the South (where the international airport is located).

How does this change if we use the crime rate that uses the workday population?

```
tm_shape(manchester) +
  tm_fill("crimr2", style="jenks", palette= "Oranges",
          title = "Crime per workday pop", alpha = 0.8,) +
  tm_basemap(leaflet::providers$OpenStreetMap)
```

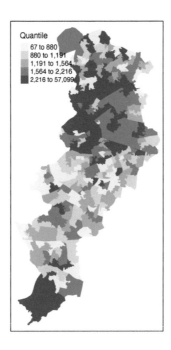

FIGURE 3.19: Interactive Jenks map of crime rate by workday population in Manchester

Things look different, don't they? For starters look at the values in the labels for the various classes. They are much less extreme. One of the reasons why we see such extreme rates in the first map is linked to the very large discrepancy between the residential population and the workday population in some parts of Manchester, like the international airport and the city centre (that attract very large volume of visitors, but have few residents). The LSOA with the highest crime rate (when using the residential population as the denominator) is E01033658. We can filter to find the number of crimes (in the count variable) that took place here.

```
manchester %>% filter(code=="E01033658") %>% select(code, count, respop, wkdpop)
```

TABLE 3.1: Number of crimes in LSOA E0103365

code	count	respop	wkdpop
E01033658	744	1303	42253

We can see in this area there were 744 crime incidents for a residential population of 1303, but the workday population is as high as 4.2253×10^4 people. So, of course, using these different denominators is bound to have an impact in the resulting rate. As noted earlier, the most appropriate denominator represents some measure of the population at risk.

Once again, we can rely on some advice from Cameron (2005):

> In the final analysis, although choropleth maps are very useful for visualizing spatial distributions, using them for hot spot analyses of crime has certain disadvantages. First, attention is often focused on the relative size of an area, so large areas tend to dominate the map. Second, choropleth maps involve the aggregation of data within statistical or administrative areas that may not correspond to the actual underlying spatial distribution of the data.

3.7 Smoothing rates: adjusting for small sample noise

In previous sections we discussed how to map rates. It seems a fairly straightforward issue: you calculate a rate by dividing your numerator (e.g., number of crimes) by an appropriately selected denominator (e.g., daytime population). You get your variable with the relevant rate and you map it using a choropleth map. However, things are not always that simple.

Rates are funny animals. Gelman and Price (1999) go so far as to suggest that all maps of rates are misleading. The problem at hand is well known in spatial epidemiology: "plotting observed rates can have serious drawbacks when sample sizes vary by area, since very high (and low) observed rates are found disproportionately in poorly-sampled areas" (Gelman and Price 1999, 3221). There is associated noise for those areas where the denominators give us a small sample size. And it is hard to solve this problem.

Let's illustrate with an example. We are going to use historic data on homicide across US counties. The dataset was used as the basis for the study by Baller et al. (2001). It contains data on homicide counts and rates for various decades across the US, as well as information on structural factors often thought to be associated with violence. The data is freely available through the webpage of Geoda, a clever point-and-click interface developed by Luc Anselin (a spatial econometrician and coauthor of the above paper) and his team, to make spatial analysis accessible. It is also available as one of the datasets in the `geodaData` package.

To read data available in a library we have loaded, we can use the `data()` function. If we check the `class()` of this object, we will see it was already stored in `geodaData` as a `sf` object.

```
data("ncovr")
class(ncovr)
```

```
## [1] "sf"          "data.frame"
```

Let's look at the "ncvor" data. We can start by looking at the homicide rate for 1960.

```
summary(ncovr$HR60)
```

```
##     Min. 1st Qu.  Median    Mean 3rd Qu.    Max.
##     0.00    0.00    2.78    4.50    6.89   92.94
```

We can see that the county with the highest homicide rate in the 1960s had a rate of 92.9368 homicides per 100,000 individuals. That is very high. Just to put it into context in the UK the homicide rate is about 0.92 per 100,000 individuals. Where is that place? I can tell you it is a place called Borden. Check it out:

```
borden <- filter(ncovr, NAME == "Borden")
```

```
## old-style crs object detected; please recreate object with a recent sf::st_crs()
```

```
borden$HR60
```

```
## [1] 92.94
```

Borden county[2] is in Texas. You may be thinking "Texas Chainsaw Massacre" perhaps? No, not really. Ed Gein, who inspired the film, was based and operated in Wisconsin. Borden's claim to fame is rather more prosaic: it was named after Gail Borden, the inventor of condensed milk. So, what's going on here? Why do we have a homicide rate in Borden that makes it look like a war zone? Is it that it is only one of the six counties where alcohol is banned in Texas?

To get to the bottom of this, we can look at the variable `HC60` which tells us about the homicide *count* in Borden:

```
borden$HC60
```

```
## [1] 1
```

What? A total homicide count of 1. How can a county with just one homicide have a rate that makes it look like the most dangerous place in the US? To answer this, let's look at the population of Borden county in the 1960s, contained in the `PO60` variable.

```
borden$PO60
```

```
## [1] 1076
```

Well, there were about 1076 people living there. It is among some of the least populous counties in our data:

```
summary(ncovr$PO60)
```

```
##     Min. 1st Qu.  Median    Mean 3rd Qu.      Max.
##      208    9417   18408   57845   39165 7781984
```

---

[2]https://en.wikipedia.org/wiki/Borden_County,_Texas

If you contrast that population count with the population of the average county in the US, that's tiny. One homicide in such a small place can end up producing a big rate. Remember that the rate is simply dividing the number of relevant events by the exposure variable (in this case, population) and multiplying by a constant (in this case, 100,000 since we expressed crime rates in those terms). Most times Borden looks like a very peaceful place:

```
borden$HR70
```

```
## [1] 0
```

```
borden$HR80
```

```
## [1] 0
```

```
borden$HR90
```

```
## [1] 0
```

It has a homicide rate of 0 in most decades. But it only takes one homicide and, bang, it goes top of the league. So a standard map of rates is bound to be noisy. There is the instability that is introduced by virtue of having areas that may be sparsely populated and in which one single event, like in this case, will produce a very noticeable change in the rate.

In fact, if you look at the counties with the highest homicide rate in the "ncovr" dataset, you will notice all of them are places like Borden, areas that are sparsely populated, not because they are that dangerous, but because of the instability of rates. Conversely, the same happens with those places with the lowest rate. They tend to be areas with a very small sample size.

This is a problem that was first noted by epidemiologists doing disease mapping. But a number of other disciplines have now noted this and used some of the approaches developed by public health researchers that confronted this problem when producing maps of disease (*aside*: techniques and approaches used by spatial epidemiologists are very similar to those used by criminologists, in case you ever think of changing careers or need inspiration for how to solve a crime analysis problem).

One way of dealing with this is by **smoothing** or **shrinking** the rates. This basically, as the word implies, aims for a smoother representation that avoids hard spikes associated with random noise. There are different ways of doing that. Some ways use a non-spatial approach to smoothing, using something called a **empirical Bayesian smoother**. How does this work? This approach takes the raw rates and tries to "shrink" them towards the overall average. What does this mean? Essentially, we compute a weighted average between the raw rate for each area and the global average across all areas, with weights proportional to the underlying population at risk. What this procedure does is adjusting considerably (brought closer to the global average) the rates of smaller areas (those with a small population at risk), whereas the rates for the larger areas will barely change.

Here we are going to introduce the approach implemented in DCluster, a package developed for epidemiological research and detection of clusters of disease. Specifically, we can implement the empbaysmooth() function which creates a smooth relative risks from a set of expected and observed number of cases using a Poisson-Gamma model as proposed by Clayton and Kaldor (1987). The function empbaysmooth() expects two parameters, the **expected** value and the **observed** value. Let's define them first.

```
#First we define the observed number of cases
ncovr$observed <- ncovr$HC60
#To compute the expected number of cases through indirect standardisation
#we need the overall incidence ratio
overall_incidence_ratio <- sum(ncovr$HC60)/sum(ncovr$PO60)
#The expected number of cases can then be obtained by multiplying the overall
#incidence rate by the population
ncovr$expected <- ncovr$PO60 * overall_incidence_ratio
```

With this parameters we can obtain the raw relative risk:

```
ncovr$raw_risk <- ncovr$observed / ncovr$expected
summary(ncovr$raw_risk)
```

```
##    Min. 1st Qu.  Median    Mean 3rd Qu.    Max.
##   0.000   0.000   0.614   0.994   1.520  20.510
```

And then estimate the smoothed relative risk:

```
res <- empbaysmooth(ncovr$observed, ncovr$expected)
```

In the new object we generated, which is a list, you have an element which contains the computed rates. We can add those to our dataset:

```
ncovr$H60EBS <- res$smthrr
summary(ncovr$H60EBS)
```

```
##    Min. 1st Qu.  Median    Mean 3rd Qu.    Max.
##   0.234   0.870   0.936   0.968   1.035   2.787
```

We can observe that the dispersion narrows significantly and that there are fewer observations with extreme values once we use this smoother.

Instead of shrinking to the global relative risk, we can shrink to a relative rate based on the neighbours of each county. Shrinking to the global risk ignores the spatial dimension of the phenomenon being mapped out and may mask existing heterogeneity. If instead of shrinking to a global risk, we shrink to a local rate, we may be able to take unobserved heterogeneity into account. Marshall (1991b) proposed a local smoother estimator in which the crude rate is shrunk towards a local, "neighbourhood", rate. To compute this, we need the list of neighbours that surround each county (we will discuss this code in Chapter 11, so for now just trust we are computing the rate of the areas that surround each country):

```
ncovr_sp <- as(ncovr, "Spatial")
w_nb <- poly2nb(ncovr_sp, row.names=ncovr_sp$FIPSNO)
eb2 <- EBlocal(ncovr$HC60, ncovr$PO60, w_nb)
ncovr$HR60EBSL <- eb2$est * 100000
```

We can now plot the maps and compare them:

```
tmap_mode("plot")
current_style <- tmap_style("col_blind")

map1<- tm_shape(ncovr) +
  tm_fill("HR60", style="quantile",
          title = "Raw rate",
          palette = "Reds") +
  tm_layout(legend.position = c("left", "bottom"),
            legend.title.size = 0.8, legend.text.size = 0.5)

map2<- tm_shape(ncovr) +
  tm_fill("HR60EBSL", style="quantile",
          title = "Local Smooth",
          palette = "Reds") +
  tm_layout(legend.position = c("left", "bottom"),
            legend.title.size = 0.8, legend.text.size = 0.5)

tmap_arrange(map1, map2)
```

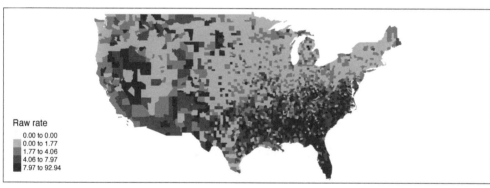

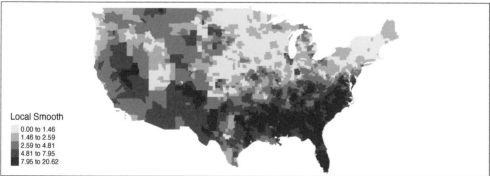

FIGURE 3.20: Compare raw and smoothed rates across counties in the USA

Notice that the quantiles are not the same, so that will make your comparison difficult. Let's look at a boxplot of these variables. In the map of raw rates, we have the most variation.

```
#Boxplots with base R graphics
boxplot(ncovr$HR60, ncovr$HR60EBSL,
        main = "Homicide per 100,000 people",
        names = c("Raw rate",
                  "Local base smoother"),
        ylab = "Rates")
```

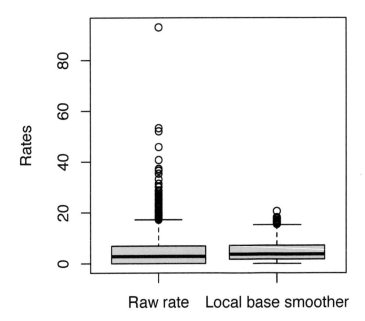

FIGURE 3.21: Boxplots to compare variation in rates

The range for the raw rates is nearly 93. Much of the variation in observed homicide rates by county is attributable to statistical noise due to the small number of (observed and expected) homicides in low-population counties. Because of this noise, a disproportionate fraction of low-population counties are observed to have extremely high (or low) homicide rates when compared to typical counties in the United States. With smoothing, we reduce this problem, and if you contrast the maps you will see how this results in a clearer and smoother spatial pattern for the rate that is estimated borrowing information from their neighbours.

So to smooth or not too smooth? Clearly we can see how smoothing stabilises the rates and removes noise. But as Gelman and Price (1999) suggests this introduces other artifacts and autocorrelation into our estimates. Some people are also not too keen on maps of statistically adjusted estimates. Yet, the conclusions one can derive from mapping raw rates (*when* the denominator varies significantly and we have areas with small sample size) means that smoothing is often a preferable alternative (Waller and Gotway 2004). The problem we have with maps of estimates is that we need information about the variability and it is hard to map this out in a convenient way (Gelman and Price 1999). Lawson (2021b), in relation to the similar problem of disease mapping, suggests that "at the minimum any map of relative risk for a disease should be accompanied with information pertaining to estimates of rates within each region as well as estimates of variability within each region" (p. 38), whereas "at the other extreme it could be recommended that such maps be only used as a presentational aid, and not as a fundamental decision-making tool" (p. 38).

3.8 Summary and further reading

This chapter introduced some basic principles of thematic maps. We learned how to make them using the `tmap` package, the importance of classification schemes, and how each one may produce a different looking map, which may tell a different story. For further reading, Brewer (2006) provides a brief condensed introduction to thematic mapping for epidemiologists but that can be generally extrapolated for crime mapping purposes. We have talked about the potential to develop misleading maps, and Monmonier (1996) "How to lie with maps" provides good guidance to avoid our negligent choices when producing a map confuse the readers. Carr and Pickle (2010) offers a more detailed treatment of small multiples and micromaps.

Aside from the issues we discussed around the computation and mapping of rates, there is a growing literature that is identifying the problem of ignoring that rates (as a standardising mechanism) using population in the denominator assumes that the relationship between crime and population is linear, when this is not generally the case. Oliveira (2021) discusses the problem and provides some solutions. Mapping rates, more generally, has been more thoroughly discussed within spatial epidemiology than in criminology. There is ample literature on disease mapping that address in more sophisticated ways some of the issues we introduce here (see Waller and Gotway (2004), Lawson (2021a), Lawson (2021b), or Lawson (2021c)). Much of this work on spatial epidemiology adopts a Bayesian framework. We will talk a bit more about this later on, but if you want a friendly introduction to Bayesian data analysis we recommend McElreath (2018).

4

Variations of thematic mapping

4.1 Introduction

In this chapter we are going to discuss some additional features around thematic maps. Specifically, we will address some of the problems we confront when we are trying to use choropleth maps, as well as some alternatives to point-based maps. We will also briefly introduce the modifiable area unit problem.

The main objectives for this chapter are that by the end you will have:

- Explored **binning** as an alternative to point maps.
- Been introduced to alternative visualisations of polygon-level data such as
    - **cartograms**, and
    - **bi-variate** thematic maps.
- Gained an insight into the **Modifiable Areal Unit Problem**

In this chapter, we will be making use of the following libraries:

```
# Packages for reading data and data carpentry
library(readr)
library(dplyr)
library(tidyr)


# Packages for handling spatial data and for geospatial carpentry
library(sf)


# Packages for mapping and visualisation
library(ggplot2)
library(ggspatial)
library(hexbin) # needs to be installed
library(cowplot)
library(cartogram)
```

4.2   Binning points

In GIS it is often difficult to present point-based data because in many instances there are several different points and data symbologies that need to be shown. As the number of different data points grows, they can become complicated to interpret and manage which can result in convoluted and sometimes inaccurate maps. This becomes an even larger problem in web maps that are able to be depicted at different scales because smaller scale maps need to show more area and more data. This makes the maps convoluted if multiple data points are included.

In many maps there are so many data points included that little can be interpreted from them. In order to reduce congestion on maps, many GIS users and cartographers have turned to a process known as binning. Binning allows researchers and practitioners a way to systematically divide a region into equal-sized portions. As well as making maps with many points easier to read, binning data into regions can help identify spatial influence of neighbourhoods, and can be an essential step in developing systematic sampling designs.

This approach to binning generates an array of repeating shapes over a user specified area. These shapes can be hexagons, squares, rectangles, triangles, circles or points, and they can be generated with any directional orientation. These are sometimes 'grids' or 'fishnets' and lots of other spatial analysis (e.g., map algebra) can be done with such files.

4.2.1   The binning process

Binning is a data modification technique that changes the way data is shown at small scales. It is done in the pre-processing stage of data analysis to convert the original data values into a range of small intervals, known as a bin. These bins are then replaced by a value that is representative of the interval to reduce the number of data points.

Spatial binning (also called *spatial discretization*) discretizes the location values into a small number of groups associated with geographical areas or shapes. This approach to binning generates an array of repeating shapes over a user-specified area. These shapes can be hexagons, squares, rectangles, triangles, circles or points, and they can be generated with any directional orientation. The assignment of a location to a group can be done by any of the following methods:

- Using the coordinates of the point to identify which "bin" it belongs to.
- Using a common variable in the attribute table of the bin and the point layers.

4.2.2   Different binning techniques

Binning itself is a general term used to describe the grouping of a dataset's values into smaller groups (Z.-F. Johnson 2011). The bins can be based on a variety of factors and attributes such as spatial and temporal and can thus be used for many different projects.

4.2.2.1 Choropleth maps

You might be thinking, "grouping points into a larger spatial unit, haven't we already done this when making choropleth maps?". In a way you are right. Choropleth maps are a type of map that uses binning. Proportional symbol and choropleth maps group similar data points together to show a range of data instead of many individual points. We've covered this extensively, and it is often the best approach to consider spatial grouping of your point variables, because the polygons (shapes) to which you are aggregating your points are *meaningful*. You can group into LSOAs, as we did in previous chapters, because you want to show variation in neighbourhoods. Or you can group into police force areas because you want to look at differences between those units of analysis. But sometimes there is just not a geography present to meet your needs.

Let's say you are conducting some days of action in Manchester city centre, focusing on antisocial behaviour. You are going to put up some information booths and staff them with officers to engage with the local population about antisocial behaviour. For these to be most effective, as an analyst you decide that they should go into the areas with the highest *count* of antisocial behaviour. You want to be very specific about where you focus your operational attentio and, thus, the LSOA level may be too large. One approach can be to split central Manchester into some smaller polygons, and just calculate the number of antisocial behaviour incidents recorded in each. That way you can then decide to put your information booths somewhere inside the top 5 highest count bins.

4.2.2.2 Rectangular binning

The aggregation of incident point data to regularly shaped grids is used for many reasons such as normalizing geography for mapping or to mitigate the issues of using irregularly shaped polygons created arbitrarily (such as county boundaries or block groups that have been created from a political process). Regularly shaped grids can only be comprised of equilateral triangles, squares, or hexagons, as these three polygon shapes are the only that can tessellate (repeating the same shape over and over again, edge to edge, to cover an area without gaps or overlaps) to create an evenly spaced grid.

Rectangular binning is the simplest binning method and as such it is heavily used. However, there are some reasons why rectangular bins are less preferable over hexagonal bins. Before we cover this, let's have a look at hexagonal bins.

4.2.2.3 Hexagonal binning

In many applications binning is done using a technique called **hexagonal binning**. This technique uses hexagon shapes to create a grid of points and develops a spatial histogram that shows different data points as a range or group of pairs with common distances and directions. In hexagonal binning the number of points falling within a particular hexagon in a gridded surface is what makes the different colors to easily visualize data (N. Smith 2014). Hexagonnal binning was first developed in 1987 and today "hexbinning" is conducted by laying a hexagonal grid on top of two-dimensional data (Z.-F. Johnson 2011). Once this is done users can conduct data point counts to determine the number of points for each hexagon (Z.-F. Johnson 2011). The bins are then symbolized differently to show meaningful patterns in the data.

So how can we use hexbinning to solve our antisocial behaviour days of action task? Well let's say we split Manchester city centre into hexagons, and count the number of antisocial behaviour instances in these. We can then identify the top hexagons, and locate our booths somewhere within these.

First, make sure you have the appropriate packages loaded. Also let's get some data. You could go and get this data yourself from police.uk, but as we've been through a few times now, a tidied set of data is ready for you within the data provided with the book (see *Preamble*). This data is one year's worth of antisocial behaviour from the police.uk[1] data, from May 2016 to May 2017, for the borough of Manchester.

```
manchester_asb <- read_csv("data/manchester_asb.csv")
```

This is currently just a text dataframe, so we need to let R know that actually this is a spatial object; the geometry can be found in its longitude and latitude coordinates. As we have long/latt we can assume it's in WGS84 projection.

```
ma_spatial <- st_as_sf(manchester_asb,
                coords = c("Longitude", "Latitude"),
                crs = 4326,
                agr = "constant")
```

Now one thing that this does is it consumes our Long and Lat columns into a geometry attribute. This is generally OK, but for the binning we will do, we would like to have them as separate coordinates. To do this, we can use the `st_coordinates()` function from the `sf` package. This function extracts the longitude and latitude from the geometry within the `sf` object, in this case "ma_spatial" object. For example, if we look at the first row:

```
ma_spatial %>%
  slice(1) %>%
  st_coordinates()
```

```
##        X     Y
## 1 -2.229 53.53
```

We have our longitude (X) and latitude (Y). We can select the first and second element of this to get only one or the other. To go through our whole dataframe, we can use the `mutate()` function, and assign each element to a longitude and latitude column, respectively:

```
ma_spatial <- ma_spatial %>%
  mutate(longitude = st_coordinates(.)[,1],
         latitude = st_coordinates(.)[,2])
```

As a first step, we can plot antisocial behaviour in the borough of Manchester using simple `ggplot`, as demonstrated in Chapter 1. We can plot our points first:

---

[1]data.police.uk

```
ggplot(ma_spatial, aes(x = longitude, y = latitude)) +
  geom_sf(size=0.001) +
  theme_minimal()
```

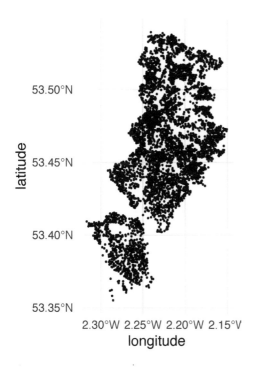

FIGURE 4.1: Point map of ASB in Manchester

We see our nice map of Manchester, as outlined by ASB incidents across the Local Authority. However, it is hard to tell from this map where these incidents might occur more frequently and where they are just a few one-off incidents. A binned map may help us better understand. To create such a map, we must first create the grid (or fishnet) of tessellating shapes. Once we have this grid, we can overlay it on top of our map, and count the number of ASB incidents (points) which fall into each grid shape (polygon). We can use the ggplot library to achieve this. It is such a great tool for building visualisations, because you can apply whatever geometry best suits your data. To map a hexbinned version of our point data of antisocial behaviour, we can use the stat_binhex() function as a layer added on to our ggplot object. We can also recreate the thematic map element, as we can use the frequency of points in each hex to shade each hexbin from white (least number of incidents) to red (most number of incidents). So let's have a go:

```
#define data and variables for x and y axes
ggplot(ma_spatial, aes(longitude, latitude)) +
  # plot geometry with geom_sf()
  geom_sf() +
  #add binhex layer (hexbin) set bin size (in degrees)
  stat_binhex(binwidth = c(.015, .01)) +
  #add shading based on number of ASB incidents
  scale_fill_gradientn(colours = c("white","red"),  name = "Frequency")  +
  theme_void()
```

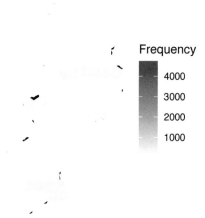

FIGURE 4.2: Hexbin map of ASB in Manchester

Note: you might get the following prompt when you run this code: "The hexbin package is required for stat_binhex() Would you like to install it?" - if you do, just choose "Yes" or quit the action and install the hexbin package separately, and then try running the code again. Similarly if you get "Warning: Computation failed in stat_binhex()" the solution might be to install this hexbin package.

Neat, but it doesn't quite tell us *where* that really dark hexbin actually is. So it would be much better if we could do this with a basemap as the background. For this we use the function annotation_map_tile() from the ggspatial package. We can also set the opacity of the binhex layer, so we can see our basemap, with the alpha parameter.

```
ggplot(ma_spatial, aes(x = longitude, y = latitude)) +
  annotation_map_tile() +
  stat_binhex(binwidth = c(.015, .01), alpha=0.7) +   # set opacity
  scale_fill_gradientn(colours = c("white","red"), name = "Frequency")   +
  theme_void()
```

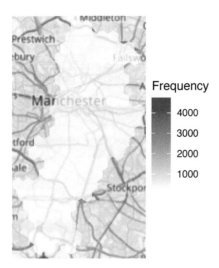

FIGURE 4.3: Hexbin map of ASB in Manchester with basemap

Adding this basemap provides us with a bit more context. And combined with the hexbin map, it is much easier to see where antisocial-behaviour concentrates (as opposed to with the point map!). Above we used a hexagon shape for our binning; however, you might choose other shapes as well. We will illustrate in a moment the approach to use rectangular binning, but first, we want to highlight why hexagon might still be your ideal choice. Here are some thoughts, summarised from (ArcGIS Pro 2021):

- Hexagons reduce sampling bias due to edge effects (Rengert and Lockwood 2009) of the grid shape.
- As hexagons are circular, they represent curves in the patterns of your data more naturally than square grids.
- When comparing polygons with equal areas, the more similar to a circle the polygon is, the closer to the centroid the points near the border are.
- Hexagons are preferable when your analysis includes aspects of connectivity or movement paths. Due to the linear nature of rectangles, fishnet grids can draw our eyes to the straight, unbroken, parallel lines which may inhibit the underlying patterns in the data.
- If you are working over a large area, a hexagon grid will suffer less distortion due to the curvature of the earth than the shape of a fishnet grid.
- You will find more neighbours with hexagons than square grids, if you're using a distance band, since the distance between centroids is the same in all six directions with hexagons (see image below).

To illustrate the different approaches. Like we did in earlier chapters, we

FIGURE 4.4: Hexagonal bins will have more neighbors included in the calculations for each feature as opposed to a fishnet grid

can assign each map to an object, and then print them all side-by-side. Let's start with rectangular binning:

```
rectangular_bin_map <- ggplot(ma_spatial, aes(x = longitude, y = latitude)) +
  annotation_map_tile() +
  stat_bin2d(binwidth = c(.015, .01), alpha=0.7) +
  scale_fill_gradientn(colours = c("white","red"),
                       name = "Frequency") +
  theme_void()
```

Now map a map using hexagonal binning:

```
hexagonal_bin_map <- ggplot(ma_spatial, aes(x = longitude, y = latitude)) +
  annotation_map_tile() +
  stat_binhex(binwidth = c(.015, .01), alpha=0.7) +
  scale_fill_gradientn(colours = c("white","red"),
                       name = "Frequency") +
  theme_void()
```

And finally for comparison a simple "heatmap" (we will discuss these more thoroughly in Chapter 7):

```
heatmap <- ggplot(ma_spatial, aes(x = longitude, y = latitude)) +
  annotation_map_tile() +
  stat_density2d(aes(fill = ..level.., # value corresponding to
                                       # discretized density estimates
                     alpha = ..level..),
                 geom = "polygon") +  # creates the bands of
                                      # different colours
  ## Configure the colours, transparency and panel
  scale_fill_gradientn(colours = c("white","red"),
                       name = "Frequency") +
  scale_alpha(guide = "none") +
  theme_void()
```

Now we can print them all side-by-side using `plot_grid()` function from the `cowplot` package to compare.

```
library(cowplot)
```

```
plot_grid(rectangular_bin_map, hexagonal_bin_map, heatmap, nrow =1,
          labels = "AUTO", scale = c(1, 1, .85))
```

```
## Warning: The dot-dot notation (`..level..`) was deprecated in ggplot2 3.4.0.
## i Please use `after_stat(level)` instead.
```

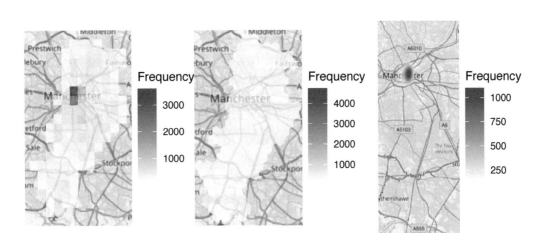

FIGURE 4.5: Comparing rectangular (A), hexagonal (B), and heat maps (C)

4.2.3 Benefits of binning

Because of the plethora of data types available and the wide variety of projects being done in GIS, binning is a popular method for mapping complex data and making it meaningful. Binning is a good option for map makers as well as users because it makes data easy to understand and it can be both static and interactive on many different map scales. If every different point were shown on a map, it would have to be a very large scale map to ensure that the data points did not overlap and were easily understood by people using the maps. According to Kenneth Field, an Esri Research Cartographer:

> Data binning is a great alternative for mapping large point-based data sets which allows us to tell a better story without interpolation. Binning is a way of converting point-based data into a regular grid of polygons so that each polygon represents the aggregation of points that fall within it (Field 2012).

Binning create maps that are easier to understand, more accurate and more visually appealing. Hexbin plots can be viewed as an alternative to scatter plots. The hexagon-shaped bins were introduced to plot densely packed sunflower plots. They can be used to plot scatter plots with high-density data.

4.3 Transforming polygons: cartograms

When you have meaningful spatial units of analysis in your polygons, for example you are interested specifically in Local Autorities, it might make sense to aggregate the points into these meaningful polygons to create thematic maps. However, while thematic maps are an accessible and visually appealing method for displaying spatial information, they can also be highly misleading. Irregularly shaped polygons and large differences in the size of areas being mapped can introduce misrepresentation. The message that analysts and researchers want to get across might be lost, or even worse, it may misdirect the viewers to erroneous conclusions. Field and Dorling (2016) provide a helpful discussion of the problem illustrating the case with UK election maps. It is worth reading.

Fortunately, there are many methods in R to enhance the legibility of geographic information and the interpretability of what it is trying to communicate. Selecting the appropriate method might depend on the research question being posed (e.g., clustering) and the data itself. Even once a method has been selected, there are different ways of operationalising them. Here we focus on **cartograms**. A cartogram is a method of distortion whereby distinct definable spatial regions on a map (e.g., countries, provinces, etc.) are distorted in some way through being weighted by some quantitative value (e.g., population count, carbon emissions, etc.) (Kirk 2016).

Let's explore this using the example of the results of the 2016 EU referendum which took place in the United Kingdom, where citizens voted to leave the European Union (in a process termed "Brexit"). We can look at the data on voting results at Local Authority level. It is well known that those who voted to remain in the EU predominantly clustered in the Local Authorities within London. A simple thematic map does not necessarily communicate this well because Local Authorities are both small and densely populated in London, compared to much larger elsewhere.

You can download the full set of EU referendum result data as a csv from the Electoral Commission website[2]. We've already done this and included in our data supplied with the book. Let's read it straight into R:

```
eu_ref <- read_csv("data/EU-referendum-result-data.csv")
```

We also need a spatial object to join our attribute table to, in order to map it. In *Appendix C: Sourcing geographical data for crime analysis*, we detail how one might go about finding such spatial data, if it is not already provided. For now, we can use the data which comes with this book from the supplementary materials. Specifically the shapefile for English Local Authorities. This file is called `england_lad_2011_gen.shp` and is found in the `England_lad_2011_gen` sub-folder.

```
las <- st_read("data/England_lad_2011_gen/england_lad_2011_gen.shp")
```

We can now join the EU referendum data using the attribute operation `left_join()`, as we have illustrated in detail in Chapter 1.

```
eu_sf <- left_join(las, eu_ref, by = c("name" = "Area"))

#make sure we are in British National Grid Projection
eu_sf <- st_transform(eu_sf, 27700)
```

Now we can have a look at these data:

---

[2]https://www.electoralcommission.org.uk/find-information-by-subject/elections-and-referendums/past-elections-and-referendums/eu-referendum/eu-referendum-result-visualisations

```
ggplot() +
  geom_sf(data = eu_sf, aes(fill = Pct_Leave)) +
  theme_void()
```

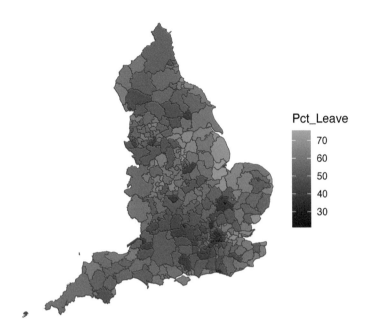

FIGURE 4.6: EU referendum results by Local Authority (no transformation)

The Local Authorities (LAs) vary greatly in their shape and size, and in the case of smaller LAs the result is barely visible. In this case, we cannot really see what was happening with the EU referendum in London, for example. This is the sort of situation where augmenting our polygons may be handy. Cartograms offer one way to achieve this.

There are different types of cartograms. **Density-Equalizing (contiguous) Cartograms** are your traditional cartograms. In density-equalizing cartograms, map features bulge out a specific variable. Even though it distorts each feature, it remains connected during its creation. On the other hand, you can have **Non-Contiguous Cartograms**, where features in non-contiguous cartograms don't have to stay connected. Finally, **Dorling Cartograms** (named after professor Danny Dorling (Dorling 1991)) use shapes like circles and rectangles to depict areas. These types of cartograms make it easy to recognize patterns.

We can explore cartograms using the `cartogram` package. Within that, we will use the `cartogram()` function. In this function, we will specify two parameters: first, the `shp =`, which asks for the shape file (it can be a SpatialPolygonDataFrame or an sf object), and second, `weight =`, which asks for the variable which it should use to distort the polygon by.

In our data set we have a variable `Electorate` which refers to the total number of registered electors in that Local Authority. It serves to give an indicator of the total number of people who were eligible to vote in the referendum. We can use this variable to distort our polygons and create our cartogram.

```
eu_cartogram <- cartogram_cont(eu_sf, "Electorate")
```

If you run this, it might take a long time. This function, while it looks nice and simple, is actually very computationally taxing for your computer. For those interested, you may like to take the time, while R works this out for you, to read up on the maths behind this transformation in Dougenik, Chrisman, and Niemeyer (1985) (it's got a fun name: a rubber sheet distortion algorithm!).

I do have a tip for you if you want to make sure the process does not take too long. You can set another parameter in the cartogram function which is the `itermax=` parameter. This specifies the maximum number of iterations we are happy to sit through for our cartogram. If you do not specify an alternative the default is set to 15. Let's set to 5 for the sake of speed:

```
# construct a cartogram using the percentage voting leave
eu_cartogram <- cartogram_cont(eu_sf, "Electorate", itermax = 5)
```

This will be faster (but may not result in the best possible cartogram output). Once your cartogram has been created, you can now plot again the referendum results, but using the electorate to change the size of the Local Authority:

```
ggplot() +
  geom_sf(data = eu_cartogram, aes(fill = Pct_Leave)) +
  theme_void()
```

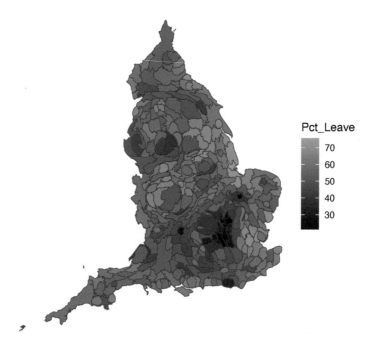

FIGURE 4.7: Cartogram weighting each polygon (Local Authority) by the size of its electorate

We can now see London much better, and see that darker-coloured cluster where a much smaller percentage of people voted leave. We can eyeball where London may be, as this continuous area cartogram tries to maintain some fidelity to our original shapes, while weighting them by some variable of interest, in our case the electorate in each Local Authority.

4.3.1 Dorling cartogram

While the continuous area cartogram we created above tries to maintain some fidelity to our original shapes, other approaches take more freedom when applying transformations. Sometimes, maintaining a resemblance to the original geometry of each polygon may not be important. In that case, you might be interested in creating a **Dorling cartogram**. We can achieve this by using the `cartogram_dorling()` function. Again we specify the sf object, and the variable which we use to scale our areas.

```
# construct a Dorling cartogram using the percentage voting leave
eu_dorling_cartogram <- cartogram_dorling(eu_sf, "Electorate")
```

Let's plot the output:

```
ggplot() +
  geom_sf(data = eu_dorling_cartogram, aes(fill = Pct_Leave)) +
  theme_void()
```

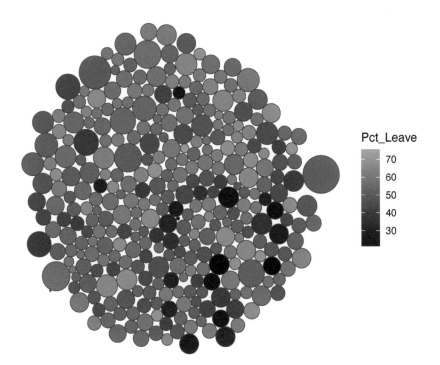

FIGURE 4.8: Dorling cartogram of LAs by electorate size

This map has transformed each Local Authority's shape into a circle, where the radius is determined as a function of the variable we supplied, which is our Electorate in each LA. The shading here is again provided by the percentage of people who voted to leave in each area, with lighter values indicating more people voting to leave, and darker values indicating fewer people voting to leave the EU. However, the relations of these Local Authorities is tough to maintain here, and we may be hard-pressed to identify London's boroughs out of this collection. However, Dorling cartograms may have their use and place. You may read up more about them in Dorling (1996).

4.4   Bivariate maps

Usually in your thematic maps you are mapping one variable. In the distortion examples above, we incorporate a second variable in order to make the map more legible. However, occasionally, you might want to use your map to illustrate the relationship between *two* variables. In this case, you may want to create a **Bivariate Choropleth Map**. Bivariate choropleths follow the same concept as the univariate ones (which display only one variable), except they show two variables at once. This is achieved through the creative use of colour.

We first came across the idea of bivariate choropleth maps from the blog of Stevens (2015). We recommend having a read as he discusses in great detail how to develop the colour schemes necessary for these maps. The idea is that there are two main colours, each one representing gradual change in one of our two variables of interest. Then, these two colours are combined to create an overlapping colour scheme.

The process itself is not too complicated. We start with binning our two variables into classes. If we bin variable 1 into n classes, and then again variable 2 into n classes, then when we compare them across one another to create our bivariate map, we will end up with n^2 classes. For example, if we bin both variables into three classes (low, medium, high), then when displaying them together, we will have nine classes to visualise. The blog by Stevens (2015) illustrates with nice visuals, so we recommend having a look. However, the practical example is in QGIS, and here we are working in R. There had been adaptations into R (e.g, see Grossenbacher (2019)) which we can borrow from here too.

Creating the map boils down to five key steps. First, we take our two variables of interest, and create bins. Second, we create a new variable, which we will use for the shading. Third we create our colour scheme. Fourth, we join this to our `sf` object. And, we map! Let's go through these steps now. Using the example of voting to leave the EU.

It may be an interesting question to look into not only how the voting outcome was distributed (`Pct_Leave`) but how this varies with voter turnout (`Pct_Turnout`). We might be interested in areas with high turnout and high percentage voting to leave, as these may be areas where people felt passionately. On the other hand, other areas may have had low turnout, which may have influenced the result, and in those places, how did people who did turn up vote? These are the kinds of questions we can answer with a bivariate map.

4.4.1 Step 1: Bin our variables of interest

We are interested in two key variables: Pct_Leave - the percentage of people who voted to leave the EU, and Pct_Turnout - the percentage of the electorate who actually voted. These are both numeric continuous variables. In order to create our bivariate choropleth map, we have to bin these values into n discrete categories. Here let's go with $n = 3$.

We can use the cut() and the quantile() functions in base R to class our variable into three quantiles. The quantile() function identifies the sample quantiles in a continuous variable. We need to include the parameters x=: the numeric vector whose sample quantiles are wanted (in this case, our variables Pct_Leave and Pct_Turnout), and probs=: numeric vector of probabilities with values in [0,1]. We can use the sequence generator function seq() to generate these from 0 to 1, by the increment of $\frac{1}{3}$ for 3 groups.

So first, let's create the breaks first for the Pct_Leave variable.

```
leave_breaks <- quantile(eu_sf$Pct_Leave,
                  probs = seq(0,1, by = 1/3), # probabilities
                  na.rm=TRUE, # remove NAs
                  names=TRUE, # keep names attribute
                  include.lowest=TRUE) # include value equal to
                                       # lowest 'breaks' value
```

And then the same again but for the turnout variable Pct_Turnout.

```
turnout_breaks <- quantile(eu_sf$Pct_Turnout,
                    probs = seq(0,1, by = 1/3),
                    na.rm=TRUE,
                    names=TRUE,
                    include.lowest=TRUE)
```

We can have a look at the output:

```
leave_breaks
```

```
##       0% 33.33333% 66.66667%      100%
##    21.38     51.72     59.12     75.56
```

The results are the cutoff values which we want to use to "cut" our numeric variable. We do this with the cut() function, where we specify again what to cut (the variables Pct_Leave and Pct_Turnout), and the breaks at which to cut (the objects we created above, leave_breaks and turnout_breaks):

```
eu_sf <- eu_sf %>%
  mutate(leave_quantiles = cut(Pct_Leave,
                        breaks = leave_breaks),
         turnout_quantiles = cut(Pct_Turnout,
                          breaks = turnout_breaks))
```

We have two resulting variables, `leave_quantiles` and `turnout_quantiles` which classify each one of our observations into one of these quartiles for both variables. In the next step, we use these to create a new variable.

4.4.2   Step 2: New variable

In this step we create a new variable. This time let's call it `group`, which tells us which quartile the specific Local Authority (each row) falls into. By applying the `as.numeric()` function, we translate the ranges of the quartile into their label (i.e., 1st, 2nd, or 3rd quartile). We do this for both variables, and paste them together using the `paste()` function, and the separator "-":

```
eu_sf <- eu_sf %>%
  mutate(group = paste(
      as.numeric(turnout_quantiles), "-",
      as.numeric(leave_quantiles))
  )
```

We now have a new column, called group, which tells us for each Local Authority, which quartile it falls into for each variable. For example, a value of "1 - 1" means the Local Authority belongs to the first quartile in both variables. This area would be considered to have low percentage voting leave, and also low turnout. On the other hand, "3 - 1" means that there was high turnout, but a low percentage voted to leave. We use this variable to assign the appropriate colour for our colour scheme for each of the 3^2 (9) combinations.

4.4.3   Step 3: Create colour scheme

Picking a colour scheme which reflects both gradual change in each individual variable, and the combined change in both is not an easy task! Luckily, Stevens (2015) has created some scale recommendations for us to choose from. We copy two of them below, but you can see the blog for another 2 options.

In this code below, we simply specify for each of the nine values of the variable created in the previous step ("1 - 1", "1 - 2", "1 - 3", "2 - 1", … "3 - 3") an associated colour using the relevant hex code.

```
library(tibble)

bivariate_color_scale_1 <- tibble(
  "3 - 3" = "#574249", # high - high
  "2 - 3" = "#985356",
  "1 - 3" = "#c85a5a", # low - high
  "3 - 2" = "#627f8c",
  "2 - 2" = "#ad9ea5", # medium - medium
  "1 - 2" = "#e4acac",
  "3 - 1" = "#64acbe", # high - low
  "2 - 1" = "#b0d5df",
```

```
  "1 - 1" = "#e8e8e8" # low - low
) %>%
  gather("group", "fill_col")


bivariate_color_scale_2 <- tibble(
  "3 - 3" = "#3b4994", # high - high
  "2 - 3" = "#5698b9",
  "1 - 3" = "#5ac8c8", # low - high
  "3 - 2" = "#8c62aa",
  "2 - 2" = "#a5add3", # medium - medium
  "1 - 2" = "#ace4e4",
  "3 - 1" = "#be64ac", # high - low
  "2 - 1" = "#dfb0d6",
  "1 - 1" = "#e8e8e8" # low - low
) %>%
  gather("group", "fill_col")
```

You can have a look at both of the colour scales we just created, and see which one you like. Feel free to use either in your future bivariate mapping adventures. Or construct your own, perhaps following the two additional ones provided by Stevens (2015), or some entirely new ones you may have constructed.

4.4.4   Step 4: Join colour scheme

Now that we have a colour scheme, we can join to the spatial object, using `left_join()`. The common element is the `group` variable, so with this approach, we join the relevant colour to each value of group in that column. Let's join `bivariate_color_scale_2`, the second of the two we created above.

```
eu_sf <- left_join(eu_sf, bivariate_color_scale_2, by = "group")
```

We now have an additional column in our `eu_sf` dataframe, called `fill_col` which we can use to shade each local authority according to the composite variable depicting the percentage who voted to leave and the percentage who turned up to vote.

4.4.5   Step 5: Create legend

The legend is a little tricky, as we need to separate out the values into separate Leave and Turnout columns. We can achieve this with the `separate()` function

```
# separate the groups
bivariate_color_scale <- bivariate_color_scale_2 %>%
  separate(group, into = c("Pct_Turnout", "Pct_Leave"), sep = " - ")
```

Then to create the legend, we actually build a `ggplot()` object. This genius bit of code is borrowed from Grossenbacher (2019) implementation of bivariate choropleth maps.

```
legend <- ggplot() +
  geom_tile( data = bivariate_color_scale,
              aes(x = Pct_Turnout, y = Pct_Leave, fill = fill_col)) +
  scale_fill_identity() +
  labs(x = "Higher % turnout ->",
       y = "Higher % voted leave ->") +
  theme(axis.title = element_text(size = 6)) + # makes text small for
                                               # adding legend to map
  coord_fixed()  # forces a specified ratio to create quadratic tiles
```

This code returns the legend as a chart itself. Have a look at what it looks like:

```
legend
```

FIGURE 4.9: Legend for bivariate map

4.4.6   Step 6: Map

Now finally we put it all on the map. For the choropleth map, we use our variable `fill_col` which contains the matched colour to the group that each observation belongs to. We pass this in the familiar `geom_sf()` geometry and use the `fill=` parameter to colour the Local Authorities according to their turnout / voted leave combination. We also have to add the `scale_fill_identity()` function, as the values in the `fill_col` variable are actually the hex codes for the colour which we use to shade the Local Authorities.

```
map <- ggplot(eu_sf) +
  geom_sf(aes( fill = fill_col)) +
  scale_fill_identity() +
  theme_void()
```

Finally, to display the legend and the map together, we can use the `ggdraw()` and `draw_plot()` functions from the `cowplot` package.

```
library(cowplot)
ggdraw() +
  draw_plot(map, 0, 0, 1, 1) +
  draw_plot(legend, 0.05, 0.075, 0.2, 0.2)
```

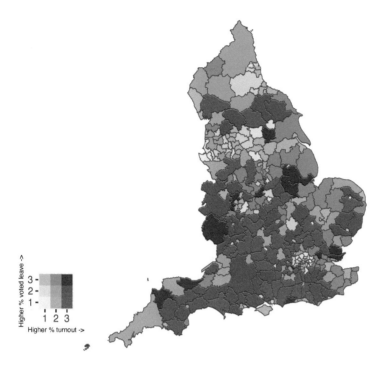

FIGURE 4.10: Bivariate choropleth map of voter turnout and percentage who voted to leave the EU

This map may now be able to provide insight into spatial patterns in turnout and voting to leave. For example, in the South you can see lots of pink, representing areas of high turnout and low percentage voting to leave the EU. You can also spot the dark blue areas; these are Local Authorities which saw high voter turnout and a high proportion voting to leave.

Overall, these maps can help visualise two variables on one map, and motivate discussion about relationships between variables in different places.

4.5 A note of caution: MAUP

Now that we've shown you how to do a lot of spatial crime analysis, we wanted to close with some words of caution. Remember that it's up to you, the researcher, the analyst, the domain expert, to apply and use the tools you are learning about with careful consideration and cautions. This discussion is very much part of spatial crime analysis, and an important field of thought.

We borrow here from George Rengert and Brian Lockwood:

> When spatial analysis of crime is conducted, the analyst should not ignore the spatial units that data are aggregated into and the impact of this choice on the interpretation

of findings. Just as several independent variables are considered to determine whether they have statistical significance, a consideration of multiple spatial units of analysis should be made as well, in order to determine whether the choice of aggregation level used in a spatial analysis can result in biased findings (Rengert and Lockwood 2009).

In particular, they highlight four main issues inherent in most studies of space:

- issues associated with politically bounded units of aggregation,
- edge effects of bounded space,
- the modifiable areal unit problem (MAUP),
- and ways in which the results of statistical analyses can be manipulated by changes in the level of aggregation.

The **scale problem** involves results that change based on data that are analyzed at higher or lower levels of aggregation (changing the number of units). For example, evaluating data at the state level vs. census tract level.

The scale problem has moved to the forefront of geographical criminology as a result of the recent interest in small-scale geographical units of analysis. It has been suggested that smaller is better since small areas can be directly perceived by individuals and are likely to be more homogenous than larger areas (Gerell 2017).

The **zonal problem** involves keeping the same scale of research (say, at the state level) but changing the actual shape and size of those areas.

The basic issue with the MAUP is that aggregate units of analysis are often arbitrarily produced by whom ever is in charge of creating the aggregate units. A classic example of this problem is known as gerrymandering. Gerrymandering involves shaping and re-shaping voting districts based on the political affiliations of the resident citizenry.

The inherent problem with the MAUP and with situations such as gerrymandering is that units of analysis are not based on geographic principles and instead are based on political and social biases. For researchers and analysts the MAUP has very important implications for research findings because it is possible that as arbitrarily defined units of analysis change shape findings based on these units will change as well.

When spatial data are derived from counting or averaging data within areal units, the form of those areal units affects the data recorded, and any statistical measures derived from the data. Modifying the areal units therefore changes the data. Two effects are involved: a zoning effect arising from the particular choice of areas at a given scale; and an aggregation effect arising from the extent to which data are aggregated over smaller or larger areas. MAUP arises in part from edge effect. If you're interested, in particular about politics and voting, you can read this interesting piece by Bycoffe et al. (2018) to learn more about gerrymandering.

The practical implications of MAUP are immense for almost all decision-making processes involving mapping technology, since with the availability of aggregated maps, policy could easily focus on issues and problems which might look different if the aggregation scheme used were changed.

All studies based on geographical areas are susceptible to MAUP. The implications of the MAUP affect potentially any area level data, whether direct measures or complex model-based estimates. Here are a few examples of situations where the MAUP is expected to make a difference; taken from Gerell (2017).

- The special case of the ecological fallacy is always present when census area data are used to formulate and evaluate policies that address problems at individual level, such as deprivation. Also, it is recognised that a potential source of error in the analysis of census data is 'the arrangement of continuous space into defined regions for purposes of data reporting'
- The MAUP has an impact on indices derived from areal data, such as measures of segregation, which can change significantly as a result of using different geographical levels of analysis to derive composite measures .
- The choice of boundaries for reporting ratios is not without consequences: when the areas are too small, the values estimated are unstable; while when the areas are too large, the values reported may be over-smoothed, i.e., meaningful variation may be lost.

Most often you will just have to remain aware of the MAUP and its possible effects. There are some techniques, that can help you address these issues, and the chapter pointed out at the beginning of this section is a great place to start to explore these. It is possible to also use an alternative, zone-free approach to mapping these crime patterns, perhaps by using kernel density estimation. Here we model the relative density of the points as a density surface — essentially a function of location (x,y) representing the relative likelihood of occurrence of an event at that point. We cover KDE in Chapter 7 of this book.

For now, it's enough that you know of, and understand the MAUP and its implications. Always be smart when choosing your appropriate spatial unit of analysis, and when you use binning of any form, make sure you consider how and if your conclusions might change compared to another possible approach.

4.6 Summary and further reading

In this chapter we explored the use of binning points into hexbins and rectangular grids, as well as transforming polygons in order to enhance the legibility of your maps. For some more information on binning, see Z.-F. Johnson (2011). For transforming polygons, read Tobler (2004) and Nusrat and Kobourov (2016). We can also suggest a read of Langton and Solymosi (2021), and check out this video from SAGE methods by Solymosi (2021).

To read up more on the Modifiable Areal Unit Problem (MAUP) in general, we recommend the original Openshaw (1981). For some criminology/ crime mapping specific reading, try Rengert and Lockwood (2009) and Gerell (2017).

5

Basics of cartographic design: elements of a map

5.1 Introduction

This chapter aims to focus on introducing good practice in map design and presentation. When putting a map together, you need to think about its intended audience (their level of expertise, whether you want them to interact with the map), purpose, and format of delivery (e.g., printed, web, projected in a screen, etc.). There are many design decisions you need to consider: fonts, labels, colour, legends, layout, etc. In this chapter we provide a general introduction to some basic design principles for map production. These themes, and the appropriate election of symbol representation, are the subject matter of cartography, the art and science of map making. Within cartography, a considerable body of research and scholarship has focused on studying the visual and psychological implications of our mapping choices. As noted in previous chapters, one of the problems with maps is that powerful as a tool as they can be, they can lead to misunderstanding. What the mapmaker chooses to emphasise and what the map reader sees may not be the same thing. We will work you through an example of a fairly basic map and the process of taking to a point where it could be ready for presentation to an audience other than yourself.

In this chapter we will be working with some data published by Hungarian police available online (Police.hu 2020). Specifically we will be looking at some statistics related to drink driving. Drink driving is one of a number of problems police confront that relate to impaired and dangerous driving. Hungary has a strict drink driving policy, with the maximum drink diving limit being 0.0 BAC. Most European countries are at 0.5 BAC, while the UK is 0.8 (except 0.5 for Scotland). We have records for each county with the number of breathalyser checks carried out, and the number of these which returned a positive result.

We have downloaded and saved these data in the data folder made available with this textbook. The two data sets we will use are this drink driving data set, called `drink_driving.csv`, and a geometry of the counties within Hungary called `hungary.geojson`.

In this chapter, we will be making use of the following libraries:

```
# Packages for reading data and data carpentry
library(readr)
library(dplyr)
```

```
# Packages for handling spatial data and for geospatial carpentry
library(sf)
library(ggspatial)
library(rnaturalearth)
```

```
# Packages for mapping and visualisation
library(ggplot2)
library(RColorBrewer)
library(ggrepel)
library(cowplot)
```

So let's read in our datasets, and join the attribute data to the geometry using `left_join()` (if you're unsure about any of the below code, revisit Chapter 1 of this book for a refresher!).

```
# read in geojson polygon for Hungary
hungary <- st_read("data/hungary.geojson")

#read in drink driving data
drink_driving <- read_csv("data/drink_driving.csv")

#join the csv (attribute) data to the polygons
hu_dd <- left_join(hungary, drink_driving, by = c("name" = "name"))
```

We can now use this example to talk through the important principles of good visualisation of spatial data. We draw specifically from two areas of research: cartography and data visualisation. Let's start with cartography.

Cartographers have always been concerned about the appearance of maps and how the display marries form with function (Field and Demaj 2012). As there is no definitive definition for what is meant by *cartographic design*, it can be challenging to evaluate what makes *good* design. However, there are themes and elements which can be used to guide the map maker, and offer points of reflection to encourage thoughtful designs.

The primary aim of maps is the communication of information in an honest and ethical way. This means each map should have a clear goal and know its audience, show all relevant data and not use the data to lie or mislead (Dent, Torguson, and Hodler 2008). It should also be reproducible, transparent, cite all data sources, and consider diversity in its audience (Dent, Torguson, and Hodler 2008). So what does that mean for specifically implementing these into practice. While a good amount of critical thought from the map-maker will be required, there are aids we can rely upon. For example, Field (2007) developed a map evaluation checklist which asks the map maker a series of questions to guide their map-making process. The questions fall into three broad categories:

- *Cartographic Requirements* such as what is the rationale for the map? who are the audience?
- *Cartographic Complication and Design* such as are all the relevant features included and the colours, symbols, and other features legible and appropriate to achieve the map's objectives?
- *Map Elements and Page Layout* which tackle some specific features such as orientation indicator, scale indicator, legend, titles and subtitles, and production notes.

We will discuss these elements in this chapter to some degree, and the recommended reading will guide the reader to further advice on these topics.

Data visualisation is a somewhat newer field; however, it seems to encompass the same guiding principles when considering what makes good design. According to Kirk (2016) three principles offer a guide when deciding what makes a *good* data visualisation. It must be: ttrustworthy, accessible, and elegant. The first principle of **trust** speaks to the integrity, accuracy, and legitimacy of any data visualisation we produce. Kirk (2016) suggests this principle to be held above all else, as our primary goal is to communicate truth (as far as we know it) and avoid at all cost to present what we know to be misleading content (see as well Cairo (2016)). **Accessibility** refers to our visualisation being useful, understandable, and unobtrusive, as well as accessible for all users. There are many things to consider in your audience such as *dynamic of need* (do they have to engage with your visualisation, or is it voluntary?), *subject-matter knowledge* (are they experts in the area, are they lay people to whom you must communicate a complex message? and many other factors) (see Kirk (2016)). Finally, **elegance** refers to aesthetics, attention to detail, and an element of *doing as little design as possible* - meaning a certain invisibility whereby the viewer of your visualisation on the content; rather than the design - that is the main point is the message that you are trying to communicate with your data. There are various schools of thought within data visualisation research. For example, the work of Tufte (2001) emphasises clean, minimalist approaches to data visualisation, emphasising a low *data-to-ink-ratio*, which means all the ink needed to print the visualisation should contribute data to the graph. However, other research has explored the usefulness of additional embellishments on charts (which Tufte calls "chart junk") — finding there may be value to these sorts of approaches as well (see Li and Moacdieh (2014)). In this chapter, we will aim to bring together the above principles, and work through a practical example of how to apply these to the maps we make.

5.2 Data representation

5.2.1 Thematic maps

We've been working with thematic maps thus far in this book. There are many decisions that go into making a thematic map, which we have explored at length in the previous chapters, such as how (and whether) to bin your data (Chapter 3) and how (or whether) to transform your polygons (Chapter 4). These are important considerations on how to represent your data to your audience, and require a technical understanding, not only an aesthetic one. So please do read over those chapters carefully when thinking about how to represent your data.

We've covered a few approaches to mapping (using ggplot2, tmap, and leaflet), but here we will continue with ggplot2 package to plot our thematic maps, specifically a choropleth map. We will map our sf objects using the geom_sf() function. To shade each polygon with the values of a specific variable, we use the fill = argument within the aes() (aesthetics) function. Most simply:

```
library(ggplot2)

map <- ggplot(data = hu_dd) + # specify data to use
  geom_sf(aes(fill = total_breath_tests)) # specify aestetics

map
```

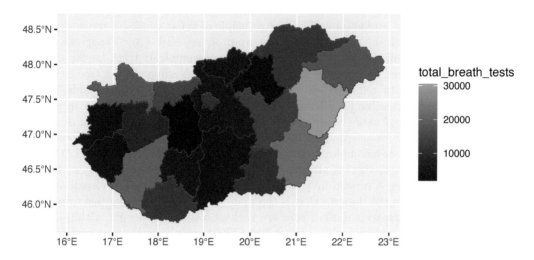

FIGURE 5.1: Quick thematic map with ggplot2

Maps made with `ggplot` are automatically placed upon a grid reference to our data (see our brief overview of building plots with `ggplot` in Chapter 1). To remove this, we can use the `theme_void()` theme, which will strip this away.

```
map <- map +
  theme_void()  # remove grid

map
```

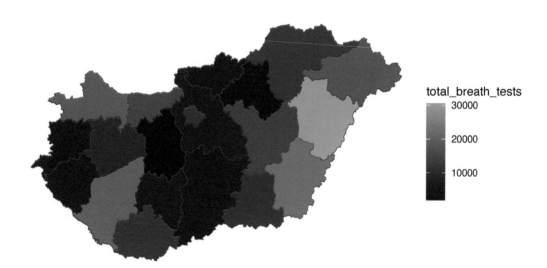

FIGURE 5.2: Add theme_void() to ggplot map

We can change the colour and size of the borders of our polygons with arguments inside the `geom_sf()` function, but outside the `aes()` function, as long as we're not using our data to define these. For example, we can change the line width (`lwd =`) to 0, eliminating bordering lines between our polygons:

```
ggplot(data = hu_dd) +
  geom_sf(aes(fill = total_breath_tests), lwd = 0) +  # specify line width
  theme_void()
```

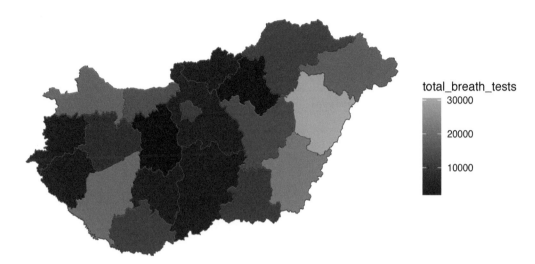

FIGURE 5.3: Thematic map with border lines removed

Or we can change the colour of the borders with the `col =` argument:

```
ggplot(data = hu_dd) +
  geom_sf(aes(fill = total_breath_tests), lwd = 0.5,
          col = "white") + # specify border colour
  theme_void()
```

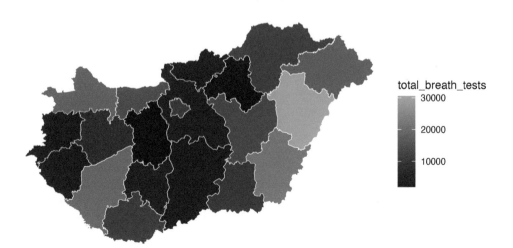

FIGURE 5.4: Thematic map with white borders

Here we have a continuous fill for our values; however, we can employ our learning from Chapter 3 and apply a classification system, such as quantiles. To do this, we might create a new variable which contains the quantiles of our numeric variable, and then use that as our fill =.

```
# create new variable for quantiles
hu_dd <- hu_dd %>%
  mutate(total_quantiles = cut(total_breath_tests,
                          breaks = round(quantile(total_breath_tests),0),
                          include.lowest = TRUE, dig.lab=10))

# plot this new variable
ggplot(data = hu_dd) +
  geom_sf(aes(fill = total_quantiles), lwd = 0.5, col = "white") +
  theme_void()
```

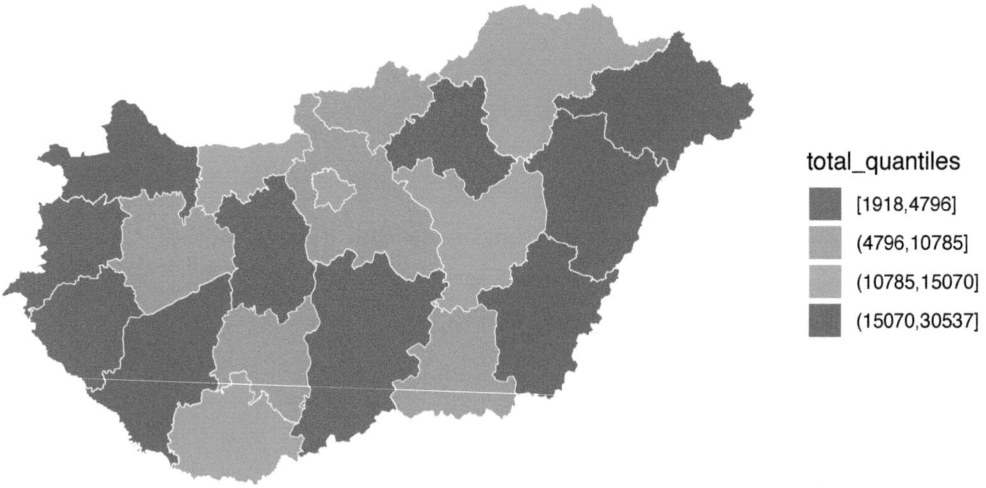

FIGURE 5.5: Map of quantiles with default colour scheme

The colour scheme is terrible, but we will talk about colour in the next section, so we can forgive that for now.

5.2.2 Symbols

You might not want to display your map as a choropleth map. You may want to use symbols. Again we explored this in Chapter 3 where you used the tmap package for this. Here is another way you can use the graduated symbol map with ggplot(). You can take the centroid of each county polygon using the st_centroid() function from the sf package, and then when mapping with geom_sf(), within the aes() function specify the size = argument to the variable you wish to visualise:

```
ggplot(data = hu_dd) +
  geom_sf() +
  geom_sf(data = st_centroid(hu_dd),  #get centroids
          aes(size = total_breath_tests)) +  # variable for size
  theme_void()
```

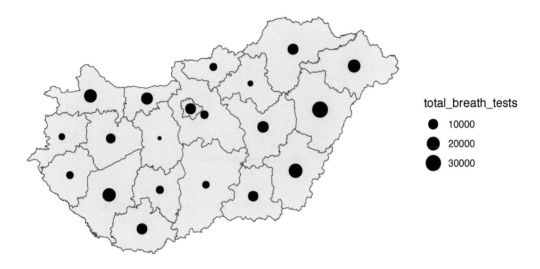

FIGURE 5.6: Graduated symbol map

Like with the thematic map, you can play around with colour and shape:

```
ggplot(data = hu_dd) +
  geom_sf(fill = "light yellow",   # specify polygon fill colour
          col = "white") +    # specify border colour
  geom_sf(data = st_centroid(hu_dd),
          aes(size = total_breath_tests),
          col = "orange") +  # specify symbol colour
  theme_void()
```

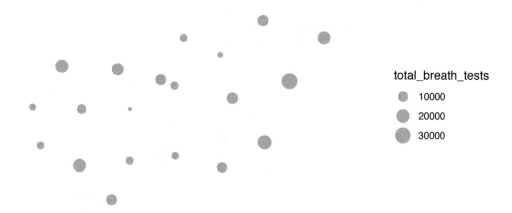

FIGURE 5.7: Graduated symbol map with adjusted colours

You can also change the symbol itself with the `shape` = parameter. For example, you could use a triangle, like below, or to any other symbol.

```
ggplot(data = hu_dd) +
  geom_sf(fill = "light yellow",
          col = "white") +
  geom_sf(data = st_centroid(hu_dd),
          aes(size = total_breath_tests),
          col = "orange",
          shape = 17) + # set shape to be a triangle
  theme_void()
```

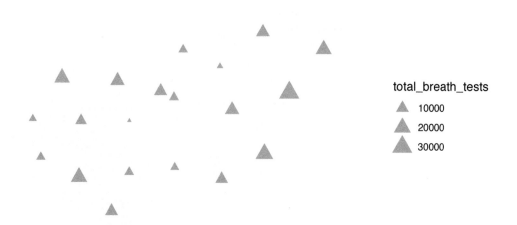

FIGURE 5.8: Graduated symbol map with triangle shape

Or to any other symbol. The possible values that you can use for the shape argument are the numbers 0 to 25, and the numbers 32 to 127. Only shapes 21 to 25 are filled (and thus are affected by the fill colour), the rest are just drawn in the outline colour. Shapes 32 to 127 correspond to the corresponding ASCII characters. For example, if we wanted to use the exclamation mark, the corresponding value is 33. How you choose to represent your data will depend on your decisions to the questions asked above about audience, message, integrity, and so on.

5.2.3   Rate vs. count

In Chapter 3 we have discussed this already in great detail, so do consult this, but it is important that your data are meaningful and easy to interpret. We might, in this case for example, want to consider the rate of positive breath tests per test carried out in each county. To compute this, we might want to consider the proportion of positive results on the breathalyser tests (where the person had been drinking and their result is over the limit). To compute this, we can simply divide the positive results by the total test, and multiply by 100. We also include the round() function in there, as we don't need much precision in this case.

```
hu_dd <- hu_dd %>%
  mutate(pos_rate = round(positive_breath_tests/total_breath_tests*100,1))
```

We can see the county with the highers proportion of test yielding drink drivers is Pest megye with 3%, while the county with the lowest is Hajdú-Bihar with 0.2%. We can visualise this rate on our thematic map in exactly the same way as the count data, but using our new variable in the fill = argument:

```
ggplot(data = hu_dd) +
  geom_sf(aes(fill = pos_rate), lwd = 0.5, col = "white") +
  theme_void()
```

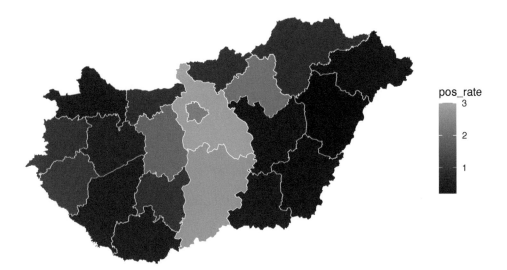

FIGURE 5.9: Map with rate instead of count

5.3   Colour

When choosing a colour palette, the first thing to consider is what kind of colour scheme we need. This will depend on the variable we are trying to visualise. Depending on the kind of variable we want to visualise, we might want a *qualitative* colour scheme (for categorical nominal variables), a *sequential* colour scheme (for categorical ordinal, or for numeric variables) or a *diverging* colour scheme (for categorical ordinal, or for numeric variables). For qualitative colour schemes, we want each category (each value for the variable) to have a perceptible difference in colour. For sequential and diverging colour schemes, we will want mappings from data to colour that are not just numerically but also perceptually uniform.

- **sequential scales** (also called gradients) go from low to high saturation of a colour.
- **diverging scales** represent a scale with a neutral mid-point (as when we are showing temperatures, for instance, or variance in either direction from a zero point or a mean value), where the steps away from the midpoint are perceptually even in both directions.
- **qualitative scales** identify as different from each other the different values of your categorical nominal variable .

For your sequential and diverging scales, the goal in each case is to generate a perceptually uniform scheme, where hops from one level to the next are seen as having the same magnitude.

Of course, perceptual uniformity matters for your qualitative scales for your unordered categorical variables as well. We often use colour to represent data for different countries, or political parties, or types of people, and so on. In those cases we want the colours in our qualitative palette to be easily distinguishable, but also have the same valence for the viewer. Unless we are doing it deliberately, we do not want one colour to perceptually dominate the others.

The main message here is that you should generally not put together your colour palettes in an ad hoc way. It is too easy to go astray. In addition to the considerations we have been discussing, we might also want to avoid producing plots that confuse people who are colour-blind. Fortunately for us, almost all of the work has been done already. Different colour spaces have been defined and standardized in ways that account for these uneven or non-linear aspects of human colour perception.

A good resource is *colorbrewer*[1]. We have come across the work of Cynthia Brewer (2006) in Chapter 3. *Colorbrewer* is a resource developed by Brewer and colleagues in order to help implement good colour practice in data visualisation and cartography (Brewer 1994). This site offers many colour schemes we can make use of for our maps, which are easily integrated into R using the Rcolorbrewer package.

```
library(RColorBrewer)
```

Once you have the package loaded, we can look at all the associated palettes with the function display.brewer.all().

---

[1]http://colorbrewer2.org/

```
display.brewer.all()
```

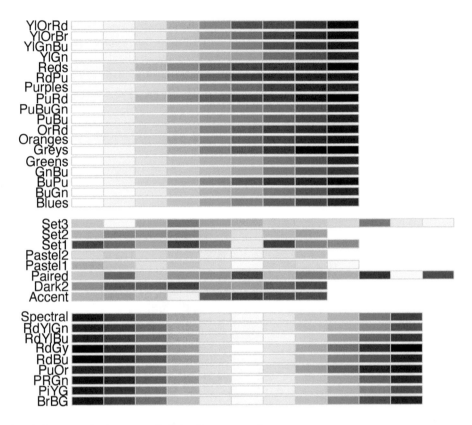

FIGURE 5.10: Palettes in RColorBrewer package

The above gives a wide choice of pallettes, and while they are applicable to all sorts of data visualisations, they were created especially for the case of thematic maps. We might use the above code to pick a palette we like. We might then want to examine the colours more closely. To do this, we can use the `display.brewer.pal()` function, and specify `n=` - the number of colours we need, as well as the palette name with `name =`:

```
display.brewer.pal(n = 5, "Spectral")
```

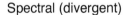

Spectral (divergent)

FIGURE 5.11: The "spectral" palette in RColorBrewer package

Let's go back to our choropleth map of the quantiles of total breath tests per county. We might be interested in this map to show distribution of policing activity, for example. We made this map earlier with the default colour scheme, which didn't really communicate to us the graduated nature of our data we were visualising. To properly do this, we may imagine using a *sequential* scale. We can use one of the sequential scales available within RColorBrewer with adding the scale_fill_brewer() function to our ggplot. In this function we can specify the type= parameter, i.e., if we want to use sequential, diverging, or qualitative colour schemes (specified as either "seq" (sequential), "div" (diverging) or "qual" (qualitative)). We can then specify our preferred palette with the palette = argument. Let's demonstrate here with the "YlOrRd" sequential palette:

```
ggplot(data = hu_dd) +
  geom_sf(aes(fill = total_quantiles),
          lwd = 0.5,
          col = "white") +
  scale_fill_brewer(type = "seq",    # pick palette type
                    palette = "YlOrRd") + # specify palette by name
  theme_void()
```

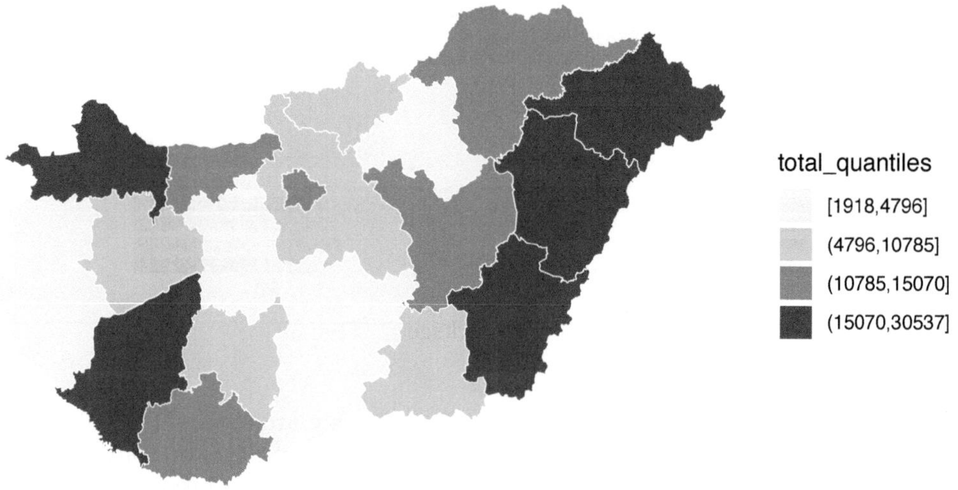

FIGURE 5.12: Thematic map with "YlOrRd" sequential palette

This looks much better, and communicates our message much more clearly. Is this accessible to our colour-blind colleagues? Earlier, when we asked to view all the palettes with the display.brewer.all() function, we did not specify any arguments. However, we can do so in order to filter only those palettes which are accessible for all audiences. We can include the parameter colorblindFriendly = to do so:

```
display.brewer.all(colorblindFriendly = TRUE)
```

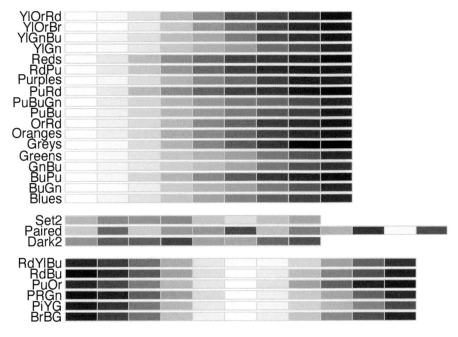

FIGURE 5.13: Colour-blind friendly palettes in RColorBrewer package

You can see that there are a few palettes missing from our earlier results, when we did not specify this requirement. Our recommendation is to always use one of these palettes.

Another way to ensure that we are making accessible maps is to use greyscale (if your map is being printed, this may also save some money). To introduce a greyscale palette, you can use the function `scale_fill_grey()` from the `ggplot2` package:

```
ggplot(data = hu_dd) +
  geom_sf(aes(fill = total_quantiles), lwd = 0.5, col = "white") +
  scale_fill_grey() + # use greyscale colour scheme for fill
  theme_void()
```

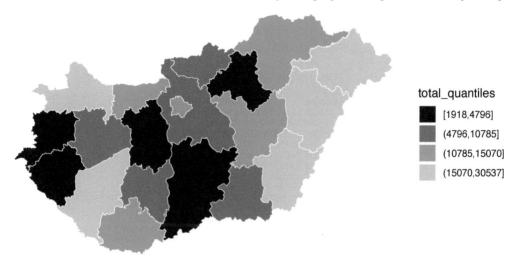

FIGURE 5.14: Map with greyscale palette

Sometimes you might prefer such a map. However, do keep in mind: a number of studies have shown the desirability of monochrome colour (over greyscale) thematic maps, as they are linked to less observer variability in interpretation (Lawson 2021b). So you might want to use something like this instead:

```
ggplot(data = hu_dd) +
  geom_sf(aes(fill = total_quantiles), lwd = 0.5, col = "white") +
  scale_fill_brewer(type = "seq", palette = "Greens") +
  theme_void()
```

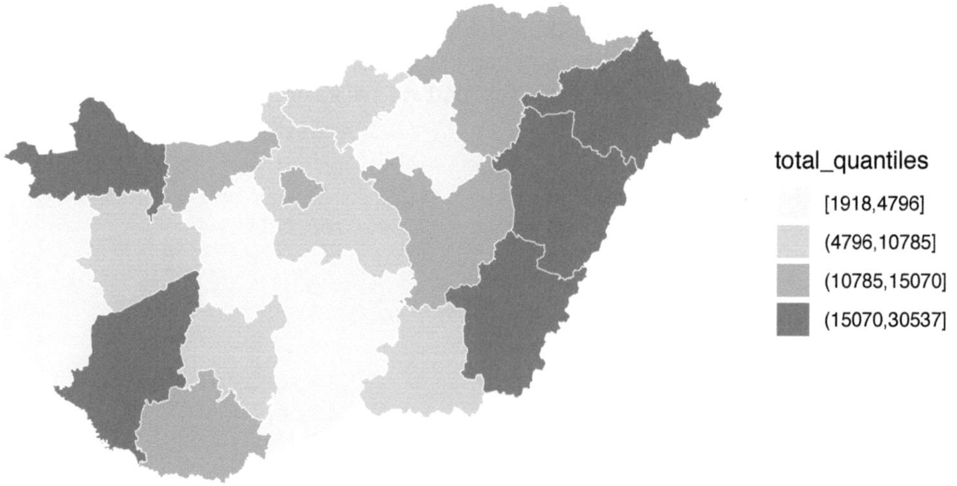

FIGURE 5.15: Map with monochrome palette

Overall, the key thing is to be conscious the colours you choose represent your data. Make sure that they are accessible for all audiences, and best represent the patterns in your data which you want to communicate.

5.4 Text

There are important pieces of information with every map which are represented by text. Titles, subtitles, legend labels, and annotations all help to make the message communicated by your map more clear and obvious to your readers. Further, important information about the underlying data can be communicated through product notes. You want to acknowledge the sources of your data (both attribute data and geometry data), as well as leave some information about yourself as the map-maker, so consumers of your map can understand who is behind this map, and leave some contact information to get in touch with any questions. In this section we go through how to add such text information to your maps.

5.4.1 Titles and subtitles

To give your map a title and subtitle, you can use the appropriate functions from the ggplot2() package. In this case we can add both within the ggtitle function. Make sure that your title is short and specific, so it is clear what your map is about. You can include a subtitle to elaborate on this, or you can add additional information such as the period for which your map represents data (in this case, January 2020).

First, let's save our map into an object, call it map. Then we can add to this object.

```
map <- ggplot(data = hu_dd) +
  geom_sf(aes(fill = total_quantiles),
          lwd = 0.5, col = "white") +
  scale_fill_brewer(type = "seq", palette = "Greens") +
  theme_void()
```

Now we can add a title, with the ggtitle() function.

```
map <- map +
  # specify both title and subtitle:
  ggtitle(label = "Number of breathalyser tests per county in Hungary",
          subtitle = "January 2020")

map
```

Number of breathalyser tests per county in Hungary
January 2020

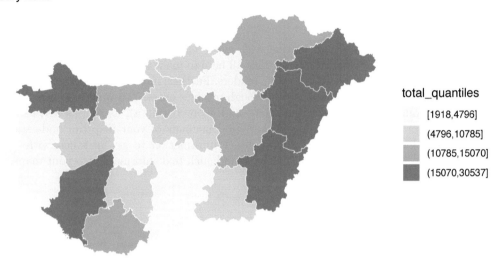

total_quantiles

[1918,4796]

(4796,10785]

(10785,15070]

(15070,30537]

FIGURE 5.16: Add title(s) to map

5.4.2 Legend

Besides an informative title/subtitle, you need your legend to be clear to your readers as well. To modify the title to your legend, you can use the `name` = parameter in your `scale_fill_brewer()` function, where we specified the colour palette.

```
map <- map +
  scale_fill_brewer(type = "seq", palette = "Greens",
                    name = "Total tests (quantiles)")  # desired legend title

map
```

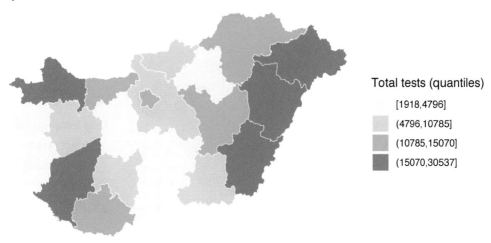

Number of breathalyser tests per county in Hungary
January 2020

Total tests (quantiles)

[1918,4796]

(4796,10785]

(10785,15070]

(15070,30537]

FIGURE 5.17: Add title to map legend

Besides the legend title, we can also change how the levels are labeled. For this, we can use text manipulation functions, such as `gsub()` which substitutes one string for another. For example, we can replace the "," with a " - " if we'd like using the `gsub()` function. We can create a new object, here `new_levels`, which has the desired labels:

```
# create object new_levels with desired labels
new_levels <- gsub(",","," - ",levels(hu_dd$total_quantiles))
```

We can then assign this new levels object in the `labels` = parameter of the `scale_fill_brewer()` function:

```
map <- map +
  scale_fill_brewer(type = "seq", palette = "Greens",
                    name = "Total tests (quantiles)",
                    labels = new_levels)   # specify our new labels

map
```

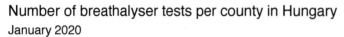

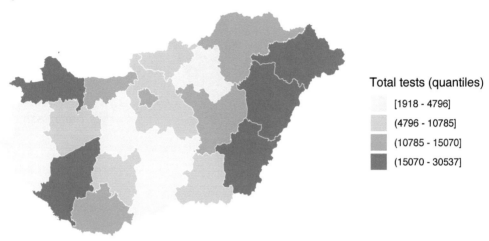

FIGURE 5.18: Edit labels on map legend

Of course it is possible that we want to completely re-write the levels, rather than just swap out one character for another. In this case, we can completely rename the levels if we liked by passing the new, desired labels into this new_levels object

```
new_levels <- c("< 4796", "4796 to < 10785", "10785  to < 15070", "> 15070")
```

And once again, we specify to use these labels in the labels = parameter of the scale_fill_brewer() function:

```
map +
  scale_fill_brewer(type = "seq", palette = "Greens",
                    name = "Total tests (quantiles)",
                    labels = new_levels) # again specify labels object
```

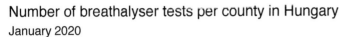

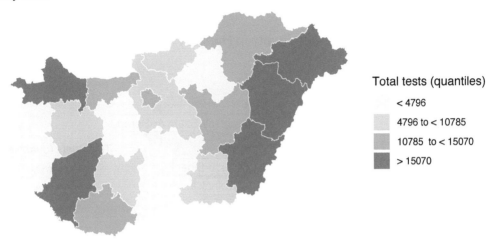

FIGURE 5.19: Re-write values on map legend

You can change the labels you would like, but do keep in mind any loss of information you may introduce. For example with this second version, we no longer know what are the minimum and maximum values on our map, as we've removed that information with our new levels. Again, there are no wrong answers here, but whatever best fits the data and the purpose of the map. Depending on the context, for example, you may use the title to draw home the key finding or lesson you want the reader to take from the map (rather than describing the plot variable).

5.4.3 Annotation

In certain cases, it might be that we want to point out something specific on our map. This would be the case if we imagine showing someone the map in person, and pointing to a specific region, or area, to highlight it. Or we might just want to label all polygons, for clarity. If we are not present to discuss our map, we might want to include some text annotation instead, which will do this for us. From `ggplot2` versions v.3.1.0, the functions `geom_sf_text()` and `geom_sf_label()` make it very smooth for us to do this.

In the below example, let's say we want to label each polygon with the name of the county which it represents. In this case, we use the `geom_sf_label()` geometry, and inside it we use `aes()` to point to which column in our dataframe we want to use (in this case, the column is called *name*).

```
map   +
  geom_sf_label(aes(label = name))   # add layer of labels from the name column
```

FIGURE 5.20: Add labels to map

You may notice, there is some overlapping here which renders some names unreadable. Well while there is work in this space to develop the function geom_sf_label_repel()[2], at the time of writing this is not yet available. However this application of the geom_label_repel() function from the ggrepel package advised by Yutani (2018) achieves the same outcome:

```
library(ggrepel)
```

```
map +
  geom_label_repel(data = hu_dd,        # add repel layer, specify dataframe
                   aes(label = name,    # specify where to find label (name column)
                       geometry = geometry),   # specify geometry
                   stat = "sf_coordinates",   # transformation to use on the data
                   min.segment.length = 0) # don't draw segments shorter than this
```

---

[2]https://github.com/slowkow/ggrepel/issues/111

FIGURE 5.21: Repel labels for legibility

This way we can display the names of all the counties without overlap, so they all become legible. While this is achievable, think back to the principles of good design. This seems busy, and it may overwhelm the reader. Not to mention, is it important that all polygons are labelled here? It might be; remember this depends on the message the map is intended to communicate! But here let's consider a different scenario, where we want to use annotation to label only those counties which meet some specific criteria. For example, you might want to label only those which are in the top quartile. One way to achieve this is to create a separate dataframe, which only includes the desired polygons, and pass this into the `geom_sf_label()` function.

```
#create new dataframe with only top counties
labs_df <- hu_dd %>% filter(total_breath_tests >= 15070)

#add to map
map +
  geom_sf_label(data = labs_df, # specify to use the labels df
                aes(label = name))
```

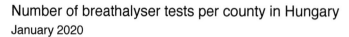

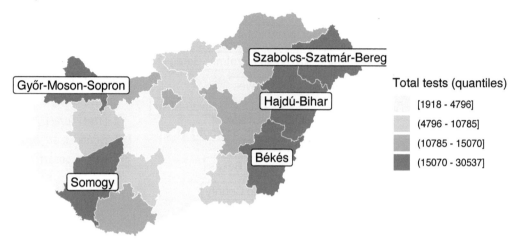

FIGURE 5.22: Label only polygons in top quartile of value on variable of interest

In another scenario, you might want to label only one region of interest. In this case, we might actually want to keep our annotation off the map, and draw an arrow onto the map pointing to where this annotation refers to. We can do this by using the `nudge_x` and `nudge_y` parameters of the `geom_sf_label()` function, to *nudge* the label position along the x and y axes respectively. In this case, let's label only Budapest.

```
#create new labels dataframe
labs_df <- hu_dd %>% filter(name == "Budapest")

map +
  geom_sf_label(data = labs_df, aes(label = name), # label from name column
                nudge_y = 0.9,   # move label on y axis
                nudge_x = -0.1)  # move label on x axis
```

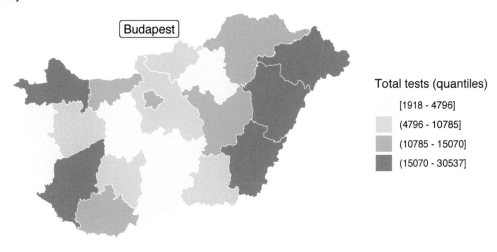

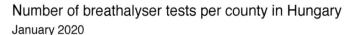

FIGURE 5.23: Label only one area of interest

But this floating label is a little ambiguous, and needs to be more explicitly connected to the map. To achieve this, we might want to use an arrow to point out Budapest on the map. To do this, we can use `geom_curve()` within `ggplot2`. We will need two sets of x and y values for this segment, the start point (x and y) and the end point (`xend` and `yend`). The end point will be the coordinates where we want the arrow pointing to. This would be some x,y pair within Budapest. We can use the `st_coordinates()` function once again the extract the centroid, this time of the Budapest polygon. Let's extract the longitude of the centroid into an object called `bp_x` for our x value, and the latitude of the centroid into an object called `bp_y` for our y value.

```
# get x coordinate
bp_x <- labs_df %>%
  mutate(cent_lng = st_coordinates(st_centroid(.))[,1]) %>% pull(cent_lng)

# get y coordinate
bp_y <- labs_df %>%
  mutate(cent_lat = st_coordinates(st_centroid(.))[,2]) %>% pull(cent_lat)
```

Great, so we have the end point for our segment! But where should it start? Well we want it pointing from our label, so we can think back to how we adjusted this label with the `nudge_x` and `nudge_y` parameters inside the `geom_sf_label()` function earlier. We can add (or subtract) these values to our bp_x and bp_y objects to determine the start points for our curve. Finally, we can also specify some characteristics of the arrow head on our curve with the `arrow =` parameter. Here we specify that we want 2 millimeters.

```
map <- map +
  geom_curve(x = bp_x - 0.1,  # starting x coordinate (the label)
             y = bp_y + 0.9, # starting y coordinate (the label)
             xend = bp_x , # ending x coordinate (BP centroid)
             yend = bp_y,  # ending y coordinate (BP centroid)
             arrow = arrow(length = unit(2, "mm"))) +
  geom_sf_label(data = labs_df,
                aes(label = name),
                nudge_y = 0.9,
                nudge_x = -0.1)

map
```

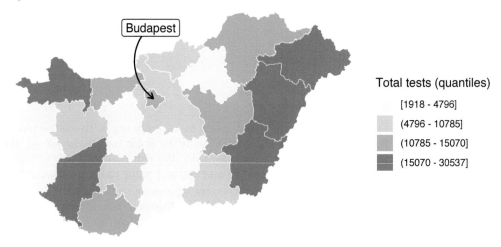

FIGURE 5.24: Add arrow to clarify label off map

This is one way to include annotation while keeping the map clear, but still using the geographic information to reference. Annotations can be useful, but think carefully about whether you need them for your map, as they can also be distracting if not used appropriately.

5.4.4 Production notes

Something that should be a key feature of all maps is the inclusion of production notes. This includes some information about who made it, as well as any attributions for data.

Here we can string together a series of information we want to include, appended with a newline character (\n), in order to keep our notes nice and legible. We save this into a new object called caption_text.

```
caption_text <- paste("Map created by Réka Solymosi (@r_solymosi)",
                      "Contains data from Police Hungary",
                      "http://www.police.hu/hu/a-rendorsegrol/",
                      "statisztikak/kozrendvedelem",
                      "Map data copyrighted OpenStreetMap contributors",
                      "available from https://www.openstreetmap.org",
                      sep = "\n")
```

Then we can include this caption_text object as a caption in the function labs(), which stands for labels.

```
map <- map + labs(caption = caption_text) # include production notes here

map
```

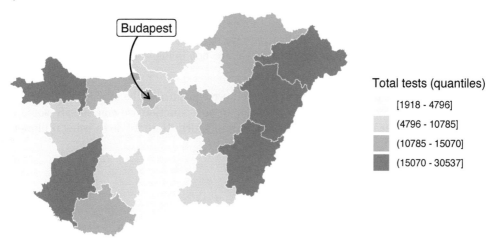

FIGURE 5.25: Include production notes in map

In this way we give credibility to our map, and we also make the proper attributions to where our data come from.

5.5   Composition

Composition of the map is the process of bringing all its elements together in order that they portray a complete image of what you are representing. Composition includes considerations of size, proportions, generalisation, simplification, and similar topics. We do not address these here, as they rely so much on the specific purpose of the map being created. Is it for the web? Is it for print? Are detailed outlines of coasts and waterways important, or is a generalised representation of the underlying geography enough? These are questions the map-maker should answer early on, and then pick geometry data, and specify output sizes and resolutions accordingly. In this section instead, we will focus on the element of composition which is concerned with the inclusion of basic map elements - information required by the map readers to make sense of our data, specifically orientation and scale indicators.

5.5.1   Orientation indicators

It used to be that no map was complete without the inclusion of an orientation indicator (known colloquially as the "North Arrow"). Readers who are geography fans may know that the issue of where North is maybe not so straightforward. True north (the direction to the North Pole) differs from magnetic north, and the latter actually moves around as the Earth's geophysical conditions change. There are reference maps which include both; however, for most crime-mapping applications we can conclude that this is overkill. Most maps are oriented to true north, anyway, so we are not being very deviant with choosing this approach.

So how to include this in our mapping in R? Well, we can turn to the `ggspatial` library, and employ the function `annotation_north_arrow()`. In this function we can specify some aesthetic properties of our arrow, such as the height and width. Here we do so using millimetres as units.

```r
library(ggspatial)

map + annotation_north_arrow(height = unit(7, "mm"), # specify arrow height
                    width = unit(5, "mm")) # specify arrow width
```

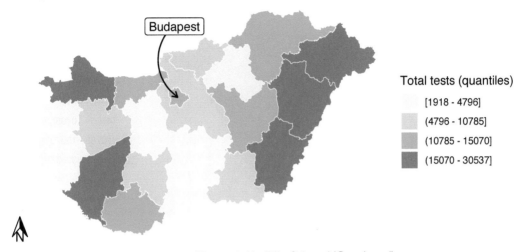

Number of breathalyser tests per county in Hungary
January 2020

FIGURE 5.26: Add north arrow

You can also change the style with the `style =` parameter, and then choose from styles such as `north_arrow_fancy_orienteering()` or `north_arrow_minimal()`:

```
map <- map + annotation_north_arrow(height = unit(7, "mm"), # specify arrow height
                    width = unit(5, "mm"),
                    style = north_arrow_minimal()) # specify arrow style

map
```

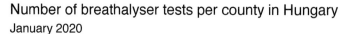

Number of breathalyser tests per county in Hungary
January 2020

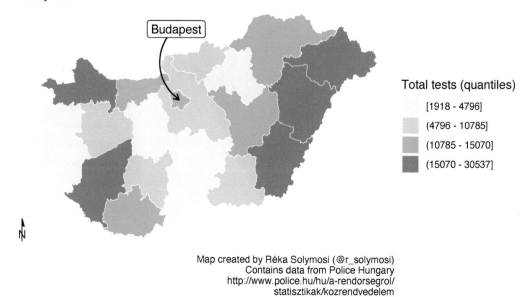

FIGURE 5.27: Change type of north arrow

5.5.2 Scale indicators

Besides the North Arrow another, key feature of maps is the scale indicator, which helps to understand distances we are presenting in our maps. Generally, scale should always be indicated or implied, unless the audience is so familiar with the map area. You could use text to indicate scale. For example you could write: "One centimeter is equal to one kilometer, or you could write 1:10000. But a common, graphical representation is to use a scale bar. Also in the ggspatial library, there is the function annotation_scale(), which helps us achieve this. To plot both the north arrow and the scale indicator, you want to think about where you place these. You can move them along the x axis using the pad_x parameter, and along the y axis with the pad_y parameter.

```
map <- map +  annotation_scale(line_width = 0.5,      # add scale and specify width
                height = unit(1, "mm"), # specify height
                pad_x = unit(6, "cm"))  # adjust on x axis


map
```

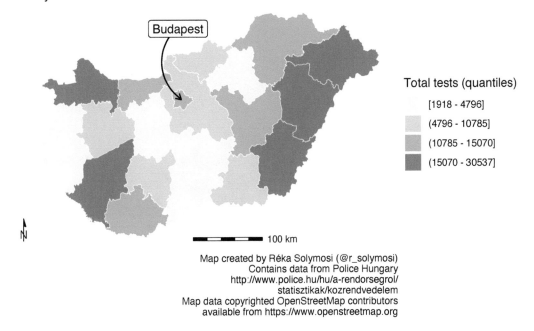

FIGURE 5.28: Include scale indicator on map

You can move these elements about to achieve your desired composition.

5.6 Context

Besides the orientation and scale indicators, there are other ways to give context to your map, that is situate it within the wider environment, and put things into perspective for your map readers. In this section we will touch on basemaps, although this is something we have already encountered in great detail in earlier chapters, here we illustrate how to add basemaps in ggplot. We also introduce inset maps, as a way of highlighting where your map sits in the wider context.

5.6.1 Basemap

As mentioned above, we have encountered and included basemaps in exercises in previous chapters. For the sake of illustration, we can add a basemap now using the `annotation_map_tile()` function also from the `ggspatial` package (like the orientation and scale indicators). Make sure that the basemap is the first layer added to the map, so that all subsequent layers are drawn on top of it. If we were to add `annotation_map_tile()` last, it would cover all the other layers.

```
ggplot(data = hu_dd) +
  annotation_map_tile() +  # add basemap layer first
  geom_sf(aes(fill = total_quantiles), lwd = 0.5, col = "white") +
  scale_fill_brewer(type = "seq", palette = "Greens", name = "Total tests")
```

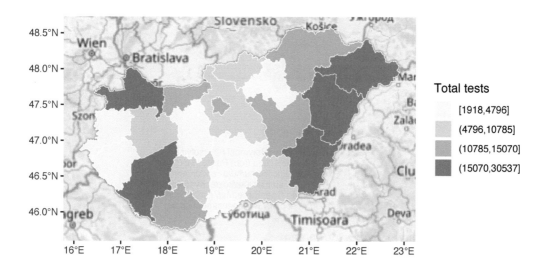

FIGURE 5.29: Add a basemap to our map

This provides one way to add context. In previous iterations, we have adjusted the opacity of our other layers, in order to aid visibility of the basemap underneath them, so this might be something to consider.

5.6.2   Inset maps

Inset maps provide another approach to situating your map in context. You might use this to show where your main map fits into the context of a larger area; for example, here we might illustrate how Hungary is situated within Europe. You might also use an inset map in another situation, where you have additional areas which you want to show which may be geographically far but politically related to your region. For example, we might want to portray a map of the United States of America and make sure to include Hawaii and Alaska on the map. The basic principles behind these maps are the same. Essentially we must create two map objects, and then bring these together. Let's illustrate how.

First, we need to create the map we will be displaying in the inset map. In this case, let's highlight the location of Hungary on a map of Europe. We can do this by creating a map of Europe (let's use the `rnaturalearth` package for this). We create a list of the countries from the world map 'countries110', and filter only Europe (we also exclude Russia, because it is so big it makes the rest of Europe hard to see on a smaller map, and Iceland as it's far, also making the map bigger than we need).

```
library(rnaturalearth)

europe_countries <- st_as_sf(countries110) %>%  # get geom for all countries
  filter(region_un=="Europe" & # select Europe
         name != "Russia" & name != "Iceland") %>%  # remove Russia and Iceland
  pull(name) # get only the names in a list

europe <- ne_countries(geounit = europe_countries, # get geoms for countries in list
                       type = 'map_units',  # country type as map_units
                       returnclass = "sf") # return sf object (not sp)
```

Now we can use the returned sf object `europe` to create a map of Europe. But this isn't necessarily enough context. We also want the inset map to highlight Hungary within this map. We can do this by creating another layer, with only Hungary, and making its border red and use a thicker line width. By layering this on top of the Europe map, we are essentially highlighting our study region.

```
inset_map <- ggplot() +  # create new ggplot
  geom_sf(data = europe,  # add europe map as first layer
          fill = "white") + # white fill
  geom_sf(data = europe %>% filter(name == "Hungary"), # new layer only Hungary
          fill = "white" ,  # white fill
          col = "red",   # make the border red
          lwd = 2) +    # make border line thick
  theme_void() +  # strip grid elements
  theme(panel.border = element_rect(colour = "black", # draw border around map
                                    fill=NA))
```

We now have this separate map, which highlights where Hungary can be found, right there in Central Europe. To display this jointly with our map of the breathalyser test, we must join the two maps. For this, we will need both as separate objects. We've already assigned our inset map to the object `inset_map`, and we also have our main map object we've been working with, called `map`.

So now we have our inset map and main map stored as two map objects. To display them together, we can use the `ggdraw()` and `draw_plot()` functions from the `cowplot` package. Let's load this package.

```
library(cowplot)
```

First, we set up an empty drawing layer for our ggplot using the `ggdraw()` function. Then we layer on the two maps both using the `draw_plot()` function. This allows us to draw plots and subplots. This function places a plot (which we specify as the first parameter of this function) somewhere onto the drawing canvas. By default, coordinates run from 0 to 1, and the point (0, 0) is in the lower left corner of the canvas. We want to therefore specify where the plots go on our canvas explicitly. Alongside position, we can also specify size. This is important, as we usually make the inset map smaller.

```
hu_dd_with_inset <- ggdraw() +  # set layer
  draw_plot(map) +  # draw the main map
  draw_plot(inset_map,  # draw inset map
            x = 0.75,   # specify location on x axis
            y = 0,      # specify location on y axis
            width = 0.35,   # specify width
            height = 0.35)  # specify height
```

We now have our final map, which we can check out now:

```
hu_dd_with_inset
```

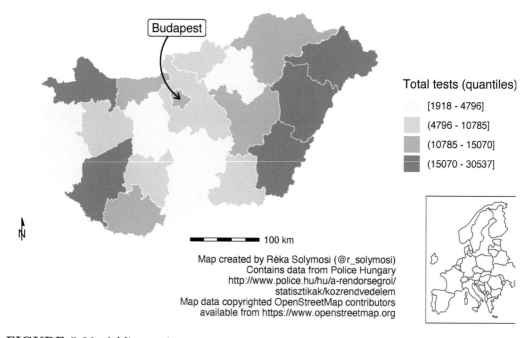

FIGURE 5.30: Adding an inset map to our map

You can play around with where you position your inset map by adjusting the x and y coordinates. You can also play around with the size of it by adjusting the parameters for height and width. And as mentioned above, you can use inset maps not only for context, but also to include geographically far away regions which belong to the same political unity, for example to include Alaska and Hawaii in maps of the United States.

5.7 Summary and further reading

In this chapter we covered some principles of data visualisation and good map design, specifically how to implement some of these using `ggplot2` library in R. We talked about symbols, colour, text, and adding context. `ggplot2` provides an incredibly flexible framework, and through the use of layers you can achieve a very beautiful and meaningful map output. We could for example consider adding topography, such as rivers, lakes, and mountains to our maps, and it would be only a case of adding another layer. To get more familiar with this, we recommend Wickham (2010) and Healy (2019).

For general readings on cartography, the work of Kenneth Field provides thorough and engaging guidance (e.g., Field and Demaj (2012) or Field (2018)). For data visualisation, while Tufte (2001) is a classic text, those looking for practical instruction may wish to turn to Kirk (2016) or Cairo (2016) for more guidance.

6

Time matters

6.1 Introduction

In this chapter we provide a brief introduction into spatio-temporal visualisation. The importance of place in criminology and crime analyses is widely recognised and is the central topic of this book. However, taking into consideration **time** is just as important as place. We often hear that crime is "going up" or "going down" over time. These variations on the levels of crime along time also vary across space. These variations across both time and place are called **spatio-temporal variations**, and are of crucial importance for crime analysis, explanation, and prevention.

Traditionally, the temporal and spatial analyses of crime are not introduced in a combined manner. Although a great deal of energy has been devoted to produce accessible training material for crime analysts on geographic information systems and spatial analysis, most criminology degrees (even at Postgraduate level) and training packages for crime analysis devote a very limited (if any) content to the appropriate visualisation and analysis of temporal and spatio-temporal data. Therefore, before we discuss the spatio-temporal, we have to introduce a few key ideas and concepts about temporal crime analysis.

In this chapter, we will therefore give a very high-level overview of **temporal crime analysis**, before moving on to ways in which we can display spatio-temporal variation in our data using maps. The key concepts covered in this chapter are:

- an introduction to temporal data in crime analysis,
- cleaning and wrangling temporal data,
- visualising time data,
- time series and its three components: trend, seasonality, and random variation,
- visualising spatio-temporal variation.

We will be making use of the following R packages:

```
# Basic reading and data carpentry
library(readr)
library(dplyr)
library(lubridate) # adds functionality for better handling of temporal data

# Packages for handling spatial data
library(sf)
library(spacetime)

# General visualisation and ggplot plugins for temporal data
```

```
library(ggplot2)
library(ggfortify)
library(ggTimeSeries)
library(ggseas)
library(gganimate)
library(tmap)
```

6.2   Temporal data in crime analysis

In this book so far we have really emphasised the role of place in presenting opportunities for crimes to occur. However, we cannot consider space without also considering time. An area might look very different during the day and during the night, on a weekend or on a weekday, and in the summer or in the winter.

On a macro-scale, the relationship between crime and the seasons is something that had been a topic of concern for researchers as long as space has (Baumer and Wright 1996). Zooming into a micro-scale, changes in routine activities with time of day or day of week will affect the profile of a place, and linking back to the idea of crime places such as *generators* or *attractors*, will have significant effect on crime rates. In conceptualising these crime places, Newton (2018) emphasised the importance to consider the measure of busyness of a place by time of day in order to understand its role as an attractor, generator, or other crime place. For example, at transit stations during school days, there is a morning peak time of work and school users combined, an afternoon school closing peak, and a secondary and slightly later end-of-workday peak time. Any calculation of crime rates needs to account for these micro-level temporal variations. At whatever unit of analysis, time is a vital variable to include in our crime analysis and criminological research. To quote Tompson and Bowers (2013):

> [I]t is important to disaggregate data into sensible temporal categories to have a real understanding of the relationship between the variables under scrutiny (p.627).

Returning to the importance of the role of place, we can introduce **spatio-temporal data** analysis. Spatio-temporal analysis is the process of utilising geo and time-referenced data in order to extract meaning and patterns from our data. In the earlier days, crime pattern analysis has tended to focus on identifying areas with higher densities of criminal activity, but not so much the monitoring of change in crime patterns over time (Ratcliffe and McCullagh 1998). However, crime hot spots display significant spatio-temporal variance, and the identification of spatio-temporal patterns of hot streets provides significant 'actionable intelligence' for police departments (Herrmann 2013). Evidently, we cannot ignore time as a variable in our analyses. And while the main focus of this book is space, we must take at least one chapter to introduce some key concepts, and provide additional resources for readers to follow up with.

6.3 Temporal data wrangling

Temporal data means that we can perform all sorts of exciting operations in our data wrangling processes. Just how we learned about spatial operations with spatial data, there are some things we can do only with temporal data. In this section we will introduce some of these.

A key R package that will help with temporal data wrangling is `lubridate`. Date-time data can be frustrating to work with and many base R commands for date-times can be unintuitive. The package `lubridate` was created to address such issues, and make it easier to do the things R does with date-times.

For crime data, we will be using crime data from New York City made available by the R package `crimedata` (Ashby 2019). You can refer to these citations to learn more about acquiring data from this fantastic resource. However for now, let's make use of the data provided with the book, where we selected a specific subset of aggravated assault in New York for a period of five years.

```
agassault_ny<-read_csv("data/agassault.csv")
```

When you read the data into R, you will see that there is a column for date called **date_single**. Let's have a look at the first value in this column:

```
agassault_ny %>%  dplyr::select(date_single) %>% head(1)
```

```
## # A tibble: 1 x 1
##    date_single
##    <dttm>
## 1 2014-01-01 00:03:00
```

We can see that the date is stored in the following format: year-month-day hour-minute-second. So the first date on there you can see is *2014-01-01 00:03:00*. What kind of variable is this?

```
class(agassault_ny$date_single)
```

```
## [1] "POSIXct" "POSIXt"
```

Our date and time variables are of class `POSIXct` and `POSIXt`. These are the two basic classes of date/times. Class "POSIXct" represents the (signed) number of seconds since the beginning of 1970[1] as a numeric vector. Class "POSIXt" is a named list of vectors representing seconds (0–61), minutes (0–59), hours (0–23), day of the month (1–31), months after the first of the year (0–11), years since 1900, day of the week, starting on Sunday (0–6), and a flag for whether it is daylight savings time or not (positive if in force, zero if not, negative if unknown). Let's plot this data:

---

[1]https://en.wikipedia.org/wiki/Unix_time

```
ggplot(agassault_ny, aes(date_single)) +
  geom_freqpoly(binwidth = 7*24*60*60) # 7 days in seconds
```

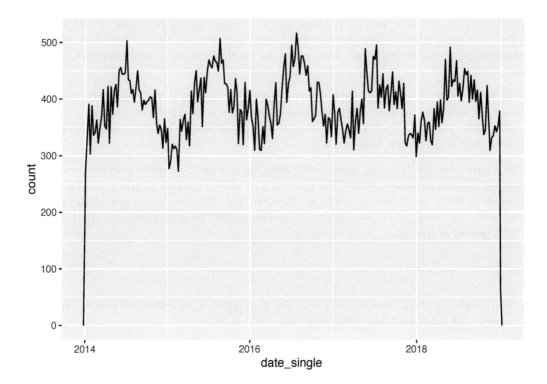

FIGURE 6.1: Number of assaults aggregated to 7 day chunks

Notice what `geom_freqpoly()` is doing. We have a dataframe with rows for each case. The data is not aggregated in any form. But this function counts on the fly the number of cases (rows) for each of the bins as we define them. It is, thus, a convenient function that saves us from having to first do that aggregation ourselves when we want to plot it.

An alternative approach to plotting individual components is to round the date to a nearby unit of time, with `floor_date()`, `round_date()`, and `ceiling_date()`. These functions live inside the lubridate package. Each function takes a vector of dates to adjust and then the name of the unit round down (floor), round up (ceiling), or round to. So to aggregate per month, we will code as:

```
agassault_ny %>%
  count(month = floor_date(date_single, "month")) %>% # use floor date function
  ggplot(aes(month, n)) +
    geom_line()
```

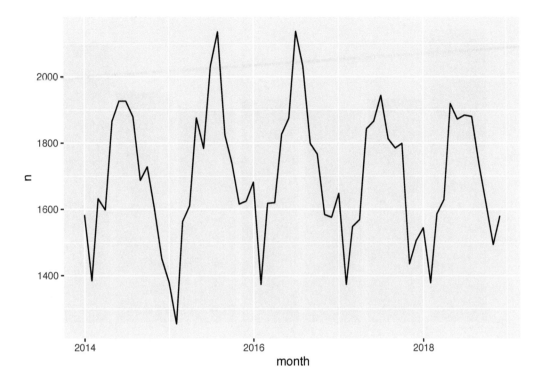

FIGURE 6.2: Number of assaults aggregated to Month

What if we ask the question: which year had the most aggravated assaults? Or what if we want to know whether aggravated assaults happen more in the weekday, when people are at work, or in the weekends, maybe when people are away for a holiday? You have the date, so you should be able to answer these questions, right?

Well you need to be able to have the right variables to answer these questions. To know what year saw the most aggravated assaults, you need to have a variable for year. To know what day of the week has the most aggravated assaults, you need to have a variable for day of the week. So how can we extract these variables from your date column? Well, luckily the `lubridate` package can help us do this. We can use the `year()`, `month()`, `day()`, and `wday()` to extract these components of a date-time.

```
agassault_ny <- agassault_ny %>%
  mutate(year = year(date_single), # extract year
         month = month(date_single, label = TRUE, abbr=FALSE), # month
         day = day(date_single), # extract day
         wday = wday(date_single, label = TRUE, abbr=FALSE) ) # extract day of week
```

We have now created a set of additional variables that have extracted information from your original time of occurrence variable. Let's consider distribution of events per day of the week.

```
ggplot(agassault_ny, aes(x = wday)) + geom_bar()
```

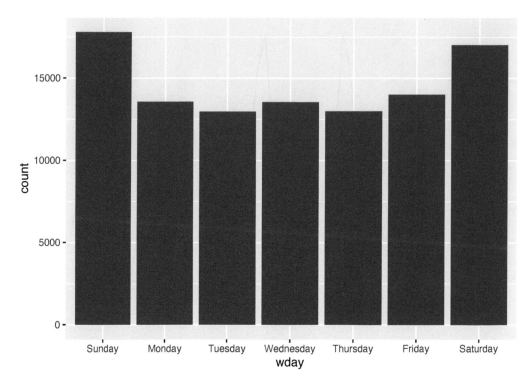

FIGURE 6.3: Distribution of events per day of week

In order to extract such date-time information from variables, we need these to be date-time objects. We saw above that in this case this assumption was met. However if it is not, you can turn text column of dates into a date-time object using `lubridate`'s functions. For example if you have a data set of day-month-year separated with "/", you can use the `dmy()` function (stands for day, month, year) to parse it as a date-time object. On the other hand, if you have some US data, and it is actually written as month-day-year, separated by "/", you can simply shuffle the order of the letters in the function, and use the function `mdy()`. They will translate into the same item. See for yourself:

```
dmy("15/3/2021") == mdy("3/15/2021")
```

```
## [1] TRUE
```

In fact, `lubridate` is so good, it can even parse text representations of months. Look at this for example:

```
dmy("15/3/2021") == mdy("March/15/2021")
```

```
## [1] TRUE
```

This is amazing stuff, which will definitely come in handy, especially if you might be a crime analyst working with some messy data. The `lubridate` package should most definitely form part of your data wrangling toolkit in this case.

6.4 Visualising time data

Once we have parsed and cleaned the temporal information that is available in our date variable, we can make use of this to visualise trends in other ways, not just using time series. For example, we might want to show where hot spots in time occur, within a year, or within a week, or some other set interval. One approach for this is to use a calendar heatmap. First, we create a column for date, removing the time (hour, minute and second). We can achieve this easily by using the `date()` function in `lubridate`:

```
agassault_ny$date <- date(agassault_ny$date_single)
```

If we have a look, we can now see that the time component has been stripped away, and we have this date object. Let's look at the value for the first row.

```
agassault_ny$date[1]
```

```
## [1] "2014-01-01"
```

That looks like what we are expecting. Now let's see the class of this variable.

```
class(agassault_ny$date)
```

```
## [1] "Date"
```

It is a date object, so we will be able to do date-time operations to it. Now what we want to do is create a heatmap, a sort of calendar, of the number of aggrevated assault incidents per date. We can use our well-known attribute operations here, specifically the `group_by()` function in the `dplyr` package to count the events per day.

```
agassault_ny_d <- agassault_ny %>%
                group_by(date) %>% summarise(assaults = n())
```

Now we have this new dataframe, the number of aggravated assaults per day. To make our temporal heatmap calendar, we will now use a `ggplot2` extension, `ggTimeSeries`, that allows us to produce calendar heat visualisations. In this package, we can use the `ggplot_calendar_heatmap()` function, which creates a calendar heatmap. This approach provides context for weeks, and day of week which makes it a better way to visualise daily data than line charts.

```
library(ggTimeSeries)
```

```
ggplot_calendar_heatmap(
    agassault_ny_d,  # our dataset of dates and number of crimes
    cDateColumnName = 'date',  # column name of the dates
    cValueColumnName = 'assaults') + # column name of the data
    xlab(NULL) + # x axis lable
```

```
   ylab(NULL) + # y axis lable
   scale_fill_continuous(low = '#f7fcfd',  # set colour for low count
                         high = '#6e016b') +   # set colour for high count
 facet_wrap(~Year, ncol = 1) + # separate out by each year
 theme_minimal()
```

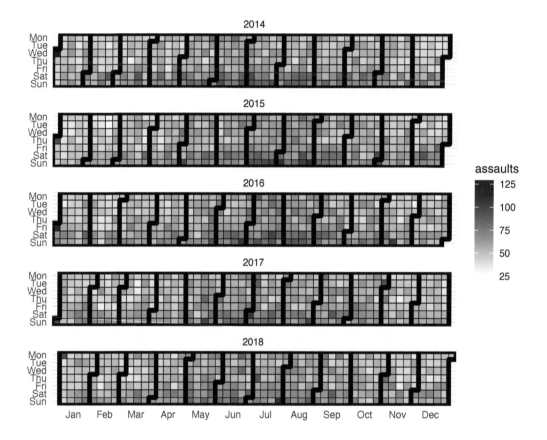

FIGURE 6.4: Calendar heat map of aggravated assaults

This sort of visualisation might be useful for example to see if certain days of the week have more incidents than others, possibly due to differences in the underlying routine activities that predominate for example weekends versus weekdays. For a bit of fun, you can read Moss (2013) for a safety rating on each day of the week.

6.4.1 How (not) to present time

In our calendar heatmap, above we presented count of crimes on each day. But there are other approaches to visualise change over time as well. One such approach is to present **percentage change** from one point in time to the next. There are many issues with this approach, which lead to misinterpreting the data, which are explored in detail by Wheeler (2016). For example, he mentions that percentage change is not symmetric: "For example, an increase from 4 to 5 crimes is a 25% increase, whereas a decrease from 5 to 4 crimes is only a 20% decrease" (p.x). He presents an alternative metric in Poisson z-scores.

Another issue is with the very popular approach of showing changes as **year-to-date**. In the post: "Why you can't identify changes in crime by comparing this month to last month," Ashby (2020) presents some great arguments for why attempts to identify changes in crime frequency that involve simply comparing the number of crimes this week/month/year to the number that occurred last week/month/year, or comparing this week, month, etc. to the same period last year is not an appropriate way to analyse change over time. For example, doing this means throwing away useful information, ignoring trends and seasonality, and leaves results vulnerable to noise. As an alternative, he promotes the use of creating a forecast based on historic data, and comparing observed values against this. It is important that when presenting temporal data, we keep in mind these notes of caution, and follow good practice recommendations such as those notes here.

6.5 Time series analysis

A key way to ensure we are analysing our time data appropriately is to deal with time series data and treat them accordingly. **Time series analysis** looks at trends in crime or incidents. A crime or incident trend is a broad direction or pattern that specific types or general crime and/or incidents are following. Three types of trend can be identified:

- **overall trend** – highlights if the problem is getting worse, better or staying the same over a period of time
- **seasonal, monthly, weekly or daily cycles of offences** – identified by comparing previous time periods with the same period being analysed
- **random fluctuations** – caused by a large number of minor influences, or a one-off event, and can include displacement of crime from neighbouring areas due to partnership activity or crime initiatives.

Decomposing these trends is an important part of what time series analysis is all about. We will see some examples.

6.5.1 Plotting time series data

The data we will work with now is called `femicidos.csv` in the companion data (see *Preamble* to this book if not sure where to find this), and is a collection of intimate partner femicides from Spain.

```
femicidios <- read_csv("data/femicidos.csv")
```

This dataframe has only two columns, `femicidos`, which is a monthly observation of the number of intimate partner femicides per month, starting in January 2003, and `month`, which is the date for each observation. So, each row represents a monthly count of these crimes. We could do something like plot the number of crimes over each month using `ggplot2`.

```
library(ggplot2)

ggplot(femicidios, aes(x = month, y = femicidios, group = 1)) +
  geom_point() +
  geom_line() +
  theme_minimal() +
  ggtitle(label = "Number of femicides per month", subtitle = "In Spain") +
  theme(axis.text.x = element_text(hjust = 1, angle = 45))
```

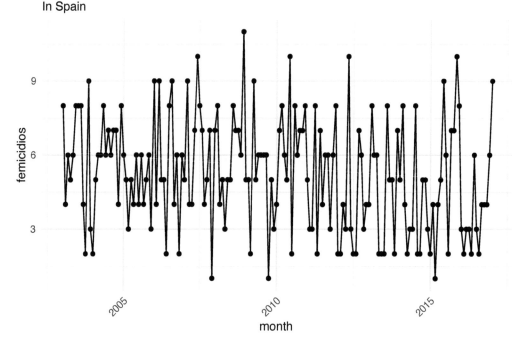

FIGURE 6.5: Plot the number of femicides per month

This visualisation gives us an insight into how the count of crimes varies between observations, but it subsumes in itself the three elements mentioned above (the overall trend, seasonal fluctuation, and random noise). This makes it difficult to isolate and discuss any one of these, and to answer the question about whether crime is going up or down, or whether there are any seasonal fluctuations present. In order to achieve this, we want to **decompose** the data into these components.

To look into decomposing into these components, we can use the functionality within R to deal with time series data. To take advantage of many of these, we will need our dataframe to be stored as a **time series** object. This enables us to apply R's many functions for analysing time series data. To store the data in a time series object, we use the ts() function. Inside this function, we pass only the column which contains the number of crimes for each month (we filter with the select() function from dplyr package).

```
fem_timeseries <- ts(femicidios %>% dplyr::select(femicidios))
```

We have taken our dataframe of observed values in our time series (monthly observations of crime counts in this case) and transformed it into a matrix with class of ts — essentially representing our data as having been sampled at set intervals (in this case, every month). Once we have the result stored in our fem_timeseries object, we can auto print to see some details:

```
fem_timeseries
```

We can see that each observation point has been numbered in order, and we have a value for each observation (the number of femicides recorded in each month). Sometimes, the time series dataset that you have may have been collected at regular intervals that were less than one year, for example, monthly or quarterly. In this case, you can specify the number of times that data was collected per year by using the frequency parameter in the ts() function. For monthly time series data, you set frequency = 12, while for quarterly time series data, you set frequency = 4. You can also specify the first year that the data was collected, and the first interval in that year by using the start parameter in the ts() function. So, in our case, we would do as follows:

```
# transform into time series
fem_timeseries <- ts(femicidios %>%
                  select(femicidios), # specify dataframe selecting column
              frequency=12,  # specify monthly frequency
              start=c(2003,1)) # specify start time (January 2003)
```

Now that we have created this ts object, we can use the ts specific functions in order to extract meaning and insight. For example, going back to plotting our data so that we can see what sort of trends might be going on with crime, we can make use of the plot.ts() function, the basic plotting method for objects that are of class ts.

```
plot.ts(fem_timeseries)
```

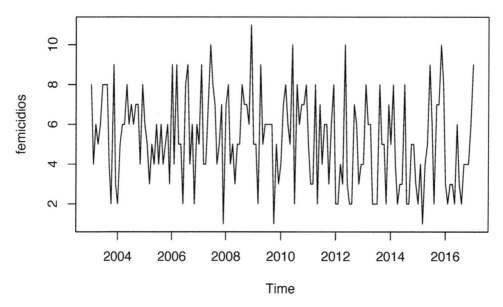

FIGURE 6.6: Basic plot of the ts object

This plot should look similar to the one we created using ggplot2 above; however, our observations are now treated as a continuous variable, labeled "Time". We can of course also use `ggplot2` to plot a time series like the one we just did but here we would need a variable encoding the date (and preferably a full date, not just month and year as here). As you can see, it is very noisy. Fortunately, the annual count for intimate partner femicides is low in Spain. There seems to be some seasonality too. But what more can we do with plotting time series objects?

A seasonal time series consists of a trend component, a seasonal component and an irregular component. Decomposing the time series means separating the time series into these three components; that is, using statistics to estimate these three components. To estimate the trend component and seasonal component of a seasonal time series that can be described using an additive model, we can use the `decompose()` function in R. This function estimates the trend, seasonal, and irregular components of a time series using moving averages. It deals with additive or multiplicative seasonal components (the default is additive). The function `decompose()` returns a list object as its result, where the estimates of the seasonal component, trend component and irregular component are stored in named elements of that list objects, called "seasonal", "trend", and "random", respectively. Let's now decompose this time series to estimate the trend, seasonal and irregular components.

```
fem_timeseriescomponents <- decompose(fem_timeseries)
```

The estimated values of the seasonal, trend and irregular components are now stored in variables `fem_timeseriescomponents$seasonal`, `fem_timeseriescomponents$trend` and `fem_timeseriescomponents$random`. For example, we can print out the estimated values of the seasonal component by typing:

```
fem_timeseriescomponents$seasonal
```

The estimated seasonal factors are given for the months from January to December and are the same for each year. The largest seasonal factor is for July (about 0.70), and the lowest is for February (about -0.76), indicating that there seems to be a peak in femicides in July and a trough in femicides in February each year. We can plot the estimated trend, seasonal, and irregular components of the time series by using the plot() function, for example:

```
plot(fem_timeseriescomponents)
```

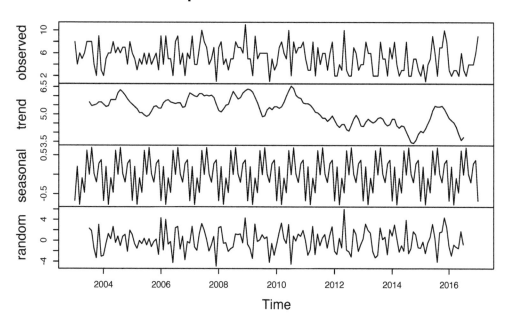

FIGURE 6.7: Decomposed time series of femicides in Spain

Once we remove the noise and the seasonal components, it becomes easier to see the estimated trend. Notice that while random and seasonal components still look messy, their scales are different and centred around zero.

We can adapt this code to decompose and estimate the trends for the aggravated assault data for NYC that we have used earlier in the chapter.

```
# select relevant column (assaults)
agassault_ny_d2 <- dplyr::select(agassault_ny_d, assaults)

#use ts() to transform to time series object
ny_timeseries <- ts(agassault_ny_d2, frequency=365, start=c(2014,1,1))

# decompose time series
ny_timeseriescomponents <- decompose(ny_timeseries)
```

```
#plot results
plot(ny_timeseriescomponents)
```

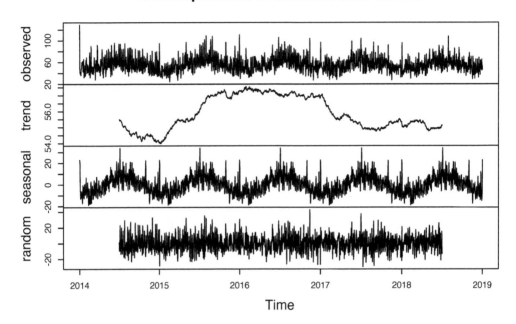

FIGURE 6.8: Decomposed time series of aggravated assaults in NYC

We can also use `ggplot2` for these purposes. In particular, we can use the `ggseas` extension which allows for seasonal decomposition within `ggplot` (see Ellis (2018) for details). First, we can use the `tsdf()` function from the `ggseas` package. This turns the `ts` object we just created into a dataframe and then plot the series.

```
library(ggseas)

ny_df <- tsdf(ny_timeseries)
```

Then we can use the `ggsdc()` function in order to create a four-facetted plot of seasonal decomposition showing observed, trend, seasonal and random components.

```
ggsdc(ny_df, aes(x = x, y = y),   method = "decompose") +
    geom_line()
```

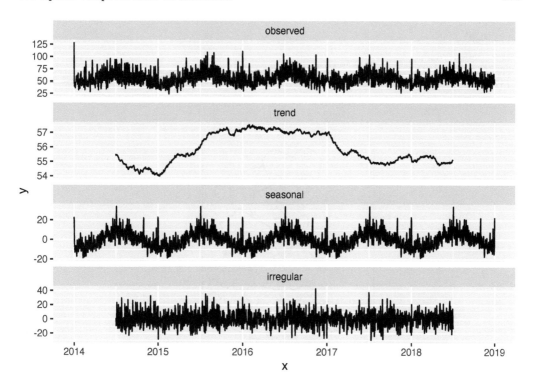

FIGURE 6.9: Decomposed time series of aggravated assaults in NYC using ggseas

The resulting graph similarly presents the components: the observed data, the trend, the seasonal component and the random fluctuation.

We have now covered a quick overview of ways of making sense of temporal data. Of course there is a lot more out there, and we urge interested readers to make use of the recommended reading section in this chapter to explore further the topic of temporal data analysis. But now let's return to the spatial focus of our book.

6.6 Spatio-temporal data visualisation

For the next set of exercises, we are going to look at temporal variations on burglary across Greater Manchester. We are going to focus on wards as the unit of analysis. Accordingly, we will need to import another data set, this time a geojson which contains the wards (administrative boundaries) for Manchester, and the monthly burglary count for each ward in 2018. To load the ward geometries for Manchester into a sf object, we use code we had already used in previous chapters.

```
mcr_burglary <- st_read("data/mcr_burglary.geojson")
```

With this data in our environment, we can plot a map using tmap.

```
library(tmap)

tm_shape(mcr_burglary) +
  tm_fill("burglaries") +
  tm_borders()
```

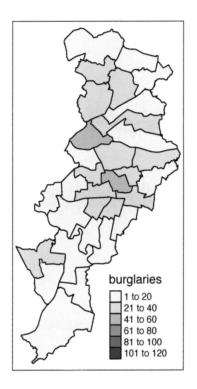

FIGURE 6.10: Burglaries in Manchester

So this is something we're already familiar with. But how can we map the temporal information? We will now cover some approaches.

6.6.1 Small multiples to show temporal variation

As noted by Pebesma (2012), spatio-temporal data often come in the form of single tables expressed in one of the three following formats:

- *time-wide* where different columns reflect different moments in time,
- *space-wide* where different columns reflect different measurement locations or areas, or
- *long formats* where each record reflects a single time and space combination.

What we have in the `mcr_burglary` object if we view the data frame is a table expressed in the long format, where each row represents a single month and ward combination. We can see as well that this is not simply a data table but a `sf` object which embeds the geographical information that allow us to place it in a map. We can now try to produce the small multiples with the `tm_facets()` function. (Note: if you wanted to use `ggplot2` instead of `tmap`, you can look up the function `facet_wrap()` to achieve the same results).

```
tm_shape(mcr_burglary) +
  tm_fill("burglaries") +
  tm_borders() +
  tm_facets("month", free.coords=FALSE)
```

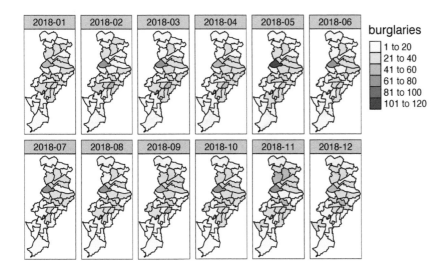

FIGURE 6.11: Small multiples of burglary across Manchester by month

So this is one way to visualise temporal variation. What are some more?

6.6.2 Spaghetti plots

In longitudinal studies and in studies looking at areas over time, sometimes researchers use spaghetti plots. On their own they are not great, but they can be used when one wants to put a trajectory within a broader context or when comparing different trajectories. You can read more about how to not use them in Nussbaumer (2013).

While we will include the ward name as a spatial component in a way, this isn't technically spatio-temporal data visualisation, as we are stripping away any sort of spatial element, and keeping ward name only as a nominal variable. It is not great, but let's show you why you might use it. The basic concept is just the same line chart we saw earlier, but with the trajectories grouped within the wards. So instead of one line, we will get 32: one for each ward in our data.

```
ggplot(mcr_burglary, aes(x = month, y = burglaries, group = ward_name)) +
  geom_line() + theme_minimal()
```

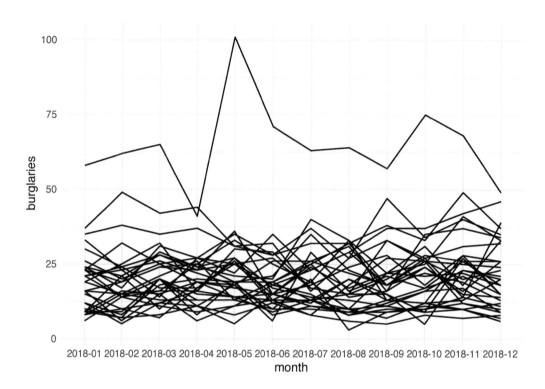

FIGURE 6.12: Spaghetti plot of burglaries by ward

This is quite the mess. So in what situation may this be useful? Well, maybe you want to compare the variation here with some central tendency. For this we can "grey out" the lines (by changing the colour) and add some summary such as the median, let's say, by using the `stat_summary()` function.

```
ggplot(mcr_burglary, aes(x = month,  y = burglaries, group = ward_name)) +
  geom_line(color="darkgrey") +
  stat_summary(aes(group = 1), # we want to summarise all the data together
               geom = "line",  # geometry to display the summary
               fun.y = median, # function to apply
               lwd = 1.5) +    # specifying thick line for attention
  theme_minimal()
```

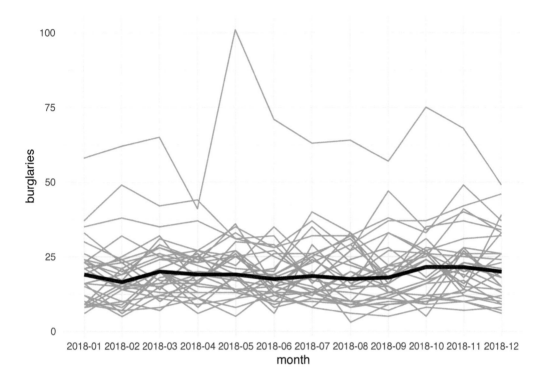

FIGURE 6.13: Spaghetti plot of burglaries by ward showing variation around central tendency measure

This way we can show something like the median, trajectory compared with the individual observations' trajectories. We could also use colour to highlight a specific area compared to the rest of the wards. For example, if we are interested in Fallowfield ward, we might highlight that as an additional layer in our spaghetti pile:

```
ggplot(mcr_burglary, aes(x = month, y = burglaries, group = ward_name)) +
  geom_line(color="darkgrey") +
  geom_line(data = mcr_burglary %>%     # pass new data in this layer
            filter(ward_name == "Fallowfield"),   #filter only Fallowfield
         aes(x = month,
         y = burglaries),
         colour = "blue",   # change colour for emphasis
         lwd = 1.5) + # change line width for emphasis
  theme_minimal()
```

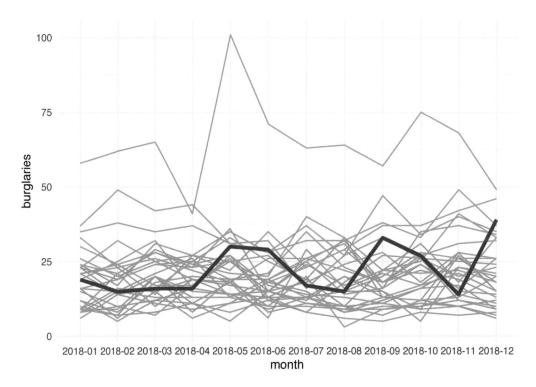

FIGURE 6.14: Spaghetti plot to highlight burglaries in Fallowfield ward

We can now maybe draw some conclusions about Fallowfield's burglary trajectory compared with the other wards in our data set. But let's move on now to the final approach, where we again make use of our spatial component.

6.6.3 Animations

The final way we will present here to visualise time and place together is through the use of animations. This feature is brought to ggplot2 thanks to the gganimate extension. The idea behind this display is really similar to that behind the small multiples, introduced earlier. A separate map is created for each month, and these are displayed near one another in order to show the change over time. However, instead of side-by-side, these maps are sequential - they appear in the same place but one after another, in order to present change. So first thing we do is to load the gganimate package:

```
library(gganimate)
```

Also, to apply gganimate to sf objects, you need to have a package called transformr installed. You don't need to load this, but make sure it is installed! If not, install with install.packages(transformr). Then, we need to make sure that our temporal variable is a date object. We can use the ymd() function, from the fantastic lubridate package (really I cannot praise this package enough, it makes handling dates so easy!) to make sure that our month variable is a date object. One thing you might notice looking at this data is that there is no date associated with each observation, only month and year. How can we use ymd() which clearly requires **y**ear, **m**onth and **d**ay! Well, one approach is to make this up, and just say that everything in our data happened on the 1st of the month. We can use the paste0() function to do this:

```
mcr_burglary$date_month <- ymd(paste0(mcr_burglary$month, "-01"))
```

Now, we can create a simple static plot, the way we already know how. Let's plot the number of burglary incidents per ward, and save this in an object called anim:

```
anim <- ggplot() +
  geom_sf(data = mcr_burglary, aes(fill = burglaries)) +
  theme_void()
```

Now, finally, we can animate this graph. Take the object of the static graph (anim) and add a form of transition, which will be used to animate the graph. In this case, we can use transition_states(). This transition splits your data into multiple states based on the levels in a given column, much like how faceting splits up the data in multiple panels. It then shifts between the defined states and pauses at each state. Layers with data without the specified column will be kept constant during the animation (again, mimicking facet_wrap). States are the unquoted name of the column holding the state levels in the data. You can then use closest_state to dynamically label the graph:

```
anim +
  transition_states(date_month,  # column holding the state levels in the data
                    transition_length = 1, # relative length of the transition
                    state_length = 2) + # elative length of the pause
  labs(title = "Month: {closest_state}")
```

In the print book, you will not see this result, so it's not very exciting, but hopefully you are following along, and in your own R environment you are now are looking at a very smooth animation of how burglary changed over the months of 2018 in Manchester.

6.7 Summary and further reading

In this chapter we had a very brief introduction to temporal data, and we considered how we can manipulate and wrangle this new type of data using packages such as lubridate. We then introduced some approaches to visualising temporal data, and finally, ways to visualise spatio-temporal variation as well. To read up on the importance of time, and specifically spatio-temporal crime analysis, we recommend Ratcliffe (2010), Wheeler (2016), and Roth et al. (2013).

While we did not get to go into any spatio-temporal analysis at this stage, visualisation is a good starting point to begin to engage with this important element of crime data. For additional details about how to visualise space-time data with R, we suggest Lamigueiro (2014) and chapter 2 of Wikle, Zammit-Mangion, and Cressie (2019).

To move into spatio-temporal analysis would require a more thorough training in temporal data analysis first, and so we do not cover it here. But we provide some resources for those interested. Notably, Hyndman and Athanasopoulos (2021) is an excellent resource for this, and we can also recommend Chatfield and Xing (2019). For those already comfortable with this, Pebesma (2012) also offers a very thoughtful introduction to how spatio-temporal data layouts appear and useful graphs for spatio-temporal data. Pebesma (2012) also introduces the R package spacetime, for handling such analyses. At the time of writing, functions in spacetime take as inputs older sp objects, but the work on the emerging package sftime, to work with sf objects is ongoing.

Specific to crime mapping, we see frequent use of the Mantel test (Mantel 1967), which provides a single index that indicates whether space-time clustering is apparent, and the Knox test (Knox and Bartlett 1964), which examines this pattern in more detail by determining whether there are more instances of events within a defined spatio-temporal area than would be expected on the basis of chance alone Adepeju and Evans (2017). To implement these in R, we recommend exploring the surveillance package, which offers statistical methods for the modelling and monitoring of time series of counts, proportions and categorical data, including the knox() function to implement Knox test Meyer, Held, and Höhle (2017), and Thioulouse et al. (2018) 'Multivariate Analysis of Ecological Data with ade4' — a package for analysis of ecological data, which includes the functions which implement Mantel's (1967) space-time interaction test. We will encounter these tests again in Chapter 11 of this book, implemented in the package NearRepeat when exploring near repeat victimisation.

As noted, there are packages that provide functionality for better handling of dates and time. It is worth to go over the details of lubridate in the official pages and the relevant chapter in Wickham and Grolemund (2017) is also very helpful. There is a brand new package clock that aims to improve on lubridate, and that is also worth exploring (for details see Vaughan (2021)).

7

Spatial point patterns of crime events

7.1 Mapping crime intensity with isarithmic maps

A key piece of information we explore in environmental criminology is the "exact" location of crimes, typically in the form of street address in which a crime is recorded as having occurred. A key goal of the analysis of this type of mapped point data is to detect patterns, in particular to detect areas where crime locations appear as clustered and reflecting an increased likelihood of occurrence.

In Chapter 1, we generated some maps with point locations representing crimes in Greater Manchester (UK), and we saw how the large volume of events made it difficult to discern any spatial pattern. In previous chapters we saw how we could deal with this via aggregation of the point data to enumeration areas such as census tracts or by virtue of binning. In this chapter we will introduce techniques that are used as an alternative visualisation for point patterns and that are based on the mathematics of density estimation. They are used to produce isarithmic maps; these are the kind of maps you often see in the weather reports displaying temperature. What we are doing is creating an interpolated surface from discrete data points. Or in simpler terms we are "effectively inventing data values for the areas on the map for which you don't have data or sample points" (Field 2015).

Before we get to the detail of how we produce these maps, we will briefly and very generally introduce the field of spatial point pattern analysis. The analysis of discrete locations are a sub-field of spatial statistics. In this chapter we will introduce an R package, called spatstat, that was developed for spatial point pattern analysis and modelling. It was written by Adrian Baddeley and Rolf Turner. There is a webpage[1] and a book (Baddeley, Rubak, and Turner 2015) dedicated to this package. In our book we are only going to provide you with an introductory practical entry into this field of techniques. If this package is not installed in your machine, make sure you install it before we carry on. In this chapter we will also be using the package crimedata developed by our colleague Matt Ashby, which we introduced in chapter 6 (Ashby 2018).

```
# Packages for reading data and data carpentry
library(readr)
library(dplyr)

# Packages for handling spatial data and for geospatial carpentry
library(sf)
library(maptools)
```

---

[1]http://spatstat.org

```
library(raster) # used for data represented in a raster model

# Specific packages for spatial point pattern analysis
library(spatstat)

# Libraries for spatial visualisation
library(tmap)
library(ggplot2)
library(leaflet)

# Libraries with relevant data for the chapter
library(crimedata)
```

7.2   Burglaries in NYC

Like in Chapter 6, our data come from the **Crime Open Database (CODE)**, a service
that makes it convenient to use crime data from multiple US cities in research on crime.
All the data in CODE are available to use for free as long as you acknowledge the source
of the data. The package crimedata was developed by Dr. Ashby to make it easier to read
into R the data from this service. In the last chapter we provided the data in a csv format,
but here we will actually make use of the package to show you how you might acquire such
data yourself.

The key function in crimedata is get_crime_data(). We will be using the crime data from
New York City police and for 2010, and we want the data to be in sf format. We will also
filter the data to only include residential burglaries in Brooklyn.

```
# Large dataset, so it will take a while
nyc_burg <- get_crime_data(
  cities = "New York", #specifies the city for which you want the data
  years = 2010,        #reads the appropriate year of data
  type = "extended",   #select extended (a fuller set of fields)
  output = "sf") %>%   #return a sf object (vs tibble)
filter(offense_type == "residential burglary/breaking & entering" &
          nyc_boro_nm == "MANHATTAN")
```

Additionally, we will need a geometry of borough boundaries. We have downloaded this
from the NYC open data portal, and it is supplied with the data included with this book.
The file is manhattan.geojson.

```
manhattan <- st_read("data/manhattan.geojson") %>%
  filter(BoroName == "Manhattan") # select only Manhattan
```

Now that we have all our data, let's plot to see how they look.

```
ggplot() +
  geom_sf(data = manhattan) + # boundary data
  geom_sf(data = nyc_burg, size = 0.5) +  # crime locations as points
  theme_void()
```

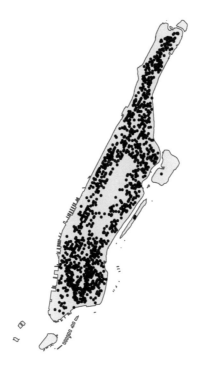

FIGURE 7.1: Plot Manhattan borough and crime locations as points

In the point pattern analysis literature each point is often referred to as an **event** and these events can have **marks**, attributes or characteristics that are also encoded in the data. In our spatial object, one of these *marks* is the type of crime (although in this case it is of little interest since we have filtered on it). Others in this object include the date and time, location category, victim sex and race, etc.

7.3 Using the spatstat package

The R package spatstat was created for spatial statistics with a strong focus on analysing spatial point patterns. It is the result of 15 years of development by leading researchers in spatial statistics. We highly recommend the book "Spatial Point Patterns: Methodology and Applications with R" to give a comprehensive overview of this package (Baddeley, Rubak, and Turner 2015).

When working within the spatstat environment, we will be working with planar point pattern objects. This means that the first thing we need to do is to transform our sf object into a ppp (planar point pattern) object. This is how spatstat likes to store its point patterns. The components of a ppp object include the coordinates of the points and the observation window (defining the study area).

It is important to realise that a point pattern is defined as a series of events in a given area, or window, of observation. Therefore, we need to precisely define this window. In spatstat the function owin() is used to set the observation window. However, the standard function takes the coordinates of a rectangle or of a polygon from a matrix, and therefore it may be a bit tricky to use with a sf object. Luckily, the package maptools provides a way to transform a SpatialPolygons into an object of class owin, using the function as.owin(). Here are the steps.

First, we transform the CRS of our Manhattan polygon into projected coordinates as opposed to the original geographic coordinates (WGS84), since only projected coordinates may be converted to spatstat class objects. As we saw in Chapter 2, measuring distance in metres is a bit easier, so since with points a good deal of the analysis we subject them to involves measuring their distance from each other, moving to a projected system makes sense. For NYC, the recommended projection is the New York State Plane Long Island Zone (EPSG 2263), which provides a high degree of accuracy and balances size and shape well. This projection, however, uses US feet (which was deprecated in 2022), so instead we are going to use EPSG 32118, as an alternative that uses metres as unit of measurement. So first, transform our object using st_transform():

```
manhattan <- st_transform(manhattan, 32118)
```

Then we use the as.owin() function to define the window based on our manhattan object.

```
window <- as.owin(st_geometry(manhattan))
```

We can plot this window to see if it looks similar in shape to our polygon of Manhattan.

```
plot(window) # owin similar to polygon
```

window

FIGURE 7.2: Plot window based on Manhattan polygon

We can see this worked and created an owin object.

```
class(window)
```

```
## [1] "owin"
```

```
window
```

```
## window: polygonal boundary
## enclosing rectangle: [295965, 307869] x [57329, 79111]
## units
```

You can see it prints the units of measurement. But it does not identify them:

```
unitname(window)
```

```
## unit / units
```

As noted above, EPGS:32118 uses metres. So we can pass this information to our owin object:

```
unitname(window) <- "Meter"
```

Now that we have created the window as an owin object, let's get the points. In the first place, we need to reproject to the same projected coordinate systems we have for our boundary data using the familiar st_transform() function from sf. Then we will extract the coordinates from our sf point data into a matrix using unlist() from base R, so that the functions from statspat can read the coordinates. If you remember sf objects have a column called geometry that stores these coordinates as a list. By unlisting we are flattening this list, taking the elements out of the list.

```
# first transform to projected CRS too to match our window
nyc_burg <- st_transform(nyc_burg, 32118)

# create matrix of coordinates
sf_nyc_burg_coords <- matrix(unlist(nyc_burg$geometry), ncol = 2, byrow = T)
```

Then we use the `ppp()` function from the `spatstat` package to create the object using the information from our matrix and the window that we created. The `check` argument is a logical that specifies whether we want to check that all points fall within the window of observation. It is recommended to set it to `TRUE`, unless you are absolutely sure that this is unnecessary.

```
bur_ppp <- ppp(x = sf_nyc_burg_coords[,1], y = sf_nyc_burg_coords[,2],
               window = window, check = T)
```

```
## Warning: data contain duplicated points
```

Notice the warning message about duplicated points? In spatial point pattern analysis an issue of significance is the presence of duplicates. The statistical methodology used for spatial point pattern processes is based largely on the assumption that processes are *simple*, that is, that the points cannot be coincident. That assumption may be unreasonable in many contexts (for example, the literature on repeat victimisation indeed suggests that we should expect the same addresses to be at a higher risk of being hit again). Even so, the point (no pun intended) is that *"when the data has coincidence points, some statistical procedures will be severely affected. So it is always strongly advisable to check for duplicate points and to decide on a strategy for dealing with them if they are present"* (Baddeley, Rubak, and Turner 2015: p. 60).

So how to perform this check? When creating the object above, we have already been warned this is an issue. But more generally we can check the duplication in a `ppp` object with the following syntax: it will return a logical indicating whether we have any duplication:

```
any(duplicated.ppp(bur_ppp))
```

```
## [1] TRUE
```

To count the number of coincidence points, we use the `multiplicity()` function from the `spatstat` package. This will return a vector of integers, with one entry for each observation in our dataset, giving the number of points that are identical to the point in question (including itself).

```
multiplicity(bur_ppp)
```

If you suspect many duplicates, it may be better to start asking how many there are. For this you can use:

```
sum(multiplicity(bur_ppp) > 1)
```

```
## [1] 915
```

That's quite something! 915 points out of 1727 burglaries share the same coordinates. And we can plot this vector if we want to see the distribution of duplicates:

```
hist(multiplicity(bur_ppp))
```

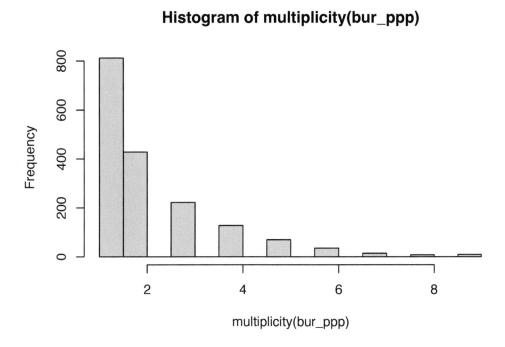

FIGURE 7.3: Histogram of incidents in overlapping coordinates (multiplicity)

In the case of crime, as we have hinted some, of this may be linked to the nature of crime itself. Hint: repeat victimisation. It is estimated that between 7 and 33% of all burglaries in several western and developed countries are repeat offenses (Chainey and Alves-da-Silva 2016). Our 52% is quite high by those standards. In this particular situation, it likely is the result of geomasking (the introduction of small inaccuracies into their geocoded location) due to privacy reasons. Most police data stored in open data portals do not include the exact address of crime events. In the case of US police data, as noted by the documentation of CODE, it relies on addresses geocoded to the nearest hundred block. So say "370 Gorse Avenue" could be rounded down and geooded to "300 Gorse Avenue". The same would happen to "307 Gorse Avenue", in the public release version of the data it would feature as "300 Gorse Avenue". In the UK there exist a pre-determined list of points so that each crime event gets "snapped" to its nearest one. So, the coordinates provided in the open data are not the exact locations of crimes. This process is likely inflating the amount of duplication we may observe, because each snap point or hundred block address might have many crimes near it, resulting in those crimes being geo-coded to the same exact location. So keep it in mind when analysing and working with this kind of data set that it may not be the same as working with the real locations. (If you are interested in the effects of this, read the paper by Tompson et al. (2015)).

What to do about duplicates in spatial point pattern analysis is not always clear. You could simply delete the duplicates, but of course that may ignore issues such as repeat victimisation. You could also use *jittering*, which will add a small perturbation to the duplicate points so that they do not occupy the exact same space. This, again, may ignore things like repeat victimisation. Another alternative is to make each point "unique" and then attach the multiplicities of the points to the patterns as *marks*, as attributes of the points. Then you would need analytic techniques that take into account these marks.

If you were to be doing this for real, you would want access to the original source data, not this public version of the data, as well as access to the data producers to check that duplicates are not artificial, and then go for the latter solution suggested above (using multiplicities as marks once we adjust for any artificial inflation). We don't have access to the source data, so for the sake of simplicity and convenience at this point and so that we can illustrate how spatstat works, we will first add some jittering to the data. For this we use rjitter().

The first argument for the function is the object, retry asks whether we want the algorithm to have another go if the jittering places a point outside the window (we want this so that we don't have to loose points), and the drop argument is used to ensure we get a ppp object as a result of running this function (which we do). The argument nsim sets the number of simulated realisations to be generated. If we just want one ppp object, we need to set this to 1. You can also specify the scale of the perturbation; the points will be randomly relocated in a uniform way in a circle within this radius. The help file notes that there is a "sensible default", but it does not say much more about it, so you may want to set the argument yourself if you want to have more control over the results. Here we are specifying 1.5 metres to minimise the effects of this random displacement.

```
set.seed(200)
jitter_bur <- rjitter(bur_ppp, retry=TRUE, nsim=1, radius = 1.5, drop=TRUE)
```

And you can check duplicates with the resulting object:

```
sum(multiplicity(jitter_bur) > 1)
```

```
## [1] 0
```

Later we will demonstrate how to use the number of incidents in the same address as a *mark* or *weight*, so let's create a new ppp object that includes this mark. We need to proceed in steps. First, we create a new ppp, just a duplicate of the one we have been using as a placeholder.

```
bur_ppp_m <- bur_ppp
```

Now we can assign to this ppp object the numerical vector created by the multiplicity function as the *mark* using the marks function.

```
marks(bur_ppp_m) <- multiplicity(bur_ppp)
```

If you examine your environment, you will be able to see the new object and that it has a 6th element, which is the mark we just assigned.

7.4 Homogeneous or spatially varying intensity?

Central to our exploratory point pattern analysis is the notion of **intensity**, that is the average number of points per area. A key question that concerns us is whether this intensity is homogeneous or uniform, or rather is spatially varying due, for example, to spatially varying risk factors. With spatstats you can produce a quick estimate of intensity for the whole study area if you assume a homogeneous distribution with intensity.ppp().

```
intensity.ppp(jitter_bur)
```

```
## [1] 2.92e-05
```

Metres are a small unit. So the value we get is very close to zero and hard to interpret. We can use the rescale() function (in spatstat package) to obtain the intensity per kilometres instead. The second argument is the number we multiply our original units by:

```
intensity.ppp(rescale(jitter_bur, 1000, "km"))
```

```
## [1] 29.2
```

We see this result that overall the intensity across our study space is about 29 burglaries per square kilometre. Does this sound right? Well, Manhattan has roughly 59.1 km^2. If we divide 1727 incidents by this area, we get around 29 burglaries per km^2. This estimate, let's remember, assumes that the intensity is homogeneous across the study window.

If we suspect the point pattern is spatially varying, we can start exploring this. One approach is to use **quadrat counting**. A **quadrat** refers to a small area - a term taken from ecology where quadrats are used to sample local distribution of plants or animals. One could divide the window of observation into quadrats and count the number of points into each of these quadrats. If the quadrats have an equal area and the process is spatially homogeneous, then one would expect roughly the same number of incidents in each quadrat.

For producing this counting, we use the quadratcount() function (in spatstat). The nx and ny argument in this function control how many breaks you want alongside the X and Y coordinates to create the quadrats. So to create 6 quadrats from top to bottom of Manhattan, we will set them to 1 and 6, respectively. The configuration along the X and Y and the number of quadrats is something you will have to choose according to what is convenient to you. Then we just use standard plotting functions from R base.

```
q <- quadratcount(jitter_bur, nx = 1, ny = 6)
plot(q, main = "Quadrat Count")
```

Quadrat Count

FIGURE 7.4: Count of burglaries in quadrats

As you can see, with its irregular shape it is difficult to divide Manhattan in equally shaped and sized quadrats, and thus it is hard to assess if regions of equal area contain a similar number of points. Notice, however, how the third quadrat from the bottom seems to have less incidents than the ones that surround it. A good portion of this quadrat is taken by Central Park, where you would not expect a burglary.

Instead, we can use other forms of tessellation, like hexagons by using the `hextess()` function (in `spatstat`). This is similar to the binning we introduced in Chapter 4, only using different functions. The numeric argument sets the size of the hexagons, you will need to play along to find an adequate value for your own applications. Rather than printing the numbers inside these hexagons (if you have many, this will make it difficult to read in a plot), we can map a colour to the intensity (the number of points per hexagon):

```
hex <- hextess(jitter_bur, 200)
q_hex <- quadratcount(jitter_bur, tess = hex)
plot(intensity(q_hex, image = TRUE), main = "Intensity map")
```

Intensity map

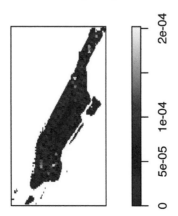

FIGURE 7.5: Intensity map of burglaries per hexagon

We can see more clearly now that areas of Manhattan, like Central Park, have understandably a very low intensity of burglaries, whereas others, like the Lower East Side and north of Washington Heights, have a higher intensity. So this map suggests to us that the process may not be homogeneous - hardly surprising when it comes to crime.

A key concept in spatial point pattern analysis is the idea of **complete spatial randomness** (CSR). When we look at a point pattern process, the first step in the process is to ask whether it has been generated in a random manner. Under CSR, points are (1) independent of each other (*independence*) and (2) have the same propensity to be found at any location (*homogeneity*).

We can generate data that conform to complete spatial randomness using the `rpoispp()` function (in `spatstat`). The `r` at the beginning is used to denote we are simulating data (you will see this is common in R), and we are using a Poisson point process, a good probability distribution for these purposes. Let's generate 300 points in a random manner and save this in the object `poisson_ppp`:

```
set.seed(1) # Use this seed to ensure you get the same results as us
poisson_ppp <- rpoispp(300)
```

We can plot this, to see how it looks

```
plot(poisson_ppp, main = "Complete Spatial Randomness")
```

Complete Spatial Randomness

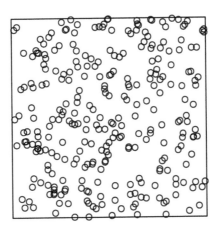

FIGURE 7.6: Simulated complete spatial randomness

You will notice that the points in a homogeneous Poisson process are not 'uniformly spread': there are empty gaps and clusters of points. This is very important. Random may look like what you may think non-random should look. There seem to be places with apparent "clustering" of points. Keep this in mind when observing maps of crime distributions. Run the previous command a few times. You will see the map generated is different each time.

In classical literature, the *homogeneous Poisson process* (CSR) is usually taken as the appropriate 'null' model for a point pattern. Our basic task in analysing a point pattern is to find evidence against CSR. We can run a Chi Square test to check this in our data. So, for example, like we did above with Manhattan, we can split our area into quadrats.

```
q2 <- quadratcount(poisson_ppp, nx = 6, ny = 6)
plot(q2, main = "Quadrat Count")
```

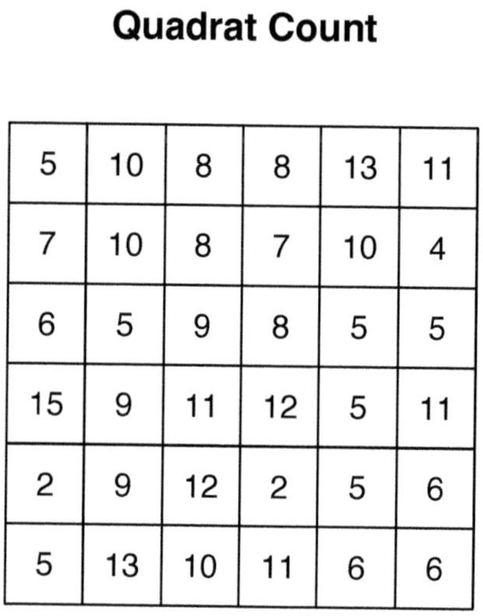

Quadrat Count

5	10	8	8	13	11
7	10	8	7	10	4
6	5	9	8	5	5
15	9	11	12	5	11
2	9	12	2	5	6
5	13	10	11	6	6

FIGURE 7.7: Count of our random points in quadrats

These numbers are looking more even than they did with our burglaries, but to really put this to the test, we run the Chi Square test using the `quadrat.test()` function.

```
quadrat.test(poisson_ppp, nx = 6, ny = 6)
```

```
##
##  Chi-squared test of CSR using quadrat counts
##
## data:  poisson_ppp
## X2 = 44, df = 35, p-value = 0.3
## alternative hypothesis: two.sided
##
## Quadrats: 6 by 6 grid of tiles
```

We can see this p-value is greater than the conventional alpha level of 0.05, and therefore we cannot reject the null hypothesis that our point patterns come from complete spatial randomness - which is great news, as they indeed do!

How about our burglaries? Let's run the `quadrat.test()` function to tell us whether burglaries in Manhattan follow spatial randomness.

```
quadrat.test(jitter_bur, nx = 1, ny = 6)
```

```
##
##  Chi-squared test of CSR using quadrat counts
##
## data:  jitter_bur
## X2 = 238, df = 5, p-value <2e-16
## alternative hypothesis: two.sided
##
## Quadrats: 6 tiles (irregular windows)
```

Observing the results, we see that the p value is well below conventional standards for rejection of the null hypothesis. Observing our data of burglary in Manhattan would be extremely rare if the null hypothesis of a homogeneous distribution was true. We can then conclude that the burglary data is not randomly distributed in the observed space. But no cop nor criminologist would really question this. They would rarely be surprised by your findings! We do know, through research, that crime is not randomly distributed in space.

Beware as well that this test would come out as significant *if the point pattern violates the property of independence between the points.* It is important to keep in mind that the null may be rejected either because the process is not homogeneous *or* because the points are not independent, but the test doesn't tell us the reason for the rejection. Equally, the power depends on the size of the quadrats (and they are not too large or too small). And the areas are not of equal size here in any case. So the test in many applications will be of limited use.

7.5   Mapping intensity estimates with spatstat

7.5.1   Basics of kernel density estimation

One of the most common methods to visualise hot spots of crime, when we are working with point pattern data, is to use kernel density estimation. **Kernel density estimation** involves applying a function (known as a "kernel") to each data point, which averages the location of that point with respect to the location of other data points. The surface that results from this model allows us to produce **isarithmic maps**. It is essentially a more sophisticated way of visualising intensity (points per area) than the tessellations we have covered so far.

Kernel density estimation maps are very popular among crime analysts. According to Chainey (2013a), 9 out of 10 intelligence professionals prefer them to other techniques for hot spot analysis. As compared to visualisations of crime that rely on point maps or thematic maps of geographic administrative units (such as LSOAs), isarithmic maps are considered best for location, size, shape and orientation of the hot spot. Chainey, Tompson, and Uhlig (2013) have also suggested that this method produces some of the best prediction accuracy. The areas identified as hot spots by KDE (using historical data) tend to be the ones that better identify the areas that will have high levels of crime in the future. Yet, producing these maps (as with any map, really) requires you to take a number of decisions that will significantly affect the resulting product and the conveyed message. Like any other data visualisation technique, they can be powerful, but they have to be handled with great care.

Essentially this method uses a statistical technique (kernel density estimation) to generate a smooth continuous surface aiming to represent the intensity or volume of crimes across the target area. The technique, in one of its implementations (quartic kernel), is described in this way by Chainey (2013b):

> *a fine grid is generated over the point distribution; a moving three-dimensional function of a specified radius visits each cell and calculates weights for each point within the kernel's radius. Points closer to the centre will receive a higher weight, and therefore contribute more to the cell's total density value; and final grid cell values are calculated by summing the values of all kernel estimates for each location*

The values that we attribute to the cells in crime mapping will typically refer to the number of crimes within the area's unit of measurement (intensity). Let's produce one of these density maps, just with the default arguments of the `density.ppp()` function:

```
ds <- density.ppp(jitter_bur)
class(ds)
```

```
## [1] "im"
```

The resulting object is of class `im`, a pixel image, a grid of spatial locations with values attached to them. You can visualise this image as a map (using `plot()`), contour plots (using `contour()`), or as a perspective plot (using `persp()`).

```
par(mfrow=c(1,3))
plot(ds, main='Pixel image')
contour(ds, main = 'Contour plot')
persp(ds, main = 'Perspective plot')
```

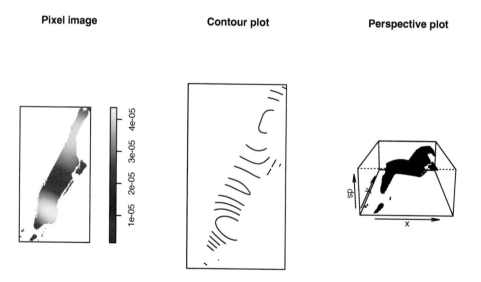

FIGURE 7.8: Visualisations of intensity of burglary in Manhattan

The density function is estimating a kernel density estimate. The result of density.ppp() *is not a probability density*. It is an estimate of the intensity function of the point process that generated the point pattern. The units of intensity are "points per unit area". Density, in this context, is nothing but the number of points per unit area. This method computes the intensity continuously across the study area, and the object returns a raster image. Depending what package you use to generate the estimation (e.g., spatstat, SpatialKDE, etc) you can set a different metric.

7.5.2   Selecting the appropriate bandwith for KDE

The defaults in density.ppp() are not necessarily the optimal ones for your application. When doing spatial density analysis, there are a number of things you have to decide upon (e.g., bandwidth, kernel type, etc.).

To perform this analysis, we need to define the **bandwidth** of the density estimation, which basically determines the area of influence of the estimation. There is no general rule to determine the correct bandwidth; generally speaking, if the bandwidth is too small the estimate is too noisy, while if bandwidth is too high the estimate may miss crucial elements of the point pattern due to oversmoothing. The key argument to pass to the density.ppp() is sigma=, which determines the bandwidth of the kernel. Sigma is the *standard deviation of the kernel*, and you can enter a numeric value or a function that computes an appropriate value for it.

A great deal of research has been developed to produce algorithms that help in selecting the bandwidth (see Baddeley, Rubak, and Turner (2015) and Davies, Marshall, and Hazelton (2017) for details). In the spatstat package the functions bw.diggle(), bw.ppl(), and bw.scott() can be used to estimate the bandwidth according to difference methods (mean square error cross-validation, likelihood cross-validation, and Scott's rule of thumb for bandwidth section in multi-dimensional smoothing, respectively). The help files recommend the use of the first two. These functions run algorithms that aim to select an appropriate bandwidth.

```
set.seed(200) # We set this seed to ensure you get this same results when running
              # these algorithms
bw.diggle(jitter_bur)
```

```
## sigma
## 2.912
```

```
bw.ppl(jitter_bur)
```

```
## sigma
## 73.06
```

```
bw.scott(jitter_bur)
```

```
## sigma.x sigma.y
##   608.4  1447.6
```

Scott's rule generally provides larger bandwidth than the other methods. Don't be surprised if your results do not match exactly the ones we provide if you do not use the same seed, the computation means that they will vary every time we run these functions. Keep in mind these values will be expressed in the unit of the projected coordinate system you use (in this case metres). We can visualise how they influence the appearance of 'hot spots':

```
set.seed(200)
par(mfrow=c(1,3))
plot(density.ppp(jitter_bur, sigma = bw.diggle(jitter_bur),edge=T),
     main = "Diggle")

plot(density.ppp(jitter_bur, sigma = bw.ppl(jitter_bur),edge=T),
     main="PPL")

plot(density.ppp(jitter_bur, sigma = bw.scott(jitter_bur) ,edge=T),
     main="Scott")
```

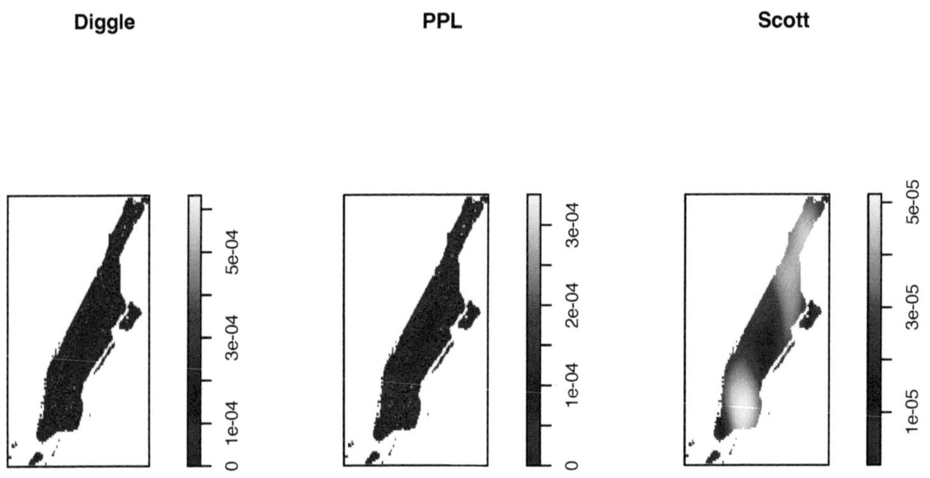

FIGURE 7.9: Kernel density maps with different bandwidth estimation algorightms

When looking for clusters, Baddeley, Rubak, and Turner (2015) suggest the use of the `bw.ppl()` algorithm because in their experience it tends to produce the more appropriate values when the pattern consists predominantly of tight clusters. But they also insist that if your purpose it is to detect a single tight cluster in the midst of random noise, then the `bw.diggle()` method seems to work best.

All these methods use a fixed bandwidth: they use the same bandwidth for all locations. As noted by Baddeley, Rubak, and Turner (2015) a "fixed smoothing bandwidth is unsatisfactory if the true intensity varies greatly across the spatial domain, because it is likely to cause oversmoothing in the high-intensity areas and undersmoothing in the low-intensity areas" (p. 174). When using these methods in areas of high point density, the resulting bandwidth will be small, while in areas of low point density the resulting bandwidth will be large. This allows us to reduce smoothing in parts of the window where there are many points so that we capture more detail in the density estimates there, whilst at the same time increasing the smoothing in areas where there are less points.

With spatstat we can use an adaptive bandwidth method, where (1) a fraction of the points are selected at random and used to construct a tessellation; (2) a quadrat counting estimate of the intensity is produced; and (3) the process is repeated a set number of times and an average is computed. The adaptive.density() function (in spatstat) takes as arguments the fraction of points to be sampled (f) and the number of samples to be taken (nrep). The logical argument verbose when set to FALSE (not the default) means we won't see a message in the console informing of the steps in the simulation.

```
plot(adaptive.density(jitter_bur, f=0.1, nrep=30, verbose=FALSE),
     main = "adaptive bandwidth method")
```

adaptive bandwidth method

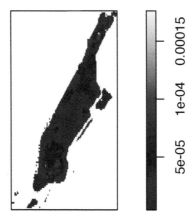

FIGURE 7.10: Adaptive bandwidth method for estimating bandwidth for KDE map

Which bandwidth to use? Chainey (2021) is critical of the various algorithms we have introduced. He suggests that "they tend to produce large bandwidths, and are considered unsuitable for exploring local spatial patterns of the density distribution of crime." As we have seen with our example this is not necessarily the case, if anything both mean square error cross-validation and likelihood cross-validation produce small bandwidths in this example. When it comes to crime, Neville (2013) suggests that the "best choice would be to produce an interpolation that fits the experience of the department and officers who travel an area. Again, experimentation and discussions with beat officers will be necessary to establish which bandwidth choice should be used in future interpolations" (p. 10.22). This is generally accepted by other authors in the field. We see no damage in using these algorithms as a starting point and then, as suggested by most authors (see as well Chainey (2021)), to engage in a discussion with the direct users (considering what the map will be used for) to set in a particular bandwidth that is not too spiky, nor oversmoothed.

7.5.3 Selecting the smoothing kernel

Apart from selecting the bandwidth, we also need to specify the particular kernel we will use. In density estimation there are different types of kernel. This relates to the type of kernel drawn around each point in the process of counting points around each point. The use of these functions will result in slightly different estimations. They relate to the way we weigh points within the radius:

> *The normal distribution weighs all points in the study area, though near points are weighted more highly than distant points. The other four techniques use a circumscribed circle around the grid cell. The uniform distribution weighs all points within the circle equally. The quartic function weighs near points more than far points, but the fall off is gradual. The triangular function weighs near points more than far points within the circle, but the fall off is more rapid. Finally, the negative exponential weighs near points much more highly than far points within the circle and the decay is very rapid.* (Levine 2013, chap. 10, p. 10).

Which one to use? Levine (2013) produces the following guidance: "*The use of any of one of these depends on how much the user wants to weigh near points relative to far points. Using a kernel function which has a big difference in the weights of near versus far points (e.g., the negative exponential or the triangular) tends to produce finer variations within the surface than functions which weight more evenly (e.g., the normal distribution, the quartic, or the uniform); these latter ones tend to smooth the distribution more.*" However, Silverman (1986) has argued that it does not make that much difference as long as the kernel is symmetrical. Chainey (2013b) suggests that in his experience most crime mappers prefer the quartic function, since it applies greater weight to crimes closer to the centre of the grid. The authors of the CrimeStat workbook S. Smith and Bruce (2008), on the other hand, suggest that the choice of the kernel should be based in our theoretical understanding of the data generating mechanisms. By this they mean that the processes behind spatial dependence may be different according to various crime patterns and that this is something that we may want to take into account when selecting a particular function. They provide a table with some examples that may help you to understand what they mean. For example, for residential burglaries they recommend 1 mile with normally dispersed quartic or uniform interpolation, while for domestic violence, 0.1 mile and the tightly focused, negative exponential method (S. Smith and Bruce 2008).

The default kernel in `density.ppp()` is `gaussian`. But there are other options. We can use the `epanechnikov`, `quartic` or `disc`. There are also further options for customisation. We can compare these kernels, whilst using the `ppl` algorithm for sigma (remember, we set seed at 200 for this particular value):

```
par(mfrow=c(2,2))
# Gaussian kernel function (the default)
plot(density.ppp(jitter_bur, sigma = bw.ppl(jitter_bur), edge=TRUE), main="Gaussian")

# Epanechnikov (parabolic) kernel function
plot(density.ppp(jitter_bur, kernel = "epanechnikov", sigma = bw.ppl(jitter_bur),
            edge=TRUE), main="epanechnikov")

# Quartic kernel function
plot(density.ppp(jitter_bur, kernel = "quartic", sigma = bw.ppl(jitter_bur), edge=TRUE),
    main="Quartic")

plot(density.ppp(jitter_bur, kernel = "disc", sigma = bw.ppl(jitter_bur), edge=TRUE),
    main="Disc")
```

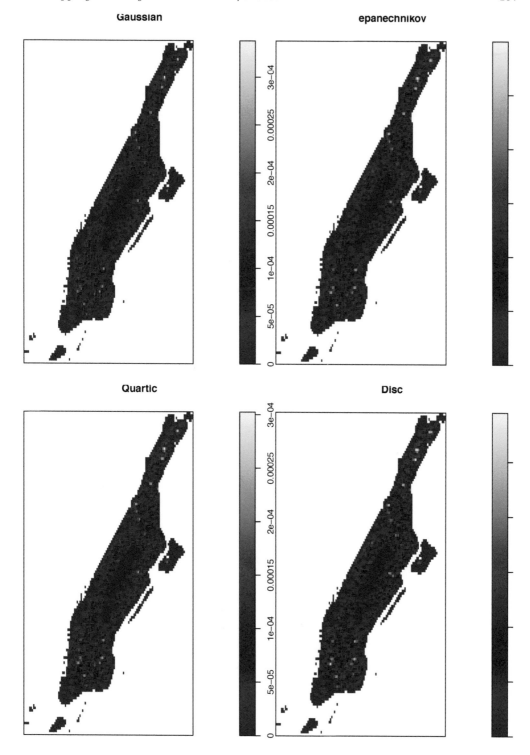

FIGURE 7.11: KDE maps created with different kernel shapes

You can see how the choice of the kernel has less of an impact than the choice of the bandwidth. In practical terms, choice of the specific kernel is of secondary importance to the bandwidth.

Equally, you may have noticed an argument in `density.ppp()` we have not discussed `edge = TRUE`. What is this? When estimating the density we are not observing points outside our window. In the case of an island like Manhattan, perhaps this is not an issue. But often when computing estimates for a given jurisdiction, we need to think about what is happening in the jurisdiction next door to us. Our algorithm does not see the crimes in jurisdictions neighbouring ours, and so the estimations around the edges of our boundaries receive fewer contributions in the computation. By setting `edge = TRUE` we are effectively applying a correction that tries to compensate for this negative bias.

7.5.4 Weighted kernel density estimates

In the past examples we used the un-weighted version of the estimation methods, but if the points are weighted (like for example by multiplicity) we can use weighted versions of these methods. We simply need to pass an argument that specifies the weights.

```
set.seed(200)
plot(density.ppp(bur_ppp_m, sigma = bw.ppl(bur_ppp_m), edge = T,
                 kernel = "quartic", weights = marks(bur_ppp_m)),
     main = "Weighted Points")
```

Weighted Points

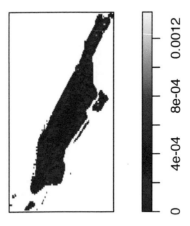

FIGURE 7.12: Weighting KDE estimates

7.5.5 Better KDE maps using `leaflet`

The plots produced by `spatstat` are good for initial exploration, but if you want to present this to an audience, you may want to use one of the various packages we have covered so far for visualising spatial data. Here we show how to use leaflet for plotting densities. Hot spot analysis often has an exploratory character, and the interactivity provided by leaflet is particularly helpful, since it allows us to zoom in and out and check the details of the areas with high intensities.

Often it is convenient to use a basemap to provide context. In order to do that, we first need to turn the image object generated by the spatstat package into a raster object, a more generic format for raster image used in R. Remember rasters from the first chapter? Now we finally get to use them a bit! First, we store the density estimates produced by density.ppp() into an object (which will be of class im, a two-dimensional pixel image), and then using the raster() function from the raster package, we generate a raster object.

```
set.seed(200)
dmap1 <- density.ppp(jitter_bur,
                     kernel = "quartic",
                     sigma = bw.ppl(jitter_bur),
                     edge=T)
# create raster object from im object
r1 <- raster(dmap1)
# plot the raster object
plot(r1, main = "Raster")
```

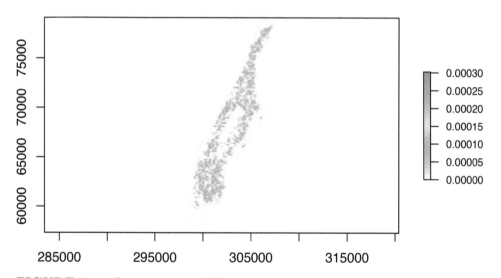

FIGURE 7.13: Raster image of KDE output

We can remove the areas with very low density of burglaries for better reading of the map:

```
#remove very low density values
r1[r1 < 0.00010 ] <- NA
```

```
#multiply by 1000 to re-express intensity in terms of burglaries per km
values(r1) <- values(r1)*1000
```

Now that we have the raster, we can add it to a basemap. Two-dimensional `RasterLayer` objects (from the `raster` package) can be turned into images and added to `Leaflet` maps using the `addRasterImage()` function. The `addRasterImage()` function works by projecting the `RasterLayer` object to EPSG:3857 and encoding each cell to an RGBA colour, to produce a PNG image. That image is then embedded in the map widget. This is only suitable for small to medium-sized rasters.

It's important that the `RasterLayer` object is tagged with a proper coordinate reference system, so that leaflet can then reproject to EPSG:3857. Many raster files contain this information, but some do not. Here is how you would tag a raster layer object `r1` which contains the EPSG:32118 data:

```
#make sure we have right CRS, which in this case is the New York 32118

crs(r1) <- "EPSG:32118"
```

The way we assign the CRS above is using the now most accepted PROJ6 format. You can find how these different formats apply to each coordinate system in the online archives.

Now we are ready to plot, although first we will create a more suitable colour palette and, since the values in the scale are quite close to 0, will multiple by a constant, to make the labels more readable. For this we use the `values()` function from the `raster` package that allows us to manipulate the values in a raster object:

```
pal <- colorNumeric("Reds", values(r1),
  na.color = "transparent") #Create colour palette

#and then make map!
leaflet() %>%
  addTiles() %>%
  addProviderTiles(providers$Stamen.TonerLite) %>%
  addRasterImage(r1, colors = pal, opacity = 0.7) %>%
  addLegend(pal = pal, values = values(r1),
    title = "Burglary map")
```

FIGURE 7.14: Leaflet map for exploring KDE layer of burglaries in Manhattan

And there you have it. Perhaps those familiar with Manhattan have some guesses as to what may be going on there?

7.5.6 Problems with KDE

There are at least two problems with these maps. First, they are just a convenient visualisation to represent the location of crime incidents. But they are not adjusting by the population at risk. If all you care, is where most crimes are taking place, that may then be fine. But if you are interested in relative risk, intensity adjusting by population at risk, then you are bound to be misled by these methods, which usually will plot very well the distribution of the population at risk. Second, the Poisson process, as we saw, creates what may appear as areas with more points than others, even though it is a random process. In other words, you may just be capturing random noise with these maps.

There are a number of ways to adjust for background population at risk when producing density estimates. Many of these methods have been developed in spatial epidemiology. In criminology, the population at risk is generally available only at more aggregated (at some census geography level) than in many of those applications (where typically you have points representing individuals without a disease being contrasted with points representing individuals with a disease). So, although one could extract the centroids of these polygons representing census tracts or districts as points and mark them with a value (say, population count or number of housing units), there isn't a great advantage to go to all this trouble. In fact by locating all the mass of the population at risk in the centroids of the polygons, you are introducing some distortion. So if you want to adjust for the population at risk, it is probably better to map out rates in the ways we described in previous chapters. We just need to be aware, then, that KDE surfaces in essence are simply mapping counts of crime. We therefore need to interpret density maps with this very significant caveat.

The other problem we had is that a Poisson process under CSR can still result in a map that one could read as identifying hot spots, whereas all we have is noise. Remember the `poisson_ppp` object we created earlier, which demonstrated complete spatial randomness? Well, if we were that way inclined, we could easily compute a KDE for this:

```
set.seed(200)
```

```
plot(density.ppp(poisson_ppp, kernel = "quartic", sigma = bw.diggle(poisson_ppp),
          edge=T), main = "KDE for random poisson point process")
```

KDE for random poisson point process

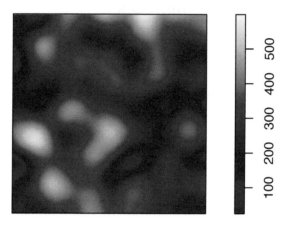

FIGURE 7.15: Kernel density map for quartic kernel

And there you go, you could come to the view looking at this density map that the intensity is driven by some underlying external risk factor when in fact the process is purely random. In subsequent chapters we will learn about methods that can be used to ensure we don't map out hot spots that may be the product of random noise.

7.5.7   Other packages for KDE

In this chapter we have focused on spatstat, for it provides a comprehensive suite of tools for working with spatial point patterns. The snippets of code and examples we have used here only provide a superficial glimpse into the full potential of this package. But spatstat is by no means the only package that can be used for generating spatial density estimates. We need to mention at least two alternatives: SpatialKDE and sparr.

SpatialKDE implements kernel density estimation for spatial data with all the required settings, including the selection of bandwidth, kernel type, and cell size (which was computed automatically by density.ppp()). It is based in an algorithm used in QGIS, the best free GIS system relying primarily on a point-and-click interface. Unlike spatstat, SpatialKDE can work with sf objects. But it also finds it more convenient to have the data on a projected coordinate systems, so that it can compute distances on metres rather than with decimal degrees.

The package `sparr` focuses on methods to compare densities and thus estimate relative risk, even at a spatio-temporal scale. It is written in such a way as to allow compatibility with `spatstat`. The basic functionality allows to estimate KDE maps through the `bivariate.density()` function. This function can take a `ppp` object and produce a KDE map. In some ways it gives the analyst more flexibility than `density.ppp()`. It allows to modulate the resolution of the grid, two different methods to control for edge effects, the type of estimate produced (intensity or density, that integrates to 1), etc. This package also introduces various methods for estimating the bandwidth different to those existing in `spatstat`. Critically, it introduces a function that allows for spatio-temporal kernel density estimation: `spattemp.density()`. For more details, see the help files for `sparr`, and the tutorial by Davies, Marshall, and Hazelton (2017).

7.6 Summary and further reading

For a deeper understanding of the issues discussed here, we recommend the following reading. The documentation manual for CrimeStat (Chapter 10) has a detailed technical discussion of density estimation that is suited for crime analysts (Levine 2013). Other manuals written for crime analysts also provide appropriate introductions, Gorr and Kurland (2012). Chapter 7 of Bivand, Pebesma, and Gómez-Rubio (2013) also provides a good introduction to spatial pattern analysis with R. For a fuller coverage of this kind of analysis with R, nothing can replace Baddeley, Rubak, and Turner (2015). This book, at 810 pages, is dedicated to `spacestat` and it provides a comprehensive discussion of spatial pattern analysis well beyond what we cover in this chapter. For a discussion of alternatives to KDE so that the method does not assign probabilities of occurrence where this is not feasible see Woodworth et al. (2014) (though beware this has not yet being implemented into R). In Chapter 8, we will discover how much of what we have learnt in this one has to be adapted when working with crime data, for this data typically only appears along street networks.

8

Crime along spatial networks

8.1 Introduction

In the previous chapter we explored how the techniques for spatial pattern analysis can be used to study the varying intensity of crime across space. We introduce the key idea of spatial randomness. These methods were developed to study locations in a continuous plane and are also referred to as **planar spatial analysis**. Point crime data, however, has an inherent structure to it. There are parts of our study region where crime points cannot appear. Since typically the data will be geocoded at the street level address, crime data almost always will appear alongside the spatial network of roads and streets. Crimes won't appear randomly anywhere in our two dimensional representation of the city; they will only appear along the street network covering this city. We can see this in the following example from Chicago using data from `spatstat`:

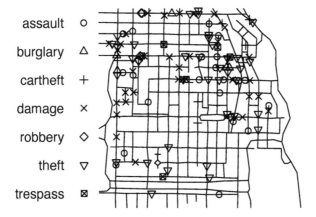

FIGURE 8.1: Crime along street network in Chicago

Clearly the offences are constrained to only occur along the street network. This is not a homogeneous Poisson process along the whole study surface, because there is an underlying structure. If we want to investigate heterogeneity on intensity, for example, we would need to account for that underlying structure, the fact that the points can only be located along the network and not in other parts of the study region. Several authors have argued that crime is best represented and predicted if we focus in the street network rather than other kind of level of aggregation and that this has clear implications for operational policing (Singleton and Brunsdon 2014; Roser et al. 2017). It is clear that if our crime occurrences are collected and represented along a spatial network, we need to ensure we use techniques of analysis and visualisation that respect this existing underlying structure in our data.

The software infrastructure for spatial analysis along networks, however, is still not as developed and consolidated. Partly, this is to do with the statistical and computational challenges when we move from planar to linear network analysis. There is some software such as SANET[1] that was originally conceived as a toolbox for ArcGIS but with a beta standalone version already available (Okabe and Sugihara 2012). R users and developers are working to expand the functionality of R in this regard. It was only in 2020 that Lucas van der Meer, a Dutch spatial data scientist, published the first release of the `sfnetworks` package to facilitate the creation and analysis of spatial networks. Two other useful packages (`SpNetPrep` and `DRHotNet`) developed by Briz-Redón and colleagues are also only a couple of years old (Briz-Redón, Martínez-Ruiz, and Montes 2019a, 2019b). Another recent package, released in 2021, is `spNetwork` by Jeremy Gelb and Philippe Apparicio that can be used for kernel density estimation along networks (Gelb 2021). Finally, the package `spatstat`, covered in the last chapter, also provides functionality for the analysis of events along spatial networks.

In this chapter we will provide an introduction to the study of crime along networks by exploring the following:

- creation of network representations from geographical vector objects
- linking data to meaningful micro-places such as **street segment** (or in a transportation example, train line segment) or **street junction** (or for transport research a train or bus station)
- introducing the idea of **hot routes**
- evaluating crime concentration in these types of micro-places using **Gini coefficient** and **Lorenz curve**
- **street profile analysis** - an alternative (non-spatial) way to visualise crime along a network
- introducing `spatstat` for the analysis of events along networks

```
# Packages for reading data and data carpentry
library(readr)
library(readxl)
library(janitor)
library(tidyr)
library(dplyr)
library(tibble)
library(lubridate)


# Packages for handling spatial data
library(sf)
library(lwgeom)
library(sfnetworks)
library(spatstat)
library(spatstat.linnet)
library(maptools)
# NOTE: maptools is still required for the as.linnet.SpatialLines() function,
```

---

[1] http://sanet.csis.u-tokyo.ac.jp/

```
# but will be retired by the end of 2023. Look out for where
# as.linnet.SpatialLines() function will migrate!

# Package for computing Lorenz curve
library(ineq)

# Packages for visualisation and mapping
library(ggplot2)
library(ggpubr)
library(raster)
library(leaflet)

# Packages providing accesss to data
library(osmdata)
library(tigris)
```

It is not only in the case of street networks that understanding spatial point patterns along a network might be more meaningful; there are other networks we might want to consider. Transport networks are a good example of this and are receiving an increasing interest in criminology and crime analysis (Newton 2008; Tompson, Partridge, and Shepherd 2009; Ceccato 2013).

8.2 Constructing and storing a spatial linear network

The first difficulty one encounters when wanting to do analysis along spatial networks is to get the data for such geographical element in the proper shape. It is fairly straightforward to get files with the administrative boundaries for census geographies, as we have seen in previous chapters. This kind of geographical data is widely available in official and open repositories (see also *Appendix C: sourcing geographical data for crime analysis*). Getting data for spatial networks used to be a little bit trickier, although it is getting better. One could, for example, use the osmdata package to access Open Street Map data for a road network (we also discuss this in greater detail in *Appendix C*). Or, if you analyse data from the United States, you could use the tigris package to extract geometries which represent road networks.

Below we show how to obtain this data for Kensington, a popular neighbourhood in Toronto city centre through osmdata and for Manhattan using tigris. As we're querying data from Open Street Map, first we draw a bounding box around the area we are interested in, then select "highways" as the feature to extract (see *Appendix C* for details on building OSM queries). As discussed by Abad, Lovelace, and Meer (2019), the osmdata package turns streets that form loops into polygons; to ensure we end up with lines we use osm_poly2line.

```
# using osm data
highways_heswall <- opq("Heswall, UK") %>%
  add_osm_feature(key = "highway") %>%
```

```
osmdata_sf()

heswall_lines <- highways_heswall$osm_lines
```

Now let's use the `tigris` package to get TIGER/Line shapefiles from the United States Census Bureau. The `roads()` function returns the content of all road shapefiles - which includes primary roads, secondary roads, local neighbourhood roads, rural roads, city streets, vehicular trails (4WD), ramps, service drives, walkways, stairways, alleys, and private roads. It returns these as an `sf` object.

```
# using tigris
manhattan_roads <- roads("NY", "New York")
```

We can now plot these two side-by-side.

```
# plot results
plot_1 <- ggplot(heswall_lines) + geom_sf() + theme_void()
plot_2 <- ggplot(manhattan_roads) + geom_sf() + theme_void()
figure_1 <- ggarrange(plot_1, plot_2, labels = c("Heswall", "Manhattan"),
                      ncol = 2, nrow = 1)
figure_1
```

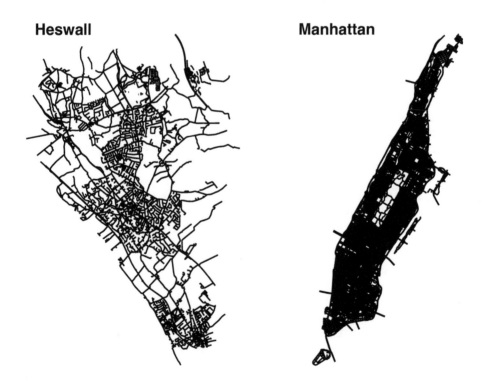

FIGURE 8.2: Street network geometries for Heswall, UK and Manhattan, USA

What we have is two `sf` object containing linestrings. Linear networks indeed can be represented in R by simple features wit `sf` as here, or with the class `SpatialLines` of package `sp`. However, for us to be able to do a particular kind of analysis and some geospatial operations on this network we often will need a clear topology: we will need the data to be stored as a network graph. In other words, we need to move from linestrings to a linear graphical network representation, with edges (for example, representing street segments) and nodes (vertices indicating where they start and end).

This can be done by virtue of manual editing, which in large networks may be prohibitive, fully automatized, or automatized with manual edits just to correct errors. Explaining how to do this in detail with R would require space we do not have. Luckily, there are some good tutorials out there that you can consult for this data pre-processing work.

- Abad, Lovelace, and Meer (2019) explain how to use the packages `rgrass7` and `link2GI` for this purpose. These packages bridge R to GRASS GIS software. So for this approach to work, you would require this software to be installed in your machine. Fortunately, GRASS GIS is freely available.
- The vignettes for `sfnetworks` also cover how to turn a linestring `sf` object into a network. The function `as_sfnetwork()` provides the basic functionality for this. This package also includes several additional functions for cleaning the resulting network, and there is a specific vignette documenting these (van der Meer et al. 2021).
- Briz-Redón, Martínez-Ruiz, and Montes (2019a) developed the `SpNetPrep` ("Spatial Network Preprocessing"). This package will open a Shiny application in RStudio that will allow for network creation and edition, network direction endowment and point pattern revision and modification. The key aim of this package is to allow for manual edits of the constructed network to correct for errors.

We quickly show below the `as_sfnetwork()` function with the road network in Kensington (which will generate an object of class `sfnetwork`) so that you get a sense for the look of a spatial network. But for further details on network construction and pre-processing, it is important you see the documentation, as well as the other materials cited above.

```
heswall_ntwk <- as_sfnetwork(heswall_lines)
plot(heswall_ntwk)
```

FIGURE 8.3: Heswall as a spatial network

Once you have a spatial network that has been properly cleaned and prepared as discussed in the references above, there are a number of things you can do with it. You can apply geospatial operations, such as blending points into the edges, segmenting the edges, running spatial filters, etc. You can also apply standard social network analysis techniques and measures (such as centrality, measuring shortest path). Illustrations of these applications are available in the vignettes for `sfnetworks` (van der Meer et al. 2021) and in Abad, Lovelace, and Meer (2019).

Another format for linear networks in R is `linnet` used by the package `spatstat`. When what we want is to do analysis not of the network itself, but of the location of points (such as crime events) along the network `spatstat` is the way to go, and so we would need to ensure we store the data in formats that are required by this package. Statistical analysis of events along these networks is of key interest for crime analysts. In the next section we will explain how to perform density estimation along networks.

8.3   Density estimation along networks

One way of estimating the intensity of crime along spatial networks is to use kernel density estimation methods. You could, as we did in Chapter 7, use standard kernel density estimation methods, but this ignores the underlying structure in the data. As Jeremy Gelb indicates in the vignette for `spNetwork`, there are a number of problems with this cruder approach:

> Calculating density values for locations outside the network is meaningless and the Euclidean distance underestimates the real distance between two objects on the network. Moreover, networks are not isotropic spaces. In other words, it is not possible to move in every direction, but only along the edges of the network (Gelb 2021).

Smoothing through kernel density estimation in linear networks is an area of active development and a number of approaches have been proposed over recent years, some of which provide a pragmatic (heuristic) solution, but "their properties are not well understood, statistical insight is lacking," and "computational cost is high" (McSwiggan, Baddeley, and Nair 2016: 325). The existing software implementation of the "traditional" algorithms that have been developed can deal with small datasets, but may become prohibitive when dealing with larger datasets. More recent algorithms aim to provide faster solutions for larger datasets without seriously compromising performance (see Rakshit et al. (2019)). You can currently do this in R with at least two separate packages `spatstat` and the newer `spNetwork`. For reasons of space, given that it provides a more general framework for point pattern analysis and a wider selection of (more principled) approaches, we will only cover `spatstat` here.

In `spatstat` a point pattern on a linear network is represented by an object of class `lpp` (stands for point pattern in a linear network). The functions `lpp()` and `as.lpp()` convert raw data into an object of class `lpp`. The two needed arguments are a matrix or data frame with the coordinates of the points (or alternatively a `ppp` object that we discussed in Chapter 7) and a liner network of class `linnet`).

The `linnet` object will represent the network of a set of line segments, containing information about each of these segments (edges) and its vertices (nodes). You can generate a `linnet` object from a SpatialLines (sp) object using as.linnet.SpatialLines(). If you need to process this network before it is ready, you can do so with some functions from spatstat or SpnetPrep as preferable (and only then create your linnet object).

We will show here how to transform a SpatialLines object into linnet and how, then, we generate a lpp object.

```
# transform to projected coordinates (British National Grid)
heswall_lines <- st_transform(heswall_lines, crs = 27700)
# make into sp object
heswall_roads_sp <- as(heswall_lines, 'Spatial')
# turn into line network (linnet)
heswall_roads_linnet <- as.linnet.SpatialLines(heswall_roads_sp)
```

Now let's get some crime data to link to this line network. We can find this from the open data portal data.police.uk and choose to get data from Merseyside police, who cover this area. For the sake of practice here, one year's worth of data is included with the datasets available with this textbook (see Preamble). The file is merseyside_asb.csv.

```
# read in asb data
shoplifting <- read_csv("data/merseyside_shoplifting.csv")
```

```
## New names:
## Rows: 7930 Columns: 14
## -- Column specification
## ---------------------------------------- Delimiter: "," chr
## (9): crime_id, month, reported_by, falls_within... dbl (4): ...1,
## ...2, longitude, latitude lgl (1): context
## i Use `spec()` to retrieve the full column specification for this
## data. i Specify the column types or set `show_col_types = FALSE` to
## quiet this message.
## * `` -> `...1`
## * `...1` -> `...2`
```

```
# transform to spatial object
shoplifting_sf <- st_as_sf(shoplifting,
                     coords = c("longitude", "latitude"),
                     crs = 4326)
# transform to match projection of lines
shoplifting_sf <- st_transform(shoplifting_sf, crs = st_crs(heswall_lines))
# plot them together to have a look
ggplot() +
  geom_sf(data = heswall_lines) +
  geom_sf(data = shoplifting_sf, colour = "red") +
  theme_void()
```

FIGURE 8.4: All shoplifting incidents in Merseyside from May 2019 to April 2020

We can see that Merseyside Police cover a lot more than just Heswall, so we can use the bounding box around the roads to subset our crime data:

```
# create a bounding box around the heswall lines geometry
heswall_bb <- st_bbox(heswall_lines) %>% st_as_sfc()
# subset using the bounding box
heswall_shoplifting <- st_intersection(shoplifting_sf, heswall_bb)
```

```
## Warning: attribute variables are assumed to be
## spatially constant throughout all geometries
```

Now that we have the line network and the shoplifting incidents in Heswall, we can create our lpp object. First we create a ppp (point pattern) object, following the steps we learned in Chapter 7 on spatial point patterns:

```
# create a window
window <- as.owin(heswall_roads_linnet)
# extract coordinates
shoplifting_coords <- matrix(unlist(heswall_shoplifting$geometry), ncol = 2, byrow = T)
# make into pppt
shoplifting_ppp <- ppp(x = shoplifting_coords[,1], y = shoplifting_coords[,2],
            window = window, check = T)
```

```
## Warning: data contain duplicated points
```

Now that we have the ppp object, we can use it to create our lpp object (notice that, as noted above, there are other ways to input your points when creating a lpp object) and then plot it. There are a 264 locations, some of which represent up to 6 incidents. This is likely the result of geomasking due to privacy concerns (Tompson et al. 2015).

```
# now use this and previously created linnet to make lpp object
shoplifting_lpp <- lpp(shoplifting_ppp, heswall_roads_linnet, window = window)
```

Now we can calculate the density and visualise. Like we did before, we can estimate the density of data points along the networks using Kernel estimation with density.lpp(). This is an area of spatstat that has seen significant development in the latest release. Depending on the arguments passed to the function, a different algorithm will be employed to estimate the density.

```
# calculate density
d100 <- density.lpp(shoplifting_lpp, 100, finespacing=FALSE)
# plot
plot(d100)
```

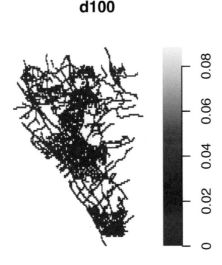

FIGURE 8.5: Network density map with defaults

```
plot(d100, style="width")
```

d100

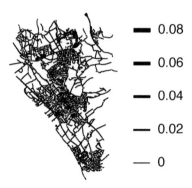

FIGURE 8.6: Network density map with defaults with line width

A `lpp` object contains the linear network information, the spatial coordinates of the data points, and any number of columns of *marks* (in this case, the mark is telling us the type of crime we are dealing with). It also contains the local coordinates `seg` and `tp` for the data points. The local coordinate `seg` is an integer identifying the particular street segment the data point is located in. A segment is each of the sections of a street between two vertices (marking the intersection with another segment). The local coordinate `tp` is a real number between 0 and 1 indicating the position of the point within the segment: `tp=0` corresponds to the first endpoint and `tp=1` correspond to the second endpoint.

The visual inspection of the map suggests that the intensity of crime along the network is not spatially uniform. Alternatively one could use the new specific functions for each of these different algorithms: `densityQuick()` (see Rakshit et al. (2019)), `densityEqualSplit()` (see Okabe and Sugihara (2012)), or `densityHeat()` (see McSwiggan, Baddeley, and Nair (2016)).

The Okabe-Suhihara method was the first to be developed for this kind of applications and can be invoked with `densityEqualSplit()`. We can specify the kernel (the default is `epanechnikov`) and the smoothing bandwidth, sigma (the standard deviation of the kernel). Above, we used 100 metres for illustration, but see *Chapter 7: Spatial Point Patterns of Crime Events* in this book for details on these parameters). There are two versions of this algorithm, the continuous that is excruciatingly slow (particularly for larger datasets) and the discontinuous that is a bit quicker but has less desirable statistical properties. We are using the number of events per point as the weight for computation, and wrapping everything on `system.time()` to provide a notion of computation time. We will only show the discontinuous version.

```
system.time(d_okabe <- densityEqualSplit(shoplifting_lpp,
                            weights = marks(shoplifting_lpp),
                            sigma = 100,
                            continuous = FALSE,
                            verbose= FALSE))
```

```
##    user  system elapsed
##   9.179   0.218   9.398
```

And to visualise:

```
plot(d_okabe, main="Okabe-Sugihara (disc.)")
```

Okabe-Sugihara (disc.)

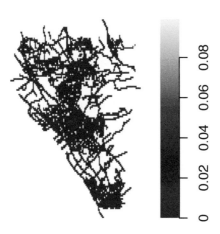

FIGURE 8.7: Network density map (Okabe-Sugihara (disc.) method)

McSwiggan, Baddeley, and Nair (2016) subsequently developed an algorithm that is equivalent to the Okabe-Sugihara method using the continuous rule applied to the Gaussian kernel, and therefore has more desirable properties than the equal split discontinuous, but that runs faster (than the Okabe-Sugihara equal split continuous version).

```
system.time(d_mcswiggan <- densityHeat.lpp(shoplifting_lpp,
                         weights = marks(shoplifting_lpp),
                         sigma = 100,
                         finespacing=FALSE))
```

```
##    user  system elapsed
##   1.053   0.015   1.068
```

```
plot(d_mcswiggan, main="McSwiggan")
```

McSwiggan

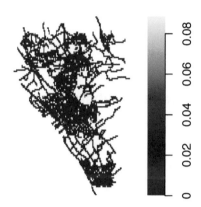

FIGURE 8.8: Network density map (McSwiggan Gaussian method)

Finally, as noted above, Rakshit et al. (2019) developed a fast algorithm, using a convolution method, that is particularly helpful with large datasets. The analysis of the authors suggest this estimator "is consistent, and its statistical efficiency is only slightly suboptimal" (Rakshit et al. 2019).

```
system.time(d_rakshit <- densityQuick.lpp(shoplifting_lpp,
                              weights = marks(shoplifting_lpp),
                              sigma = 100,
                              edge2D = TRUE))
```

```
##    user  system elapsed
##   0.247   0.001   0.248
```

We can now plot the data and compare the results:

```
plot(d_rakshit, main="Rakshit")
```

Rakshit

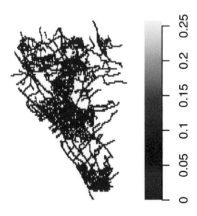

FIGURE 8.9: Network density map (Rakshit method)

So these are some ways in which we can create a density map of crime along networks. You can compare the three maps to see how choice of method affects conclusions. This is something we must always reflect on, when applying these techniques.

8.3.1 Adding context

Another thing you might like to do is add some context to explore where these hot spots are. Similar to how we did this with our KDE in continuous space in Chapter 7, we can use the raster and leaflet packages here too. One difference, we must also use the as.im() function to make sure our heatmap is of class "im".

```
# convert to raster wrapped inside as.im() function
r1 <- raster(as.im(d_mcswiggan))
#remove very low density values
r1[r1 < 0.0001 ] <- NA
# set CRS
crs(r1) <- "EPSG:27700"

#create a colour palet
pal <- colorNumeric(c("#0C2C84", "#41B6C4", "#FFFFCC"),
                values(r1), na.color = "transparent")


#and then make map!
leaflet() %>%
  addTiles() %>%
  addRasterImage(r1, colors = pal, opacity = 0.8) %>%
  addLegend(pal = pal, values = values(r1),
    title = "Shoplifting map")
```

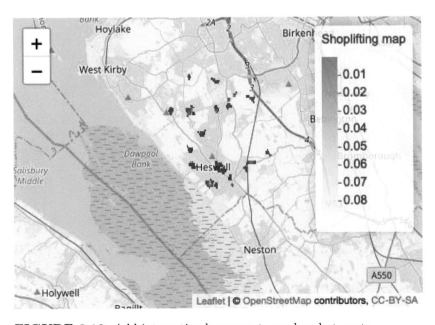

FIGURE 8.10: Add interactive basemap to explore hot spots

8.4   Visualising crime distribution along networks

To demonstrate some simpler approaches to visualising crime distribution along networks, we can zoom in a little bit on Heswall town centre. The coordinates for that are: 53.3281041, -3.1025683. Let's build a buffer.

```r
# create a df of one central point
heswall_tc <- data.frame(longitude = -3.1025683,
                         latitude = 53.3281041)

# create a buffer by making point into sf object
heswall_tc_buffer <- st_as_sf(heswall_tc,
                    coords = c("longitude", "latitude"),
                    crs = 4326) %>%
  st_transform(., 27700) %>% # project to BNG (for metres)
  st_buffer(., 1000)  # build 1km buffer


# select roads that go through town centre
heswall_tc <- st_intersects(heswall_tc_buffer, heswall_lines)
heswall_tc <- heswall_lines[unlist(heswall_tc),]
# select shoplifting incidents in town centre
tc_shoplifting <- st_intersects(heswall_tc_buffer, heswall_shoplifting)
tc_shoplifting <- heswall_shoplifting[unlist(tc_shoplifting),]

ggplot() +
  geom_sf(data = heswall_tc) +
  geom_sf(data = tc_shoplifting, col = "red") +
  geom_sf(data = heswall_tc_buffer, col = "blue", fill = NA) +
  theme_void()
```

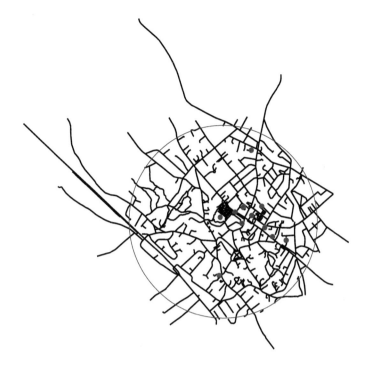

FIGURE 8.11: Narrow focus on Heswall town centre

8.4.1 Hot Routes

In order to map hot spots along a network, we can use a technique called *hot routes*. Hot routes were devised to be a straightforward spatial technique that analyses crime patterns that are associated with a linear network (e.g., streets and other transportation networks). It allows an analyst to map crime concentrations along different segments of the network and visualise this through colour. We can create a hot routes visualisation by thematically shading each street segment with a colour (and line thickness if desired) that corresponds to the range of the rate of crime per metre. It was used by Newton (2008) to map crime and disorder on the bus network in Merseyside. A useful how-to guide on using hot routes was later produced by Tompson, Partridge, and Shepherd (2009). In it they break down the process into four stages:

```
- Step 1. Preparing the network layer
- Step 2. Linking crime events to street segments
- Step 3. Calculating a rate
- Step 4. Visualising the results
```

We can follow those steps here.

8.4.1.1 Preparing the network layer

The prerequisite to visualising hot routes is to have a network layer which meets certain conditions. For example, Tompson, Partridge, and Shepherd (2009) advise that since network layers typically contain streets of unequal length, longer segments might show up as hot simply because they have more space to contain more crimes. Therefore in such cases, it is advisable in this analysis to use equal length street segments, where possible. To address this, they segment their study area into 1 metre length segments using a grid overlay. Another approach is to calculate a rate per metre, instead of count, so that's the approach we will take here.

Another requirement is that the network layer has a unique identifier for each segment. One quick way to do this, if your data don't already contain such a column is to create a column that just numbers each row from 1 to the number of observations (rows) in the data frame. Even if you have such a unique identifier, having the row name to a numeric column is required for joining the crimes to the street segments in the next step.

```
# create numeric unique id column
heswall_tc$unique_id <- 1:nrow(heswall_tc)
```

8.4.1.2   Link crime events to street segments

One approach to solve this is to snap each shoplifting incident to the nearest street segment. We can use the `st_nearest_feature()` function that will create a list of the ID number of the nearest street segment for each crime incident in our `tc_shoplifting` object.

```
nearest_segments <- st_nearest_feature(tc_shoplifting, heswall_tc)
```

This new object is simply a list of the ID numbers of matched line segments for each of the ASB incidents. For example, here is the ID number of the nearest street segment for the first five ASB incidents in our data:

```
nearest_segments[1:5]
```

```
## [1] 315 315  65 317 317
```

We can use this to create a frequency table which counts the number of ASB incidents linked to each street segment (by virtue of being nearest to it).

```
#make list of nearest into df of frequency
nearest_freq <- as.data.frame(table(nearest_segments))
#make sure id is numeric
nearest_freq$nearest_segments <- as.numeric(
  as.character(
    nearest_freq$nearest_segments
    )
  )
nearest_freq[1:3,]
```

```
##    nearest_segments Freq
## 1                 1    1
## 2                17    3
## 3                27    3
```

We can see here some examples of the number of shoplifting incidents snapped to each segment. Now we can join this to our spatial layer:

```
#join to sections object and replace NAs with 0s
heswall_tc <- left_join(heswall_tc,
                        nearest_freq,
                        by = c("unique_id" = "nearest_segments")) %>%
  mutate(Freq = replace_na(Freq, 0))
```

We can produce a quick visualisation to check that everything linked OK before moving on to compute a rate normalising for length of street segment.

```
ggplot() +
  geom_sf(data = heswall_tc,
          aes(colour = Freq), lwd = 0.5) +
  theme_void() +
  scale_colour_gradient2(name = "Number of shoplifting incidents",
                         midpoint = 1,
                         low = "#2166ac",
                         mid = "#d1e5f0",
                         high = "#b2182b")
```

FIGURE 8.12: Check that crimes are linked to street segments

Looks like we joined the shoplifting incidents to our network layer successfully. Now let's normalise for the varying length of the street segments.

8.4.1.3 Calculating a rate

In order to calculate a rate we need the numerator, number of shoplifting incidents, which we joined above, and a denominator to divide by. To control for varying length of the street segments, we might want to consider this length as a suitable denominator. So we need to calculate this for each segment. To achieve this, we can use the `st_length()` function.

```
heswall_tc$length <- st_length(heswall_tc)
```

Now we have a column in the data which shows the length of each segment in metres (as we are working with projected coordinates). The class of this variable is units. We can see this if we print out one of the values:

```
heswall_tc$length[1]
```

```
## 729.8 [m]
```

To use it in our calculation, we need this to be a numeric variable, so let's transform it, and then use it as our denominator to calculate our rate of crimes per metre of street segment length:

```
heswall_tc$length <- as.numeric(heswall_tc$length)
heswall_tc$shoplifting_rate <- heswall_tc$Freq /heswall_tc$length
```

Now we have a measure of incidents per metre. Finally, we can move to visualise the results.

8.4.1.4 Visualise the results

This is very similar to how we carried out a visual check earlier, but this time using our rate as the value to present.

```
ggplot() +
  geom_sf(data = heswall_tc, aes(colour = shoplifting_rate),
          lwd = 0.5, alpha = 0.8) +
  theme_void() +
  scale_colour_gradient2(name = "Shoplifting incidents per metre",
                          midpoint = mean(heswall_tc$shoplifting_rate),
                          low = "#2166ac", mid = "#d1e5f0",
                          high = "#b2182b")
```

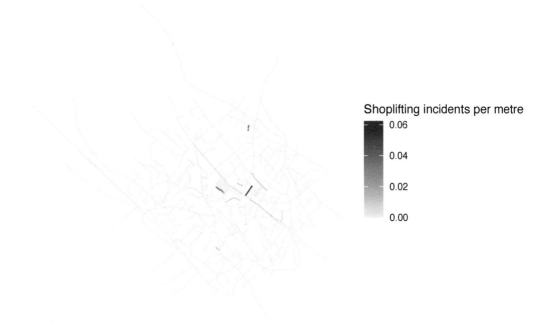

FIGURE 8.13: Hot routes map displaying rate of crimes per metre for each segment

We can now see that the longer road there (Telegraph Road) is no longer so dominant once we've controlled for length.

Something to consider is that the longer the road, the larger the denominator, so, since there are not many crimes on each road, we might artificially dilute the crime rate. If you suspect this might be the case, then segmentation might be a better approach, and then refer back to the paper by Tompson, Partridge, and Shepherd (2009) for guidance on that.

In any case, we have now created a hot routes visualisation, that helps us map where crime rates are higher along a network. We can add some variations to this if we like; for example we could adjust the width of the segments as well as the colour, although this doesn't always look the best as it distorts the street network:

```
ggplot() +
  geom_sf(data = heswall_tc,
          aes(colour = shoplifting_rate, size = shoplifting_rate),
          alpha = 0.8) +
  theme_void() +
  scale_colour_gradient2(name = "Shoplifting incidents per metre",
                         midpoint = mean(heswall_tc$shoplifting_rate),
                         low = "#2166ac", mid = "#d1e5f0",
                         high = "#b2182b") +
  scale_size_continuous(name = "Rate of crimes per metre (width)")
```

FIGURE 8.14: Hot routes map displaying rate of crimes per metre for each segment with line width as well as colour

8.4.2   Street Profile Analysis

Visualising crime along networks presents an important first step in identifying crime concentration. It may not always be necessary to include the geographic component of a spatial network for the purposes of crime analysis in these networks. Instead, another approach could be to make use of street profile analysis introduced by Spicer et al. (2016). Street profile analysis was initially developed as a method for visualising temporal and spatial crime patterns along major roadways in metropolitan areas. The idea is that the geographical location of the street may not actually be as important as the ability to visualise data from multiple years in a comparable way. So the approach is to treat the street in question as the x-axis of a graph, with a start point (A) and end point (B) breaking it into some interval. Then, you can visualise how crime is distributed along this street by plotting count or rate of crimes on the y-axis, and have multiple lines for different years for example. You can add context by re-introducing some key intersections, or other points of interest. In their paper, Spicer and colleagues demonstrate this using the example of Kingsway Avenue in Vancouver, BC. In this section we will go through how we can apply street profile analysis in R. We will follow the same four steps we did for the hot routes tutorial:

```
- Step 1: Prepare the network layer
- Step 2: Link crime events to points of interest
- Step 3: Calculate a rate
- Step 4: Visualise the results
```

8.4.2.1 Prepare the network layer

The strength of this approach is to focus on one area of interest. Here let's choose Telegraph Road. It is a long street which runs through the town centre. To subset it, we can use the names provided by the volunteers of Open Street Map, available in the `name` column:

```
telegraph_rd <- heswall_tc %>% filter(name == "Telegraph Road")
```

The geometry we have is already separated into seven segments. This is just how we downloaded from OSM. We see there are seven segments, but they don't necessarily correspond to anything specific. In Street Profile Analysis we want to be conscious about how we divide up the segment. In their paper, Spicer et al. (2016) just segment into 100-metre lengths. Another thing we might be interested in is to slice into segments wherever there is an intersection with another road. Here let's use the intersections as a way to divide our street into segments.

To do this, we create an object of the intersecting streets; let's call this object `tr_intersects`.

```
# and select also the cross streets into separate object
tr_intersects <- st_intersects(telegraph_rd, heswall_tc)
# subsetting
tr_intersects <- heswall_tc[unlist(tr_intersects),]
# remove Telegraph Road itself
tr_intersects <- tr_intersects %>% filter(name != "Telegraph Road")
```

We can plot our segments to see whether they correspond to these intersections:

```
ggplot() +
  geom_sf(data = tr_intersects, lwd = 0.5) +
  geom_sf(data = telegraph_rd, aes(colour = osm_id), lwd = 1.5, alpha = 0.8) +
  theme_void() + theme(legend.position = "none")
```

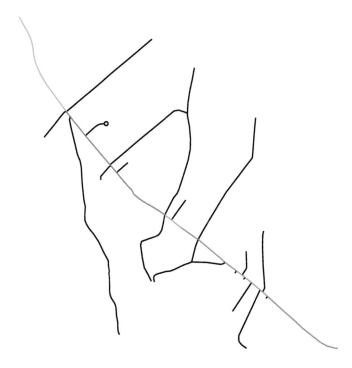

FIGURE 8.15: Default OSM segments of Telegraph Road versus intersections

We can see that they do not correspond to all the intersections. Unfortunately, we will have to carry out some data cleaning. To do this, first we unite the entire street into one, and then use the intersections to divide into the appropriate segments.

```
# unite the segments into one long street
telegraph_rd_u <- telegraph_rd %>%
  group_by(name) %>%
  st_union() %>%
  st_cast("LINESTRING")


# use st_split and st_combine to split into parts by intersecting roads
library(lwgeom)
parts <- st_split(telegraph_rd_u, st_combine(tr_intersects$geometry)) %>%
  st_collection_extract("LINESTRING")

# combine the parts into one shapefile
datalist = list()
for (i in 1:length(parts)) {

  datalist[[i]] <- st_as_sf(data.frame(section = i),
                            geometry = st_geometry(parts[i]))

}
tr_sections <- do.call(rbind, datalist)
```

In the above code we do a lot, but it means we end up with a geometry where the segments were defined by us, with the rationale that we should segment at each intersection. Whether you use this method or divide at some set interval like every 100 metres is up to you, but a decision made and justified is usually better than relying on defaults, such as the segments from OSM.

Let's plot the new segments to see if they correspond to intersections.

```
ggplot() +
  geom_sf(data = tr_intersects, lwd = 0.5) +
  geom_sf(data = tr_sections,
          aes(colour = as.character(section)),
          lwd = 1.5, alpha = 0.8) +
  theme_void() + theme(legend.position = "none")
```

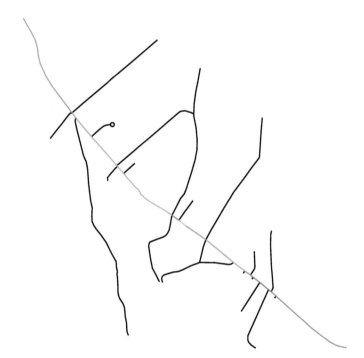

FIGURE 8.16: Telegraph Road segmented at each intersection

This looks good. Now we can move to the next step and link our shoplifting incidents to these segments.

8.4.2.2 Link events to street segments

To link the shoplifting incidents, first we should filter those crimes which will be relevant. For this we build a 400-metre buffer around the road, and include only those shoplifting incidents which fall within this buffer (for more about buffers, see *Chapter 2 Basic Geospatial Operations in R*).

```
tr_buffer <- st_buffer(telegraph_rd_u, 400)

# select shoplifting incidents in town centre
tr_shoplifting <- st_intersects(tr_buffer, tc_shoplifting)
tr_shoplifting <- tc_shoplifting[unlist(tr_shoplifting),]
```

The strength of street profiles comes out in comparing different sources of data' so let's split our shoplifting incidents by season. Let's compare summer and winter.

```
# split seasonally (to compare later) let's focus on winter and summer
summer_sl <- tr_shoplifting %>%
  filter(month %in% c("2019-06", "2019-07", "2019-08"))
winter_sl <- tr_shoplifting %>%
  filter(month %in% c("2019-12", "2020-01", "2020-02"))
```

Then, the steps for linking the crimes are the same as we followed for hot routes above, using the `st_nearest_feature()` function.

```
# attach summer shoplifting incidents
nearest_segments_summer <- st_nearest_feature(summer_sl, tr_sections)
nearest_freq_summer <- as.data.frame(table(nearest_segments_summer))
nearest_freq_summer <- nearest_freq_summer %>%
  mutate(nearest_segments_summer = as.numeric(
    as.character(
      nearest_freq_summer$nearest_segments_summer
      )
    )) %>%
  rename(summer_sl = Freq)
tr_sections <- left_join(tr_sections,
                      nearest_freq_summer,
                      by = c("section" = "nearest_segments_summer")) %>%
  mutate(summer_sl = replace_na(summer_sl, 0))



# repeat for the winter
nearest_segments_winter <- st_nearest_feature(winter_sl, tr_sections)
nearest_freq_winter <- as.data.frame(table(nearest_segments_winter))
nearest_freq_winter <- nearest_freq_winter %>%
  mutate(nearest_segments_winter = as.numeric(
    as.character(
      nearest_freq_winter$nearest_segments_winter
      )
    )) %>%
  rename(winter_sl = Freq)
tr_sections <- left_join(tr_sections,
                      nearest_freq_winter,
                      by = c("section" = "nearest_segments_winter")) %>%
  mutate(winter_sl = replace_na(winter_sl, 0))
```

Similar to how we did with hot routes, we can carry out a visual check to make sure they linked OK. We can also try to compare the two seasons on these side-by-side maps.

```
p1 <- ggplot() +
  geom_sf(data = tr_sections, aes(colour = summer_sl), lwd = 0.5) +
  theme_void() +
  scale_colour_gradient2(name = "Shoplifting in summer",
                         midpoint = 1,
                         low = "#2166ac", mid = "#d1e5f0", high = "#b2182b")
p2 <- ggplot() +
  geom_sf(data = tr_sections, aes(colour = winter_sl), lwd = 0.5) +
  theme_void() +
  scale_colour_gradient2(name = "Shoplifting in winter",
                         midpoint = 1,
                         low = "#2166ac", mid = "#d1e5f0", high = "#b2182b")

gridExtra::grid.arrange(p1,p2, nrow = 1)
```

FIGURE 8.17: Shoplifting in summer and winter months

We can see there may be some differences between winter and summer, but it is unclear how much.

8.4.2.3 Calculate a rate

This is the same process we used in the hot routes example above, in that we get a length for each segment, and then we use this to calculate a rate for summer and winter:

```
tr_sections$length <- st_length(tr_sections)

tr_sections$summer_rate <- as.numeric(tr_sections$summer_sl / tr_sections$length)
tr_sections$winter_rate <- as.numeric(tr_sections$winter_sl / tr_sections$length)
```

8.4.2.4   Visualise the results

Finally, we reach the visualisation stage. To do this, we need to find the "Point A" location, which will serve as our starting point to travel down Telegraph Road. Most likely you will have to do this manually. In the case of a bus or railway route, you know the terminus stations, so you can use these. In the case of street segments, it might be more arbitrary. Here I have identified "section 3" as our starting point. Note that it's not section 1 or 20, which means that the segments are not numbered sequentially. This means we will have to assign an order.

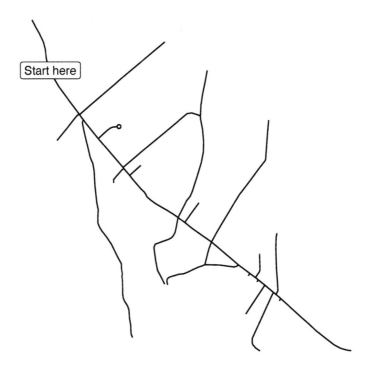

FIGURE 8.18: Find starting point "Point A" on the street

To assign this order, we can take our starting point (labelled "Start here") and for each line segment we can calculate the distance to this. The idea is that we can then use this distance to order them from nearest to furthest.

```
datalist <- list()
i <- 1
for(segment_id in tr_sections$section){
  datalist[[i]] <- data.frame(section = segment_id,
                              dist_from_a = st_distance(
                        tr_sections %>% filter(section == 3),
                        tr_sections %>%  filter(section == segment_id)))
i <- i + 1
}
```

```
dist_matrix <- do.call(rbind, datalist)
tr_sections <- left_join(tr_sections, dist_matrix)
```

```
## Joining, by = "section"
```

```
# finally give our anchor point "a" a value of -1
tr_sections$dist_from_a[tr_sections$section == 3] <- -1
```

Now we can arrange the segments in order from Point A to Point B.

Let's plot the summer shoplifting incidents using the street profile approach. This is a ggplot where the x-axis is the sequence of segments, and the y-axis is the crime rate we want to portray.

```
ggplot(tr_sections, aes(x = reorder(as.character(section),
                             dist_from_a),
                   y = summer_rate,
                   group = 1)) +
  geom_point() +
  geom_line() +
  xlab("Telegraph Road") +
  ylab("Shoplifting incidents") +
  theme_bw() +
  theme(axis.text.x = element_text(angle = 60, hjust = 1))
```

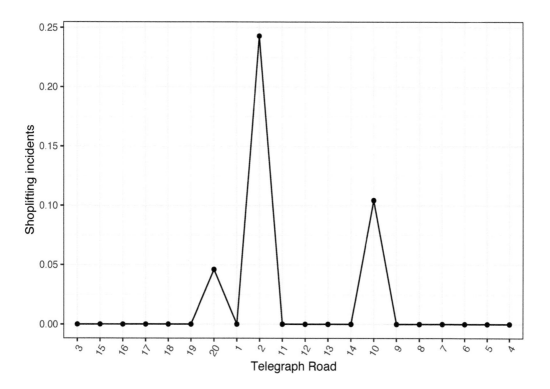

FIGURE 8.19: Street profile with unlabelled segments

You might notice here that the labels are not very meaningful. We could manually recode each one to something better, or another approach is to try to get the name of the intersecting road in order to give some meaning. To do this, we can use a little loop below which will cycle through each segment, and we can find the nearest intersecting road from the `tr_intersects` object.

```
# label the cross streets from 1 to n
x_streets <- tr_intersects %>%
   mutate(xst_id = 1:nrow(.))

# loop through each segment and find the nearest intersecting road
datalist <- list()
i <- 1
for(segment_id in tr_sections$section){
  datalist[[i]] <- data.frame(section = segment_id,
                   x_street = st_nearest_feature(
                     tr_sections %>%
                       filter(section == segment_id), x_streets))
i <- i + 1
}
# bind into a df
nearest_matrix <- do.call(rbind, datalist)

# join the intersections to the street data frame
nearest_matrix <- left_join(nearest_matrix,
                            x_streets %>%
                              dplyr::select(xst_id, name) %>%
                              st_drop_geometry(),
                            by = c('x_street' = 'xst_id'))
tr_sections <- left_join(tr_sections, nearest_matrix,
                       by = c("section" = "section")) %>%
  mutate(name = make.unique(name))
```

Once this is done, we can use this label on the x-axis instead of the segment ID, to provide a little more context.

```
ggplot(tr_sections, aes(x = reorder(name,
                              dist_from_a),
                    y = summer_rate,
                    group = 1)) +
  geom_point() +
  geom_line() +
  xlab("Telegraph Road") +
  ylab("Shoplifting incidents") +
  theme_bw() +
  theme(axis.text.x = element_text(angle = 60, hjust = 1))
```

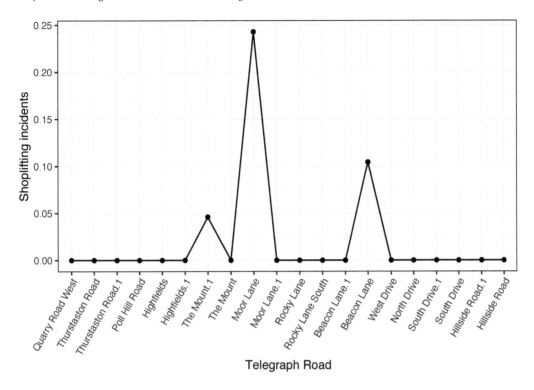

FIGURE 8.20: Street profile using intersecting roads on the x-axis

Now you may notice that there are some repeated names (e.g., "Thurstaston Road" and "Thurstaston Road.1") where the same intersection was snapped as "nearest" to two segments. In these cases the only solution is to go back to manual data cleaning, unfortunately. However, this is already a nice point from which to start.

Finally, the power of the street profile analysis comes from the ease of comparison of data across the route. For example, here we can compare shoplifting between winter and summer:

```
ggplot() +
  geom_point(data = tr_sections, aes(x = reorder(name,
                              dist_from_a),
                  y = summer_rate, group = 1), col = "#2166ac") +
  geom_line(data = tr_sections, aes(x = reorder(name,
                              dist_from_a),
                  y = summer_rate, group = 1), col = "#2166ac") +
    geom_point(data = tr_sections, aes(x = reorder(name,
                              dist_from_a),
                  y = winter_rate, group = 1), col = "#b2182b") +
  geom_line(data =  tr_sections, aes(x = reorder(name,
                              dist_from_a),
                  y = winter_rate, group = 1), col = "#b2182b") +
  scale_colour_manual(values = c("#2166ac", "#b2182b"),
                  labels = c("Summer", "Winter")) +
  xlab("Telegraph Road") +  ylab("Shoplifting incidents") + theme_bw() +
```

```
theme(axis.text.x = element_text(angle = 60, hjust = 1)) +
guides(colour = guide_legend(title = "Season"))
```

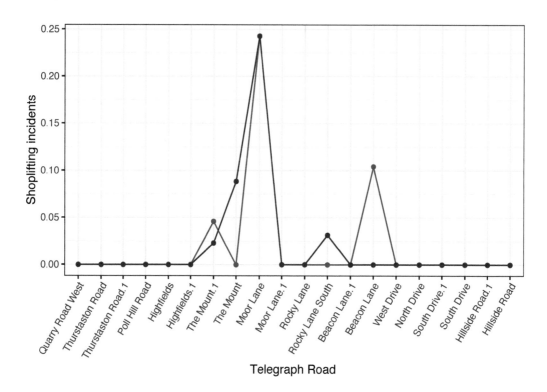

FIGURE 8.21: Street profile comparing summer and winter months

The analyst could also add the other two seasons of apring and autumn, or compare multiple sources of data about the same thing to identify where there might be areas of underreporting in certain sources. For example, I used this to map fare evasion along London bus routes from three different data sources. Similarly, we could compare different crime types along the road, to see if robbery also peaks in these areas, or maybe elsewhere. The possibilities are many, and this is where the strength of this method is evident, over a map, where only one variable at one time can be displayed.

8.5   Quantifying crime concentration at micro-places

So far we have shown ways to visualise crime along a network, but what if we want to actually quantify the extent to which crime concentrates in these micro-places. In crime and place literature we explore the concentration of crime at micro-places. "Perhaps the first and most important empirical observation in the criminology of place is that crime concentrates at very small units of geography" (Weisburd 2015, 135). A significant proportion of crime concentrates around a small proportion of micro places.

One way to measure inequality in the distribution of a quantitative variable is to use the Lorenz curve, and associated Gini coefficient. The Lorenz curve is a probability plot (P–P plot) comparing the distribution of a variable against a hypothetical uniform distribution of that variable. It can usually be represented by a function $L(F)$, where F, the cumulative portion of the population, is represented by the horizontal axis, and L, the cumulative portion of the variable of interest (e.g., crime), is represented by the vertical axis. While Lorenz curves are used typically to graph inequality of distribution of wealth, they can be applied in this case to explore unequal distribution of crimes between micro-places. A perfectly equal distribution would be depicted by the straight line $y = x$. Zeileis, Kleiber, and Zeileis (2012). The corresponding Gini coefficient represents the ratio of the area between the line of perfect equality and the observed Lorenz curve to the area between the line of perfect equality and the line of perfect inequality (Gastwirth 1972). The closer the coefficient is to 1, the more unequal the distribution is (Zeileis, Kleiber, and Zeileis 2012).

In R we can implement these tests using the functions in the `ineq` package. To obtain a Lorenz curve, we can use the `Lc()` function. `Lc()` computes the (empirical) ordinary Lorenz curve of a vector x (in this case, our `crimes_per_m` variable). The function also computes a generalised Lorenz curve (= ordinary Lorenz curve $* mean(x)$). The result can be interpreted like this: $p * 100 \%$ account for $L(p) * 100 \%$ of x.

Let's illustrate this again with the segments of Telegraph Road.

```
tr_lorenz <- Lc(tr_sections$summer_rate)
```

Our resulting `tr_lorenz` object has three elements. First, the `p` represents the cumulative percent of crimes (per metre) for each line segment. Then, `L` contains the values of the ordinary Lorenz curve, while the `L.general` element the values for the generalised Lorenz curve. We can plot the Lorenz curve with the `plot()` function from base R.

```
plot(tr_lorenz)
```

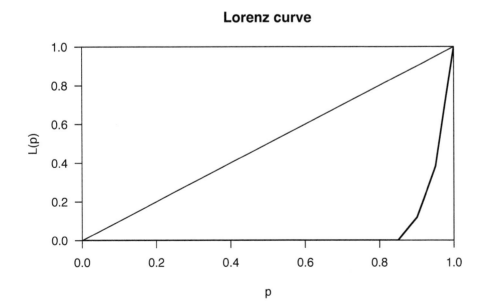

FIGURE 8.22: Lorenz curve for crimes in Heswall

Upon seeing this, we can consider that many of the segments of the Telegraph Road contribute very little to overall crimes, and it is instead the top few (less than the top 20%) which contribute to all the shoplifting incidents. From a visual inspection, it appears that the Telegraph Road very precisely fits (and exceeds) the Pareto Principle, whereby 20% of the segments seem to account for 80% of the crimes (per metre length of the segment). We can quantify this further using the Gini coefficient.

```
ineq(tr_sections$summer_rate, type="Gini")
```

```
## [1] 0.9
```

This score of 90% is quite high. The Gini Index is calculated from the Lorenz curve, by taking the area between the line of equality and the Lorenz curve, and dividing this by the total area under the line of equality. This number is bounded between 0 (perfect equality where the Lorenz curve sits right on top of the line of equality) and 1 (perfect inequality, where the Lorenz curve sits right on top of the x-axis and bends at right angle); so the closer we get to 1, the higher the inequality in the distribution in our value of interest, in this case crimes per metre. Clearly we see crime concentrate in certain segments of Telegraph Road.

8.6 Summary and further reading

The focus in this chapter has been to introduce the issue of crime events along a network. There has been increasing recognition in recent years that the spatial existence of many phenomena is constrained by networks. We have discussed some of the issues associated with the storage, handling, and visualisation of this kind of data. Chapter 1 of Okabe and Sugihara (2012) provides a fairly accessible introduction to the relevance of point pattern data along a network, whereas Chapter 3 offers a slightly more technical discussion of the computational issues associated with the statistical analysis of points along a network. The first few sections of chapter 17 of Baddeley, Rubak, and Turner (2015) provide background and very useful detail on the way that spatstat handles and stores point pattern data in a network. Baddeley et al. (2021) offers an excellent overview of the challenges of analysing this kind of data and the available methods we have. This overview can also offers an excellent framework to understand the issues of clustering and detection of local clusters along networks to which we will return in the next two chapters.

But spatial networks analysis is a more general area, with various other applications that we could not explore in detail here. O'Sullivan (2014) provides an introduction to spatial network analysis, basically a translation of social network metrics and concepts applied to spatial networks. This kind of perspective is particularly helpful within transport, routing, and similar applications. See, for example, the kind of functionality and applications provided by stplanr. We are also witnessing cross-fertilisation between social network analysis and spatial analysis. In epidemiology, for example, we see how this is done for studying the transmission of disease through personal and geographic networks (see, for example, Emch et al. (2010)). Whereas Radil, Flint, and Tita (2010) and Tita and Radil (2011) offer some criminological examples (focused on the study of gangs) linking social network analysis and spatial analysis. Along these lines, the package spnet facilitates the rendering of social network data into a geographical space. Some of these applications can be relevant for investigative crime analysis and other criminological uses. Finally, another way in which social networks are being used within the context of environmental criminology is in contributing to the alternative specifications of neighbourhoods that aim to move beyond the traditional use of census geographies (see Hipp, Faris, and Boessen (2012) and Hipp and Boessen (2013)). Finally, it is also worth exploring the literature on **space syntax** and crime, and the various analyses that are looking at the structural characteristics of the street network (such as betweenness) and how these characteristics influence crime (Davies and Bowers 2018; Kim and Hipp 2019).

9

Spatial dependence and autocorrelation

9.1 Introduction

This chapter begins to explore the analysis of spatial autocorrelation statistics. It has been long known that attribute values that are close together in space are unlikely to be independent. This is often referred to as Tobler's first law of geography: "everything is related to everything else, but near things are more related than distant things". For example, if area n_i has a high level of violence, we often observe that the areas geographically close to n_i tend to also exhibit high levels of violence. Spatial autocorrelation is the measure of this *correlation* between near things.

The analysis of spatial dependence and global clustering is generally a first step before we explore the location of clusters. In spatial statistics we generally distinguish global clustering from the detection of local clusters. Some techniques, the focus of this chapter, are based on the global view and are appropriate for assessing the existence of dependence in the whole distribution, or *clustering*. It is important to first evaluate the presence of clustering before one explores the location of *clusters*, which we will cover in Chapter 10.

In this chapter we are going to discuss ways in which you can quantify the answer to this question. We will start introducing methods to assess spatial clustering with point pattern data, which typically start from exploring continuous distance between the location of the points. Then, we will discuss measures of global spatial autocorrelation for lattice data, which essentially aim to answer the degree to which areas (census tracts, police precincts, etc.) that are near each other tend to be more alike.

We will be making use of the following packages:

```
# Packages for reading data and data carpentry
library(readr)
library(dplyr)
library(lubridate)
# Packages for handling spatial data and for geospatial carpentry
library(sf)
library(sp)
# Specific packages for spatial point pattern analysis
library(spatstat)
library(spatstat.Knet)
# Packages to generate spatial weight matrix and compute measures of dependence
library(spdep)
library(rgeoda)
# Packages for mapping and visualisation
```

```
library(tmap)
library(ggplot2)
library(ggspatial)
```

9.2 Exploring spatial dependence in point pattern data

9.2.1 The empirical K-function

In Chapter 7 we assessed homogeneity, or lack of, in point pattern data. Another important aspect of point location is whether they show some kind of interpoint dependence. In observing a point pattern, one is interested in what are referred to as first order properties (varying intensity across the study surface vs. homogeneity) and second order properties (independence vs. correlation among the points). As Bivand, Pebesma, and Gómez-Rubio (2013) explain:

> Spatial variation occurs when the risk is not homogeneous in the study region (i.e. all individuals do not have the same risk) but cases appear independently of each other according to this risk surface, whilst clustering occurs when the occurrence of cases is not at random and the presence of a case increases the probability of other cases appearing nearby. (p. 204)

In Chapter 7 we covered the Chi Square test and we very briefly discussed how it can only help us to reject the notion of spatial randomness in our point pattern, but it cannot help us to determine whether this is due to varying intensity or to interpoint dependence.

Both processes may lead to apparent observed clustering:

- **First order clustering** near particular locations in the region of study takes place when we observe more points in certain neighbourhoods or street segments or whatever our unit of interest is;
- **Second order clustering** is present when we observe more points around a given location in a way that is associated with the appearance of points around this given location.

This may sound subtle and in practice it may be difficult, nearly impossible, to distinguish "first order inhomogeneity" and "second order clustering due to interactions between points" based on a single point pattern[1].

A technique commonly used to assess whether points are independent is Ripley's K-function. For exploring global clustering in a point pattern, we use the K-function. Baddeley, Rubak, and Turner (2015) define the K-function as the "cumulative average number of data points lying within a distance r of a typical data point, corrected for edge effects, and standardised by dividing by the intensity" (p. 204). This is a test that, as we will see, assumes that the process is spatially homogeneous. As Lu and Chen (2007) indicate:

---

[1]https://stats.stackexchange.com/questions/411402/spatial-point-process-does-an-inhomogeneous-first-order-intensity-function-affe

"The basic idea of K-function is to examine whether the observed point distribution is different from random distributions. In a random distribution, each point event has the equal probability to occur at any location in space; the presence of one point event at a location does not impact the occurrence possibilities of other point events." (p. 613)

Imagine we want to know if the locations of crime events are independent. Ripley's K-function evaluates distances across crime events to assess this. In this case, it would count the number of neighbouring crime events represented by the points found within a given distance of each individual crime location. The number of observed neighbouring crime events is then traditionally compared to the number of crime events one would expect to find based on a completely spatially random point pattern. If the number of crimes found within a given distance of each individual crime is greater than that for a random distribution, the distribution is then considered to be clustered. If the number is smaller, the distribution is considered not to be clustered.

Typically, what we do to study correlation in a point pattern data is to plot the empirical K-function estimated from the data and compare this with the theoretical K-function of the homogeneous Poisson process. So, for this to work, we need to assume that the process is spatially homogeneous. If we suspect this is not the case, then we cannot use the standard Ripley's K-methods. Let's illustrate how all this works with the example of some homicide data in Buenos Aires.

9.2.2 Homicide in Buenos Aires

For this exercise we travel to Buenos Aires, the capital and largest city in Argentina. The government of Buenos Aires provides open geocoded crime data from 2016 onwards. The data is freely available from the official source[2]. But we have added it already as supplementary material available with this book. Specifically, we are after the file delitos.csv.

We will filter on "homicidios dolosos" (intentional homicides) and we are using all homicides from 2016 and 2017. After reading the .csv file, we turn it into a sf object and plot it. We are also reading a GeoJSON file with the administrative boundary of Buenos Aires. The code below uses functions in ways we have demonstrated in earlier chapters.

```
delitos <- read_csv("data/delitos.csv") %>%    # read csv
  filter(tipo_delito == "Homicidio Doloso") %>% # filter international homicides
  mutate(year = year(fecha)) %>%    # create new variable for 'year'
  filter(year == "2016" | year == "2017")    # filter only 2016 and 2017


# create sf object of crimes as points
delitos_sf <- st_as_sf(delitos,
                coords = c("longitud", "latitud"),
                crs = 4326)


# transform to projected coods
delitos_sf <- st_transform(delitos_sf, 5343)
```

[2]http://mapa.seguridadciudad.gob.ar/

```
# read in boundary geojson
buenos_aires <- st_read("data/buenos_aires.geojson", quiet = TRUE)
```

After reading the files and turning them into sf objects, we can plot them. Let's use the tmap package introduced in Chapter 3.

```
tm_shape(buenos_aires) +
  tm_polygons() +
  tm_shape(delitos_sf) +
  tm_dots(size = 0.08, col = "red") +
  tm_layout(main.title = "Homicides in Buenos Aires, 2016",
            main.title.size = 1)
```

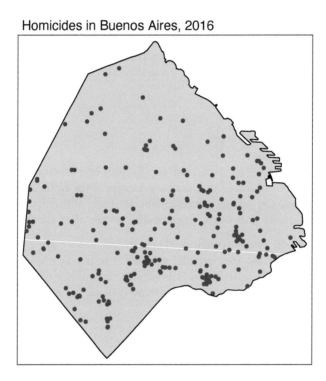

FIGURE 9.1: Map of homicides in Buenos Aires

We will be using the spatstat functionality; thus, we need to turn this spatial point pattern data into a ppp object, using the boundaries of Buenos Aires as an owin object. We are using code already explained in Chapter 7. As we will see, there are duplicate points. Since the available documentation does not mention any masking of the locations for privacy purposes, we cannot assume that's the reason for it. For simplicity we will do as in Chapter 7 and introduce a small degree of jittering with a rather narrow radius.

```
# create window from geometry of buenos_aires object
window <- as.owin(st_geometry(buenos_aires))
```

```
# name the unit in window
unitname(window) <- "Metre"

# create matrix of coordinates of homicides
delitos_coords <- matrix(unlist(delitos_sf$geometry), ncol = 2, byrow = T)

# convert to ppp object
delitos_ppp <- ppp(x = delitos_coords[,1], y = delitos_coords[,2],
                    window = window, check = T)

# set seed for reproducibility and jitter homicide points because of overlap
set.seed(200)
delitos_jit <- rjitter(delitos_ppp, retry=TRUE, nsim=1,
                       radius = 2, drop=TRUE)
```

9.2.3 Estimating and plotting the K-function

For obtaining the observed K, we use the `Kest` function from the `spatstat` package. `Kest` takes two key inputs, the `ppp` object and a character vector indicating the edge *corrections* we are using when estimating K. Edge corrections are necessary because we do not have data on homicide outside our window of observation and, thus, cannot consider dependence with homicide events around the city of Buenos Aires. If we don't apply edge correction, we are ignoring the possibility of dependence between those events outside the window and those in the window close to the edge. There is also a convenient argument (`nlarge`) if your data-set is very large (not the case here) that uses a faster algorithm.

As noted by Baddeley, Rubak, and Turner (2015), "most users... will not need to know the theory behind edge correction, the details of the technique, or their relative merits" and "so long as some kind of edge correction is performed (which happens automatically in `spatstat`), the particular choice of edge correction technique is usually not critical" (p. 212). As a rule of thumb, they suggest using translation or isotropic in smaller data-sets, the border method in medium-sized data-sets (1000 to 10000 points), and that there is no need for correction with larger data-sets. Since this is a smallish data-set, we show how we use two different corrections: *isotropic* and *translation* (for more details, see Baddeley, Rubak, and Turner (2015)).

```
k1 <- Kest(delitos_jit, correction = c("isotropic", "translate"))
plot(k1)
```

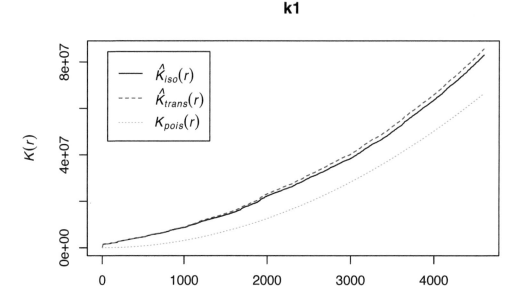

FIGURE 9.2: Isotropic and translation correction techniques for edge effects

The graphic shows the K-function using each of the corrections and the theoretical expected curve under a Poisson process of independence. When there is clustering, we expect the empirical K-function estimated from the observed data to lie above the empirical K-function for a completely random pattern. This indicates that a point has more "neighbouring" points than it would be expected under a random process.

It is possible to use Monte Carlo simulations to draw the confidence interval around the theoretical expectations. This allow us to draw *envelope* plots that can be interpreted as a statistical significance test. If the homicide data were generated at random, we would expect the observed K-function to overlap with the envelope of the theoretical expectation. Baddeley, Rubak, and Turner (2015), in particular, suggest the use of global envelopes as a way to avoid data snooping. For this we use the envelop function from spatstat. As arguments we use the data input, the function we are using for generating K (Kest), the number of simulations, and the parameters needed to obtain the global envelope.

```
set.seed(200)
E_1 <- envelope(delitos_jit, Kest, nsim=39, rank = 1, global = TRUE,
            verbose = FALSE) # We set verbose to false for sparser reporting
plot(E_1)
```

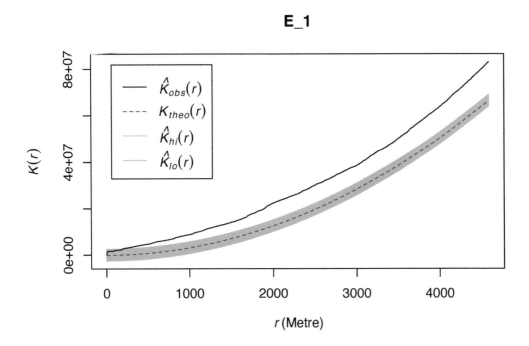

FIGURE 9.3: Global envelope plot for clustering in Buenos Aires homicide data

This envelope suggests significant clustering above the 200 metre mark. The problem, though, is that (as noted above) this test only works if we have a homogeneous process. The envelope will also fail to overlap with the observed K-function if there are variations in intensity. We also saw previously the reverse, that it is hard to assess homogeneity without assuming independence. In Chapter 7 we saw how if we suspect the process is not homogeneous, we can estimate it with non-parametric techniques such as kernel density estimation. We can apply this here (see also *Chapter 7: Spatial Point Patterns of Crime Events* in this book).

```
set.seed(200)
plot(density.ppp(delitos_jit, sigma = bw.scott(delitos_jit),edge=T),
     main=paste("Density estimate of homicides in Buenos Aires"))
```

Density estimate of homicides in Buenos Aires

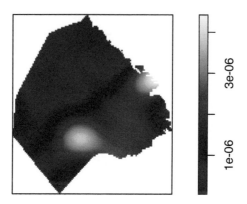

FIGURE 9.4: KDE map of homicides in Buenos Aires

Baddeley, Rubak, and Turner (2015) discuss how we can adjust our test for independence when we suspect inhomogeneity, as long as we are willing to make other assumptions (e.g., the correlation between any two given points only depend on their relative location). This adjusted K-function is implemented in the Kinhom() function in spatstat. This function estimates the inhomogeneous K-function of a non-stationary point pattern. It requires as necessary inputs the ppp object with the point pattern we are analysing and a second argument lambda which should provide the values of the estimated intensity function. The function provides flexibility in how lambda can be entered. It could be a vector with the values, an object of class im like the ones we generate with density.ppp(), or it could be omitted. If lambda is not provided, we can provide the parameters as we did in the density.ppp() function, and the relevant values will be generated for us that way. We could, for example, input this parameter as below:

```
set.seed(200)
inhk_hom <- Kinhom(delitos_jit,
                   correction = c("isotropic", "translate"),
                   sigma = bw.scott)
plot(inhk_hom)
```

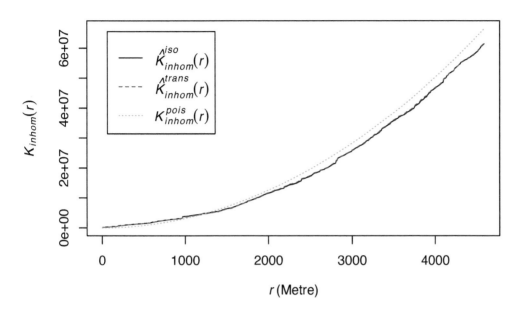

FIGURE 9.5: Inhomogeneous (adjusted) K-function for non-stationary point pattern

Once again we can build an envelope to use in the manner of confidence intervals to interpret whether there is any significant clustering.

```
set.seed(200)
E_2 <- envelope(delitos_jit, Kinhom, nsim=39, rank = 1, global = TRUE,
                verbose = FALSE) # We use this for sparser reporting
plot(E_2)
```

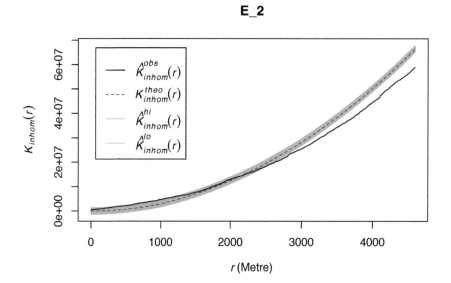

FIGURE 9.6: Global envelope plot for clustering in case of inhomogeneity

In this example, once we allow for varying intensity, the spatial dependence no longer holds for a range of distances.

9.2.4   Network constrained K-function

We already warned in Chapter 8 about the dangers of using planar point pattern analysis to data that lies along a network. Not recognising this network structure can lead to errors. Lu and Chen (2007), for example, have shown how the standard K-functions are problematic when studying crime in an urban environment. They discuss how the planar K-function to analyze the spatial autocorrelation patterns of crimes (that are typically distributed along street networks) can result in false alarm problems, which depending on the nature of the street network and the particular point pattern, could be positive or negative. As Baddeley, Rubak, and Turner (2015) also indicate when the point locations are constrained to only exist in a network we are likely to find "spurious evidence of short-range clustering (due to concentrations of points in a road) and long-range regularity (due to spatial separation of different roads)" (p. 736).

Lu and Chen (2007) recommend to use a network K-function that examines distance along the network, by street distance. They argued that this should be "more relevant for crime control policing and police man-power dispatching purpose, considering that police patrol and many other crime management activities are commonly conducted following street networks" (p. 627). This recommendation is being followed in recent studies (see, for example, Khalid et al. (2018)), although it is still much more common in the professional practice of crime analysis to continue using, incorrectly, the planar K-function.

The network K-function was developed by Okabe and Yamada (2001) and was first implemented in SANET. Chapter 7 of Okabe and Sugihara (2012) also provides a thorough discussion of this adaptation and its computation. We can use functionality implemented in the `spatstat` package to use the network constrained K. For that we need a `lpp` object, which we discussed in Chapter 8. For convenience we will use a `lpp` data-set about crime around University of Chicago that ships with the `spatstat` package. We could get the standard K as above.

```
set.seed(200)    # set seed for reproducibility
data(chicago)    # load chicago data (in spatstat package)

# Estimate Ripley's K with envelope
E_3 <- envelope(as.ppp(chicago), Kest, nsim=99, rank = 1, global = TRUE, verbose = FALSE)
plot(E_3) # plot result
```

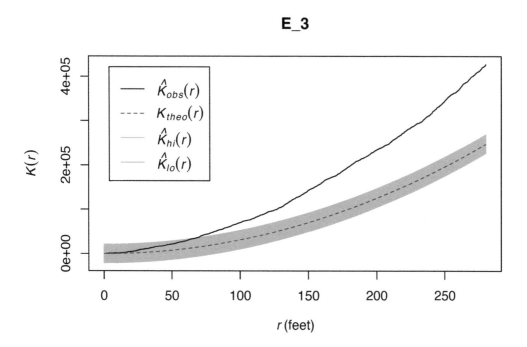

FIGURE 9.7: Ripley's K for crime around the University of Chicago with envelope

How would we do this now taking the network into consideration? We can use the methods introduced by Okabe and Yamada (2001) with the function `linearK()`. The function `linearK()` computes an estimate of the linear K-function for a point pattern on a linear network. For this we need to specify the parameter for `correction=` to be "none". If correction="none", the calculations do not include any correction for the geometry of the linear network. The result is the network K-function as defined by Okabe and Yamada (2001).

We can wrap this within the `envelope()` function for inference purposes. This will generate an envelope assuming a homogeneous Poisson process. We also limit the number of simulations (`nsim =`) to 99, and we suppress a verbose output.

```
set.seed(200)    # set seed for reproducibility

E_4 <- envelope(chicago, linearK, correction = "none", nsim= 99, verbose = FALSE)

plot(E_4) # plot result
```

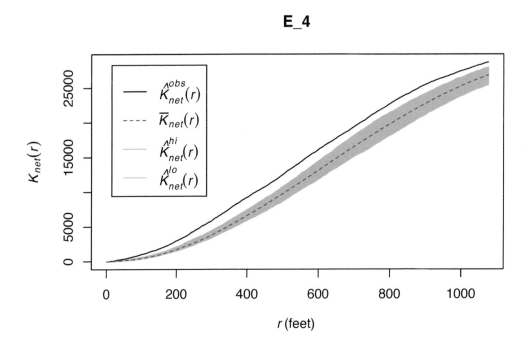

FIGURE 9.8: Estimate linear K for a point pattern (crimes) on a linear network (University of Chicago network)

As noted by Baddeley, Rubak, and Turner (2015), there is a difference between what we get with Kest and with linearK. Whereas the two planar "K-function is a *theoretical expectation* for the point process that is assumed to have generated our data... the empirical K-function of point pattern data" (that we get with linearK) "is an *empirical estimator* of the K-function of the point process" (p. 739, the emphasis is ours).

A limitation of the Okabe-Yamada network K-function is that, unlike the general Ripley's K which is normalised, you cannot use it for "comparison between different point processes, with different intensities, observed in different windows... because it depends on the network geometry" (p. 739). This is a problem that was approached by Ang, Baddeley, and Nair (2012) who proposed a "geometrically corrected" K-function. You can see the mathematical details for this function in either Baddeley, Rubak, and Turner (2015) or in greater depth in Ang, Baddeley, and Nair (2012). This geometrically corrected K-function is implemented as well in the linearK() function. If we want to obtain this solution, we need to specify as an argument correction="Ang" or simply not include this argument, for this is the default in linearK():

```
set.seed(200)
E_5 <- envelope(chicago, linearK, correction = "Ang", nsim= 99, verbose = FALSE)
plot(E_5)
```

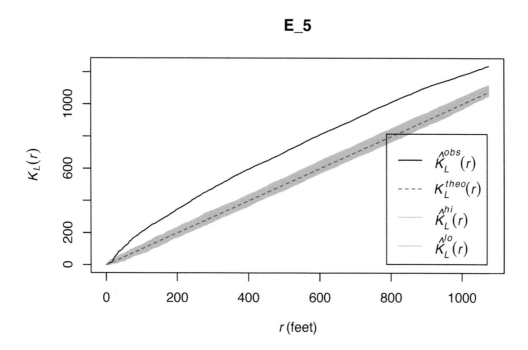

FIGURE 9.9: Geometrically corrected linear K-function for a point pattern (crimes) on a linear network (University of Chicago network)

Finally, there is also an equivalent to the geometrically corrected K-function that allows for an inhomogeneous point process that can be obtained using linearKinhom() (see Baddeley and Turner (2005) for details).

These methods all help discuss spatial autocorrelation in point pattern data. In the next section, we will consider how we might find out whether such clustering exists in data measured at polygon (rather than point) level.

9.3 Exploring spatial autocorrelation in lattice data

9.3.1 Burglary in Manchester

To illustrate spatial autocorrelation in lattica data, we are going to hop across the globe to Manchester, UK. To follow, read in the file burglary_manchester_lsoa.geojson which contains burglaries per Lower Layer Super Output Area (LSOA), which is the data with which we will be working here.

```
burglary <- st_read("data/burglary_manchester_lsoa.geojson",
                    quiet=TRUE)
```

To have a look at the data, we can plot it quickly with tmap:

```
tm_shape(burglary) +
  tm_bubbles("burglary", border.lwd=NA, perceptual = TRUE) +
  tm_borders(alpha=0.1) +
  tm_layout(legend.position = c("right", "bottom"),
            legend.title.size = 0.8,
            legend.text.size = 0.5)
```

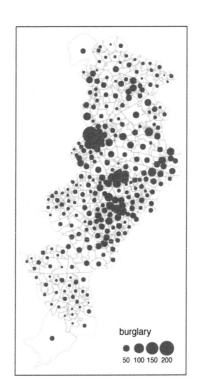

FIGURE 9.10: A quick look at burglary in Manchester

Do you see any patterns? Are burglaries randomly spread around the map? Or would you say that areas that are closer to each other tend to be more alike? Is there evidence of clustering? Do burglaries seem to appear in certain pockets of the map? To answer these questions, we will need to consider whether we see similar values in *neighbouring* areas.

9.3.2 What is a neighbour?

Previously we asked whether areas are similar to their neighbours or to areas that are close. But what is a neighbour? Or what do we mean by close? How can one define a set of neighbours for each area? If we want to know if what we measure in a particular area is similar to what happens on its neighbouring areas, we first need to establish what we mean by a neighbour.

9.3.2.1 Contiguity-based neighbours

There are various ways of defining a neighbour. Most rely on topological or geometrical relationships among the areas. First, we can say that two areas are neighbours if they share boundaries, if they are next to each other. In this case we talk of neighbours by **contiguity**. Contiguous can, at the same time, mean all areas that share common boundaries (what we call contiguity using the **rook** criteria, like in chess) or areas that share common boundaries and common *corners*, that is, that have any point in common (and we call this contiguity using the **queen** criteria). When we use this criteria we can refine our definition by defining the intensity of neighbourliness "as a function of the length of the common border as a proportion of the total border" (Haining and Li 2020, 89). Given that in criminology we mostly work with irregular lattice data (such as police districts or census geographies), the queen criteria makes more sense. There is little theoretical justification for why, say, a police district would only exhibit dependence according to the rook criteria.

When defining neighbours by contiguity, we may also specify the *order* of contiguity. **First order contiguity** means that we are focusing on areas immediately contiguous (those in dark blue in the figure below). **Second order** means that we consider neighbours only those areas that are immediately contiguous to our first order neighbours (the light blue areas in the figure below) and you could go on and on. Look at the graphic below for clarification:

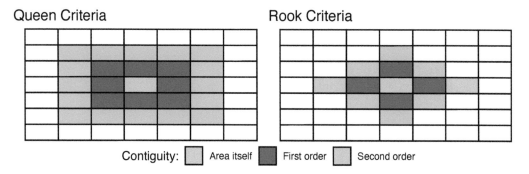

FIGURE 9.11: Defining neighbours by contiguity

9.3.2.2 Distance-based neighbours

We may also define neighbours by **geographical distance**. You can consider neighbours those areas that distance-wise are close to each other (regardless of whether boundaries are shared). The distance metric often will be Euclidean, but other commonly used distance metrics could also be employed (for example, Manhattan). Often one takes the centroid of the polygon as the location to take the distance from. More sophisticated approaches may involve taking a population weighted centroid, where the location of the centroid depends on the distribution of the population within an area.

Other approaches to define neighbours are graph-based, attribute-based, or interaction-based (e.g., flow of goods or people between areas). Contiguity and distance are the two most commonly used methods, though.

You will come across the term **spatial weight matrix** at some point or, using mathematical notation, W. Essentially the spatial weight matrix is a n by n matrix with, often, ones and zeroes (in the case of contiguity-based definitions) identifying if any two observations are neighbours. So you can think of the spatial weight matrix as the new data table that we are constructing with our definition of neighbours (whichever definition that is).

When working with contiguity measures, the problem of "islands" sometimes arises. The spatial weight matrix in this case would have zeroes for the row representing this "island". This is not permitted in the subsequent statistical use of this kind of matrix. A common solution is to "join" islands to the spatial units in the "mainland" that are closer.

So how do you build such a matrix with R? Well, let's turn to that. But to make things a bit simpler, let's focus not on the whole of Manchester, but just in the LSOAs within the city centre. Calculating a spatial weights matrix is a computationally intensive process, which means it takes a long time. The larger area you have (which will have more LSOAs), the longer this will take.

We will use code we covered in Chapter 2 to clip the spatial object with the counts of burglaries to only those that intersect with the City Centre ward. If any of the following are unclear, visit *Chapter 2: Basic Geospatial Operations in R* in this book.

```
# Read a geojson file with Manchester wards and select only the city centre ward
city_centre <- st_read("data/manchester_wards.geojson", quiet = TRUE) %>%
  filter(wd16nm == "City Centre")


# Reproject the burglary data
burglary <- st_transform(burglary, 27700)
# Intersect
cc_intersects <- st_intersects(city_centre, burglary)
cc_burglary <- burglary[unlist(cc_intersects),]
# Remove redundant objects
rm(city_centre)
rm(cc_intersects)
```

We can now have a look at our data. Let's use the view mode of tmap to enable us to interact with the map.

```
# Plot with tmap
tmap_mode("view")
tm_shape(cc_burglary) +
  tm_fill("burglary", style = "quantile", palette = "Reds", id="code") +
  tm_borders() +
  tm_layout(main.title = "Burglary counts", main.title.size = 0.7 ,
            legend.position = c("right", "top"), legend.title.size = 0.8)
```

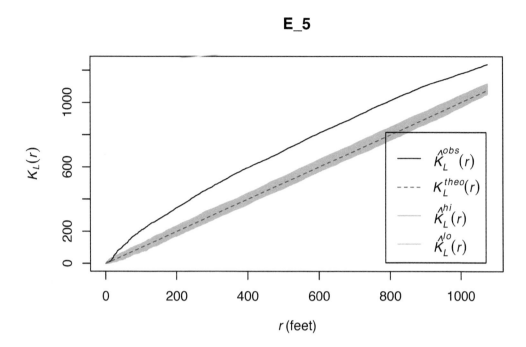

FIGURE 9.12: Interactive plot of data for lattice spatial dependence

So now we have a new spatial object "cc_burglary" with the 23 LSOA units that compose the City Centre of Manchester. By focusing in a smaller subset of areas, we can visualise perhaps a bit better what comes next.

9.3.3 Creating a list of neighbours based on contiguity

Since it is an interactive map, if you are following along, you should be able to click on each LSOA, and see its unique ID. You could, manually go through and find the neighbours for each one this way. However, it would be very, very tedious having to identify the neighbours of all the areas in our study area by hand. That's why we love computers. We can automate tedious work so that they do it, and we have more time to do fun stuff.

To calculate our neighbours, we are first going to turn our sf object into spatial (sp) objects, so we can make use of the functions from the sp package that allow us to create a list of neighbours.

```
#We coerce the sf object into a new sp object
cc_burglary_sp <- as(cc_burglary, "Spatial")
```

In order to identify neighbours, we will use the poly2nb() function from the spdep package that we loaded at the beginning of our session. The spdep package provides basic functions for building neighbour lists and spatial weights and tests for spatial autocorrelation for areal data like Moran's I (more on this below).

The poly2nb() function builds a neighbours list based on regions with contiguous boundaries. If you look at the documentation, you will see that you can pass a queen argument that takes TRUE or FALSE as options. If you do not specify this argument the default is set to true; that is, if you don't specify queen = FALSE, this function will return a list of first order neighbours using the Queen criteria.

```
nb_queen <- poly2nb(cc_burglary_sp, row.names=cc_burglary_sp$code)
class(nb_queen)
```

```
## [1] "nb"
```

We see that we have created a nb, neighbour list object called nb_queen. We can get some idea of what's there if we ask for a summary.

```
summary(nb_queen)
```

```
## Neighbour list object:
## Number of regions: 23
## Number of nonzero links: 100
## Percentage nonzero weights: 18.9
## Average number of links: 4.348
## Link number distribution:
##
## 2 3 4 5 6 7 8
## 3 4 6 5 3 1 1
## 3 least connected regions:
## E01033673 E01033674 E01033684 with 2 links
## 1 most connected region:
## E01033658 with 8 links
```

This is basically telling us that using this criteria each LSOA polygon has an average of 4.3 neighbours (when we just focus on the city centre) and that all areas have some neighbours (there are no islands). The link number distribution gives you the number of links (neighbours) per area. So here we have 3 polygons with 2 neighbours, 3 with 3, 6 with 4, and so on. The summary function here also identifies the areas sitting at both extreme of the distribution. We can graphically represent the links using the following code:

```
# We first plot the boundaries
plot(cc_burglary_sp, col='white', border='blue', lwd=2)
# Then we use the coordinates function to obtain
# the coordinates of the polygon centroids
xy <- coordinates(cc_burglary_sp)
#Then we draw lines between the polygon centroids
# for neighbours that are listed as linked in nb_bur
plot(nb_queen, xy, col='red', lwd=2, add=TRUE)
```

FIGURE 9.13: Neighbouring relationships between LSOAs with Queen contiguity criteria

9.3.4 Generating the weight matrix

We can transform the object nb_queen, the list of neighbours, into a spatial weights matrix. A spatial weights matrix reflects the intensity of the geographic relationship between observations. For this we use the spdep function nb2mat() (make a matrix out of a list of neighbours). The argument style sets what kind of matrix we want. B stands for the basic binary coding, you get a 1 for neighbours, and a 0 if the areas are not neighbours.

```
wm_queen <- nb2mat(nb_queen, style='B')
```

You can view this matrix by autoprinting the object wm_queen which we just created:

```
wm_queen
```

This matrix has values of 0 or 1 indicating whether the elements listed in the rows are adjacent (using our definition, which in this case was the Queen criteria) with each other. The diagonal is full of zeroes. An area cannot be a neighbour of itself. So, if you look at the first two and the second column, you see a 1. That means that the LSOA with the code E01005065 is a neighbour of the second LSOA (as listed in the rows) which is E01005066. You will have zeroes for many of the other columns because this LSOA only has 4 neighbours.

In many computations we will see that the matrix is **row standardised**. We can obtain a row standardise matrix changing the code, specifically setting the style to W (the indicator for row standardised in this function):

```
wm_queen_rs <- nb2mat(nb_queen, style='W')
```

Row standardisation of a matrix ensures that the sum of the rows adds up to 1. So, for example, if you have 4 neighbours and that has to add up to 4, you need to divide 1 by 4, which gives you 0.25. So in the columns for a polygon with 4 neighbours you will see 0.25 in the column representing each of the neighbours.

9.3.5 Spatial weight matrix based on distance

9.3.5.1 Defining neighbours based on a critical threshold of distance

As noted earlier, we can also build the spatial weight matrix using some metric of distance. Remember that metrics such as Euclidean distance only makes sense if you have projected data. This is so because it does not take into account the curvature of the earth. If you have non-projected data, you can either transform the data into a projected system (what we have been doing in earlier examples) or you need to use distance metrics that work with the curvature of the Earth (e.g., arc-distance or great-circle distance).

The simplest spatial distance-based weights matrix is obtained defining a given radius within each area and only considers as neighbours other areas that fall within such radius. So that we avoid "islands", the radius needs to be chosen in a way that guarantees that each location has at least one neighbour. So the first step would be to find an appropriate radius for establishing neighbourliness.

Previous to this, we need the centroids for the polygons. We already showed above how to obtain this using `coordinates()` function in the `sp` package.

```
xy <- coordinates(cc_burglary_sp)
```

Then we use the `knearneigh()` function from the `spdep` package to find the nearest neighbour to each centroid. This function will just do that, which is to look for the centroid closest to each of the other centroids. Then we turn the created object with this information into a `nb` object as defined earlier using `knn2nb()` also in `spdep` package:

```
# find nearest neighbours for each centroid coordinate pair
nearest_n <- knearneigh(xy)
# transform into nb class object
nb_nearest_n <- knn2nb(nearest_n)
```

Now we have a `nb` list that for each area defines its nearest neighbour. What we need to do is to compute the distance from each centroid to its nearest neighbour (`nbdists()` will do this for us) and then find out what is the maximum distance (we can use `max()`), so that we use this as the critical threshold to define the radius we will use (to avoid the appearance of isolates in our matrix).

```
radius <- max(unlist(nbdists(nb_nearest_n, xy)))
radius
```

```
## [1] 690.9
```

Now that we know the maximum radius, we can use it to avoid isolates. The function `dnearneigh()` will generate a list of neigbhours within the radius we will specify. Notice the difference between `knearneigh()`, that we used earlier, and `dnearneigh()`. The first function generates a `knn` class object with the single nearest neighbour to each area i, whereas `dnearneigh()` generates a `nb` object with *all* the areas j within a given radius of area i being defined as neighbours.

For dnearneigh() we need to define as parameters the object with the coordinates, the lower distance bound and the upper distance bound (the "radius" we just calculated). If you were to be working with non-projected data, you would need to specify longlat as TRUE. The function assumes we have projected data and, thus, we don't need to specify this argument in our case, since our data are indeed using the British Grid projection. As we did with the nb objects generated through contiguity, we can use summary() to get a sense of the resulting list of neighbours.

```
nb_distance <- dnearneigh(xy, 0, radius)
summary(nb_distance)
```

```
## Neighbour list object:
## Number of regions: 23
## Number of nonzero links: 88
## Percentage nonzero weights: 16.64
## Average number of links: 3.826
## Link number distribution:
##
## 1 2 3 4 5 6 7
## 3 4 4 4 2 3 3
## 3 least connected regions:
## 1 17 23 with 1 link
## 3 most connected regions:
## 5 12 18 with 7 links
```

And as earlier, we can also plot the results:

```
# We first plot the boundaries
plot(cc_burglary_sp, col='white', border='blue', lwd=2)
# Then we plot the distance-based neighbours
plot(nb_distance, xy, lwd=2, col="red", add=TRUE)
```

FIGURE 9.14: Neighbouring relationships between LSOAs with distance-based criteria

We could now use this `nb` object to generate a spatial weight matrix in the same way as earlier:

```
wm_distance <- nb2mat(nb_distance, style='B')
```

If we we autoprint the object, we will see this is a matrix with 1 and 0 just as the contiguity-based matrix we generated earlier.

9.3.5.2   Using the inverse distance weights

Instead of just using a critical threshold of distance to define a neighbour, we could use a function to define the weight assigned to the distance between two areas. Typical functions used are the inverse distance, the inverse distance raised to a power, and the negative exponential. All these functions essentially do the same; they decrease the weight as the distance increases. They impose a **distance decay** effect. Sometimes these functions are combined with a distance threshold (Haining and Li 2020).

The first steps in the procedure simply reproduce what we did earlier up to the point where we generated the `bur_dist_1` object, so we won't repeat them here and simply invoke this object. We adapt code and instructions provided by the tutorials of the Center for Spatial Data Science of the University of Chicago prepared by Anselin et al. (2021). Once we get to this point, we need to generate the inverse distances. This requires us to take the following steps:

- Calculate the distance between all neighbours
- Compute the inverse distance for this calculated distances
- Assign them as weight values

Let's go one step at the time. First we use `nbdists()` from the `spdep` package to generate all the distances:

```
distances <- nbdists(nb_distance, xy)
```

Now that we have an object with all the distances, we need to compute the inverse distances. Taking the inverse simply involves dividing 1 by each of the distances. We can easily do that with the following code:

```
inv_distances <- lapply(distances, function(x) (1/x))
```

If you look inside the generated object, you will see all the inverse distances:

```
View(inv_distances)
```

You will notice that the values are rather small. The projection we use measures distance in metres. Considering our polygons are census geographies "close" to the ideas of neighbourhoods you will have distances between their centroids that are large. If we take the inverse, as we did, we will unavoidably produce small values very close to zero. We could rescale this to obtain values that are not this small. We can change our units to something larger, like kilometres, or if we want, we can just apply some maths here:

```
inv_distances_r <- lapply(distances, function(x) (1/(x/100)))
```

Once we have generated the inverse distance, we can move to generate the weights based on these using nb2mat().

```
wm_inverse_dist <- nb2mat(nb_distance,
                glist = inv_distances_r,
                style='B')
```

The `glist` argument identifies the inverse distances, and the `style` set to "B" ensures we are not using row standardisation. If we view this object, we will see that rather than zeroes and ones we see the inverse distances used as weights. In this case, all areas neighbour each other (but not themselves!) and their "neighbouringness" is defined with this distance measure.

9.3.6 Spatial weight matrix based on k-neighbours

Chi and Zhu (2020) suggest that using the distance-based spatial weight matrix tend to result in too few neighbours for rural areas and many neighbours for urban areas. In this case they point out that the k-nearest neighbourhood structure may be more appropriate. K-neighbours also avoids the problem of isolates without having to specify a radius for a critical threshold.

Computing the list of neighbours is pretty straightforward and uses functions we are familiar with by now. In the first place we use knearneigh() to get an oject of class knn. In this instance, though, we pass a parameter k identifying the number of neighbours we want for each area. We will set this to 3. Then we use knn2nb() to generate the list of neighbours.

```
nb_k3 <- knn2nb(knearneigh(xy, k = 3))
```

As before, we can use this list of neighbours to create a matrix and we can visualise the relationships with a plot:

```
# Get the matrix
wm_k3 <- nb2mat(nb_k3, style='B')
# We first plot the boundaries
plot(cc_burglary_sp, col='white', border='blue', lwd=2)
# Then we plot the distance-based neighbours
plot(nb_k3, xy, lwd=2, col="red", add=TRUE)
```

FIGURE 9.15: Neighbouring relationships between LSOAs with distance-based criteria using k-nearest neighbours approach (k=3)

9.4 Choosing the correct matrix

So what is the best approach to creating your spatial weights matrix? As we will see later, the spatial weight matrix not only is used for exploring spatial autocorrelation, it also plays a key role in the context of spatial regression. How we specify the matrix is important.

As noted above, contiguity and distance are the two most commonly used methods. In earlier applications researchers would select a particular matrix without a strong rationale for it. You should be guided by theory in how you define the neighbours in your study; unfortunately "substantive theory regarding spatial effects remain an underdeveloped research area in most of the social sciences" (Darmofal 2015, 23) or as Chi and Zhu (2020) put it "there is limited theory available regarding how to select which spatial matrix to use."

In this context, it may make sense to explore various definitions and see if our findings are robust to these specifications. Aldstadt and Getis (2006) developed an algorithm (AMOEBA) which is a design for the construction of a spatial weights matrix using empirical data that can also simultaneously identify the geometric form of spatial clusters. Chi and Zhu (2020) propose to use this as a possible starting point in the comparisons, though the procedure is not yet fully implemented in R (there is an AMOEBA package, but it only generates a vector with the spatial clusters, not the matrix). They also suggest using the spatial weight matrix corresponding to high spatial autocorrelation and high statistical significance as more appropriate for exploratory data analysis of a given variable, and for the purpose of regression to use the matrix with the highest correlation and significance when exploring the residuals of the model.

In any case, the choice of spatial weights should be made carefully, as it will affect the results of subsequent analyses based on it. But now, that we know how to build one, we can move to implementing it in order to measure whether or not we can observes spatial autocorrelation in our data.

9.4.1 Measuring global autocorrelation

9.4.1.1 Moran's I

The most well-known measure of spatial autocorrelation is Moran's I. It was developed by Patrick Alfred Pierce Moran, an Australian statistician. Like the more familiar Pearson's r correlation coefficient, which is used to measure the dependency between a pair of variables, Moran's I aims to measure dependency. This statistic measures how one area is similar to others surrounding it in a given attribute or variable. When the values of an attribute in areas defined as neighbours in our spatial weight matrix are similar, Moran's I will be large and positive (closer to 1). When the values of an attribute in neighbouring areas are dissimilar Moran's I will be negative, tending to -1. In the first case, when there is positive spatial dependence, we speak of clustering. In the latter, what we observe is spatial dispersion (areas alike, in an attribute, "repel" each other). In social science applications, positive spatial autocorrelation is more likely to be expected. There are fewer social processes that would lead us to expect negative spatial autocorrelation. When Moran's I is close to zero, we fail to find evidence for either dispersion or clustering. The null hypothesis in a test of spatial autocorrelation is that the values of the attribute in question are randomly distributed across space.

Moran's I is a weighted product-moment correlation coefficient, where the weights reflect geographic proximity. Unlike autocorrelation in timeseries (which is unidimensional: the future cannot affect the past), spatial autocorrelation is multi-dimensional, the dependence between an area and its neighbours is either simultaneous or reciprocal. Moran's I is defined as:

$$I = \frac{N}{W} * \frac{\sum_i \sum_j w_{ij}(x_i - \overline{x})(x_j - \overline{x})}{\sum_i (x_i - \overline{x})^2}$$

where N is the number of spatial areas indexed by i and j; x is our variable of interest; $\overline{x}$ is the mean of x; w_{ij} is the spatial weight matrix with zeroes in the diagonal; and W is the sum of all of w_{ij}.

We can easily compute this statistic with the `moran.test()` function in spdep package. This function does not take as an input the spatial weight matrix that we create using `nb2mat()`. Instead it prefers to read this information from a `listw` class object. This kind of object includes the same information than the matrix, but it does it in a more efficient way that helps with the computation (see Brunsdon and Comber (2015), p. 230 for details). So the first step to use `moran.test()` is to turn our matrix into a `listw` class object.

Before we can use the functions from spdep to compute the global Moran's I, we need to create a `listw` type spatial weights object (instead of the matrix we used above). Let's use the objects we created using the various criteria. To get the same value as above, we use `style='B'` to use binary (TRUE/FALSE) weights.

```
lw_queen <-  nb2listw(nb_queen, style='B')
lw_distance <- nb2listw(nb_distance, style='B')
lw_k3 <- nb2listw(nb_k3, style='B')
lw_inverse_d <- nb2listw(nb_distance,
               glist = inv_distances_r,
               style='B')
```

Now we can apply the `moran.test()` function.

```
moran.test(cc_burglary_sp$burglary, lw_queen)
```

```
##
##   Moran I test under randomisation
##
## data:  cc_burglary_sp$burglary
## weights: lw_queen
##
## Moran I statistic standard deviate = 1.6,
## p-value = 0.06
## alternative hypothesis: greater
## sample estimates:
## Moran I statistic        Expectation           Variance
##            0.12033           -0.04545            0.01111
```

So the Moran's I here is 0.12, which does not look very large. With this statistic, we have a benchmarking problem. Whereas with standard correlation the value is always restricted to lie within the range of -1 to 1, the range of Moran's I varies with the W matrix. How can we establish the range for our particular matrix? It has been shown that the maximum and minimum values of I are the maximum and minimum values of the eigenvalues of $\frac{(W+W^T)}{2}$ (see Brunsdon and Comber (2015), p. 232 for details). Don't worry if you don't know what an eigenvalue is at this point. We can use this understanding to calculate Moran's range using the following code:

```
# Adapted from Brunsdon and Comber (2015)
moran_range <- function(lw) {
  wmat <- listw2mat(lw)
  return(range(eigen((wmat + t(wmat))/2)$values))
}
moran_range(lw_queen)
```

```
## [1] -2.541  5.023
```

On absolute scale, thus, this does not look like a very large value of spatial autocorrelation.

The second issue is whether this statistic is significant. If we assume a null hypothesis of independence, what is the probability of obtaining a value as extreme as 0.12. The spatial autocorrelation (Global Moran's I) test is an inferential statistic, which means that the results of the analysis are always interpreted within the context of its null hypothesis. For Global Moran's I statistic, the null hypothesis states that the attribute being analysed is randomly distributed among the features in your study area; said another way, the spatial process promoting the observed pattern of values is random chance. Imagine that you could pick up the values for the attribute you are analysing and throw them down onto your features, letting each value fall where it may. This process (picking up and throwing down the values) is an example of a random chance spatial process. When the p-value returned by this tool is statistically significant, you can reject the null hypothesis of complete spatial randomness.

The `moran.test()` function can compute the variance of Moran's I under the assumption of normallity or under the randomisation assumption (which is the default at least we set the argument `randomisation` to `FALSE`). We can also use a Monte Carlo procedure. The way Monte Carlo works is that the values of burglary are randomly assigned to the polygons, and Moran's I is computed. This is repeated several times to establish a distribution of expected values under the null hypothesis. The observed value of Moran's I is then compared with the simulated distribution to see how likely it is that the observed values could be considered a random draw.

We use the function `moran.mc()` to run a permutation test for Moran's I statistic calculated by using some number of random permutations of our numeric vector, for the given spatial weighting scheme, to establish the rank of the observed statistic in relation to the simulated values. Pebesma and Bivand (2021) argue that if we compare the Monte Carlo and analytical variances of I under randomisation, we typically see few differences, arguably rendering Monte Carlo testing unnecessary.

In `moran.mc()` we need to specify our variable of interest (`burglary`), the `listw` object we created earlier (`ww`), and the number of permutations we want to run (here we choose 99).

```
set.seed(1234)

burg_moranmc_results <- moran.mc(cc_burglary_sp$burglary, lw_queen, nsim=99)

burg_moranmc_results
```

```
##
##   Monte-Carlo simulation of Moran I
##
## data:  cc_burglary_sp$burglary
## weights: lw_queen
## number of simulations + 1: 100
##
## statistic = 0.12, observed rank = 92, p-value =
## 0.08
## alternative hypothesis: greater
```

So, the probability of observing this Moran's I if the null hypothesis was true is 0.08. This is higher than our alpha level of 0.05. In this case, we cannot conclude that there is significant global spatial autocorrelation. We can use this test to check for our various specifications of the spatial weight matrix:

```
moran.test(cc_burglary_sp$burglary, lw_distance)
```

```
##
##   Moran I test under randomisation
##
## data:  cc_burglary_sp$burglary
## weights: lw_distance
##
## Moran I statistic standard deviate = 3, p-value
## = 0.002
```

```
## alternative hypothesis: greater
## sample estimates:
## Moran I statistic        Expectation          Variance
##           0.28763           -0.04545           0.01266
```

```
moran.test(cc_burglary_sp$burglary, lw_k3)
```

```
##
##   Moran I test under randomisation
##
## data:  cc_burglary_sp$burglary
## weights: lw_k3
##
## Moran I statistic standard deviate = 1.9,
## p-value = 0.03
## alternative hypothesis: greater
## sample estimates:
## Moran I statistic        Expectation          Variance
##           0.18567           -0.04545           0.01423
```

```
moran.test(cc_burglary_sp$burglary, lw_inverse_d)
```

```
##
##   Moran I test under randomisation
##
## data:  cc_burglary_sp$burglary
## weights: lw_inverse_d
##
## Moran I statistic standard deviate = 2.4,
## p-value = 0.008
## alternative hypothesis: greater
## sample estimates:
## Moran I statistic        Expectation          Variance
##           0.23700           -0.04545           0.01380
```

As we can see, using the different specifications of neighbours produces results that are more suggestive of dependence. Such results further highlight the importance of picking appropriate and justified spatial weights for your analysis.

We can also make a "Moran scatterplot" to visualize spatial autocorrelation. This type of scatterplot was introduced by Anselin (1996) as a tool to visualise the relationship between the value of interest and its spatial lag, but also to explore local clusters and areas of non-stationarity (we will discuss this in greater detail in Chapter 10). Note the row standardisation of the weights matrix.

```
wm_queen_rs <- mat2listw(wm_queen, style='W')
# Checking if rows add up to 1 (they should)
mat <- listw2mat(wm_queen_rs)

#This code simply sums each row to see if they sum to 1
```

```
# we are only displaying the first 15 rows in the matrix
apply(mat, 1, sum)[1:15]
```

```
## E01005065 E01005066 E01005128 E01005212 E01033653
##         1         1         1         1         1
## E01033654 E01033655 E01033656 E01033658 E01033659
##         1         1         1         1         1
## E01033661 E01033662 E01033664 E01033667 E01033672
##         1         1         1         1         1
```

Now we can plot. Anselin (1996) proposes to center the variables around their mean, which is not the default taken by `moran.plot()`, so we need to specify this ourselves when plotting by virtue of using `scale()`. This simply subtracts the mean from each value, so that the mean of the resulting variable will be 0.

```
moran.plot(as.vector(scale(cc_burglary_sp$burglary)), wm_queen_rs)
```

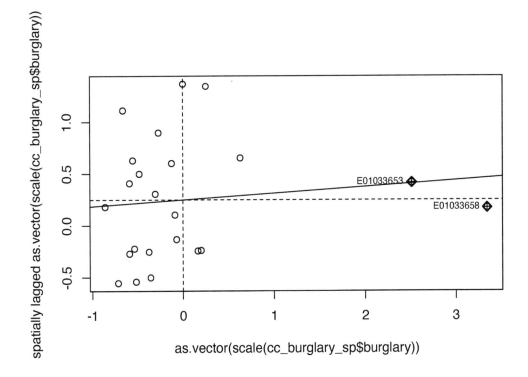

FIGURE 9.16: Moran scatterplot of burglary

The x-axis represents the values of our burglary variable in each unit (each LSOA) and the y-axis represents a spatial lag of this variable. A spatial lag in this context is simply the average value of the burglary count in the areas that are considered neighbours of each LSOA. So we are plotting the value of burglary against the average value of burglary in the neighbours. The slope of the fitted line represents Moran's I. And you can see the correlation is almost flat here. We will further explain this scatterplot in Chapter 10. Anselin (1996) suggests that the effective interpretation of this scatterplot should focus on the extent to which the fitted regression line that we see plotted reflects well the overall pattern of association between the variable of interest and its spatial lag: "the indication of observations that do not follow the overall trend represents useful information on local instability or non-stationarity" (p. 116-117).

As with any correlation measure, you could get **positive spatial autocorrelation**, which would mean that as you move further to the right in the x-axis, you have higher levels of burglary in the surrounding area. This is what we see here. But the correlation is fairly low and as we saw is not statistically significant. You can also obtain **negative spatial autocorrelation**. That would mean that areas with high level of crime *tend* (it's all about the global average effect!) to be surrounded by areas with low levels of crime. This is clearly not what we see here.

9.4.1.2 Geary's C

Another measure of global autocorrelation is Geary's C. Unlike Moran's I, which measures similarity, Geary's C measures *dissimilarity*. The value is achieved from considering the squared difference for attribute similarity. In the equation

$$C = \frac{(N-1) * \sum_i \sum_j w_{ij} * (x_i - x_j)^2}{2 * \sum_i \sum_j w_{ij} * \sum_i (x_i - \bar{x})^2}$$

the $(x_i - x_j)^2$ measures the *dissimilarity* between observations x_i and neighbours x_j.

This means that the interpretation for Geary's C is essentially the inverse of that of Moran's I. Positive autocorrelation is seen when $C < 1$ (or $z < 0$), while negative autocorrelation is seen when $C > 1$ (or $z > 0$). The nice thing about Geary's C is that it doesn't need any assumptions about linearity to be made (unlike Moran's I).

So how do we go about implementing this in R? For obtaining this statistic, we can use `geary.test()` function from the `spdep` package. As with Moran's I, we specify the variable of interest (`cc_burglary_sp$burglary`) and the spatial weights matrix (`lw_queen`):

```
geary.test(cc_burglary_sp$burglary, lw_queen)
```

```
##
##  Geary C test under randomisation
##
## data:  cc_burglary_sp$burglary
## weights: lw_queen
##
## Geary C statistic standard deviate = -1.7,
## p-value = 1
## alternative hypothesis: Expectation greater than statistic
## sample estimates:
## Geary C statistic       Expectation          Variance
##           1.39340           1.00000           0.05395
```

Note that for Geary's C, the test assumes that the weights matrix is *symmetric*. For inherently non-symmetric matrices, such as k-nearest neighbour matrices discussed above, you can make use of the listw2U() helper function which constructs a weights list object to make the matrix symmetric.

It is very important to understand that global statistics like the spatial autocorrelation (Global Moran's I, Geary's C) tool assess the overall pattern and trend of your data. They are most effective when the spatial pattern is consistent across the study area. In other words, you may have clusters (local pockets of autocorrelation), without having clustering (global autocorrelation). This may happen if the signs of the clusters negate each other. In general, Moran's I is a better measure of global correlation, as Geary's C is more sensitive to local correlations.

9.4.2 The `rgeoda` and `sfweight` packages

Traditionally, the generation of a weight matrix was done with functions included in the spdep program, and this is what we had introduced above. We do want to note, however, that more recently the functionality of **GeoDa**, a very well-known point and click interface for spatial statistics originally developed by Luc Anselin (a key figure in the field of spatial statistics), has been brought to R via the rgeoda package. This package is fast, though "less flexible when modifications or enhancements are desired" (Pebesma and Bivand 2021). The nomenclature of the functions is slightly more intuitive, and there are subtle variations in some of the formulas. It works with sf objects, and so may be useful for integrating our analysis to these objects. It is enough now that you are aware of the content of this chapter, and one path to implementing what we learned with the sp package. This should equip you to be able to explore these other packages as well, which will have different ways of achieving these outcomes.

For example, to create queen contiguity weights, you can use the queen_weights() function from the rgeoda package, and as an argument you only need to pass the name of your polygon object (cc_burglary).

```
queen_w_geoda <- queen_weights(cc_burglary)
summary(queen_w_geoda)
```

```
##                        name              value
## 1 number of observations:                 23
## 2            is symmetric:               TRUE
## 3                sparsity: 0.189035916824197
## 4        # min neighbors:                  2
## 5        # max neighbors:                  8
## 6       # mean neighbors: 4.34782608695652
## 7     # median neighbors:                  4
## 8           has isolates:              FALSE
```

There are similar functions for other types of weights:

```
# For example:
# For our k 3 nearest neighbours
knn_weights(cc_burglary, 3)
# For distance
distance_weights(cc_burglary,  min_distthreshold(cc_burglary))
```

More details are available in the documentation for rgeoda.

It is also worth mentioning `sfweight`. The `sfweight` package, still only available from GitHub at the time of this writing, and therefore we could consider it as "an ongoing development" that aims "to provide a more streamlined interface to the `spdep`" package. The goal of this package is to create a simpler workflow for creating neighbours, spatial weights, and spatially lagged variables. You can read more about it in its GitHub repository[3].

9.5 Further Reading

For a deeper understanding of the study of spatial dependence on point patterns, we recommend Baddeley, Rubak, and Turner (2015). Haining and Li (2020) and Darmofal (2015) both contain useful chapters on the spatial weight matrix that provide a bit more detail than we do here, whilst maintaining an accessible level. We highly recommend too the series of video lectures provided by Luc Anselin in the YouTube channel of the Center for Spatial Data Science (University of Chicago). Prof. Anselin has been one of the key authors in the field of spatial econometrics during the last 25 years. His lectures are both high quality and accessible for someone starting in the field of spatial statistics. You will find lectures on the spatial weight matrix, global autocorrelation, and local indicators of spatial autocorrelation (which we cover in Chapter 10) in Anselin (2021). In the website for the Center for Spatial Data Science[4] you will also find R tutorials that provide instructions on how to specify the spatial weight matrix using other approaches we have not covered here (Anselin et al. 2021).

---

[3]https://github.com/JosiahParry/sfweight
[4]spatial.uchicago.edu

10

Detecting hot spots and repeats

10.1 Introduction

10.1.1 Clusters, hot spots of crime, and near repeat victimisation

There has been a great interest within criminology and crime analysis in the study of local clusters of crime or hot spots over the last 30 years. "Hot spots" is the term typically used within criminology and crime analysis to refer to small geographical areas with a high concentration of crime. Weisburd (2015) argues for a *law of concentration* of crime that postulates that a large proportion of crime events occur at relatively few places (such as specific addresses, street intersections, street blocks, or other small micro-places) within larger geographical areas such as cities. In one of the earlier contributions to this field of research, Sherman, Gartin, and Buerger (1989) noted how roughly 3% of all addresses in Minneapolis (USA) generated about 50% of all calls to police services. A number of subsequent studies have confirmed this pattern elsewhere. Steenbeek and Weisburd (2016) argue that, in fact, most of the geographical variability of crime (58% to 69% in their data from the Hague) can be attributed to micro-geographic units, with a very limited contribution from the neighbourhood level (see, however, Ramos et al. (2021)). This literature also argues that many crime hot spots are relatively stable over time (Andresen and Malleson 2011; Andresen, Curman, and Linning 2017; Andresen, Linning, and Malleson 2017).

There is also now a considerable body of research evaluating police interventions that take this insight as key for articulating responses to crime. **Hot spots policing**, or place-based policing, assumes that police can reduce crime by focusing their resources on the small number of places that generate a majority of crime problems (Braga and Weisburd 2010). Policing crime hot spots has become a popular strategy in the US and efforts to adopt this approach have also taken place elsewhere. Theoretically, one could use all sorts of proactive creative approaches to solve crime in these locations (using for example conceptual and tactical tools from Goldstein (1990) problem oriented policing approach); though too often in practice directed patrol becomes the default response. Recent reviews of the literature (dominated by the US experience) suggest the approach can have a small effect size (Braga et al. 2019), though some authors contend alternative measures of effect size (which are suggestive of a moderate effect) should be used instead (Braga and Weisburd 2020). Whether these findings would replicate well in other contexts is still an open question (see, for example, Collazos et al. (2020)).

Thus, figuring out the location of hot spots and how to detect clusters of crime is of great significance for crime analysis. In previous chapters we explored ways to visualise the concentration of crime incidents in particular locations as a way to explore places with an elevated intensity of crime. We used tesselation and kernel density estimation as a way to do this. But, as noted by Chainey (2014):

Maps generated using KDE and the other common hot spot mapping methods are useful for showing where crime concentrates but may fail to unambiguously determine what is hot (in hot spot analysis terms) from what is not hot. That is, they may fail in separating the significant spatial concentrations of crime from those spatial distributions of less interest or from random variation.

Apparent clusters may indeed appear by chance alone (Marshall 1991b). There are a number of tools that have been developed to detect local clusters among random variation. It is not exaggerated to say that this has been a key concern in the field of spatial statistics with practical applications in many fields of study. In crime analysis, for example, it is common to use the local Gi^* statistic, which is a type of a **local indicator of spatial association** (LISA). Another popular LISA is the local Moran's I, developed by Anselin (1995).

These LISA statistics serve two purposes. On one hand, they may be interpreted as indicators of local pockets of non-stationarity, to identify outliers. On the other hand, they may be used to assess the influence of individual locations on the magnitude of the global statistic (Anselin 1995). Anselin (1995) suggests that any LISA satisfies two requirements (p. 94):

- the LISA for each observation gives an indication of the extent of significant spatial clustering of similar values around that observation;
- the sum of LISAs for all observations is proportional to a global indicator of spatial association.

These measures of local spatial dependence are available in tools often used by crime analysts (CrimeStat, GeoDa, ArcGIS, and also R). But it is important to realise that there are some subtle differences in implementation that reflect design decisions by the programmers. These differences in design may result on apparent differences in numerical findings generated with these different programs. The defaults used by different software may give the impression different results are achieved. Bivand and Wong (2018) offer an excellent thorough review of these different implementations (and steps needed for ensuring comparability). Our focus here will be in introducing the functionality of the spdep packages.

Aside from these LISAs, in spatial epidemiology a number of techniques have been developed for detecting areas with an unusually high risk of disease (Marshall 1991a). Although originally developed in the context of clusters of disease, these techniques can also be applied to the study of crime (Gomez-Rubio, Ferrandiz, and Lopez 2005). Most of the methods in spatial epidemiology are based on the idea of "moving windows", such as the spatial scan statistic developed by Kulldorff (1997). Several packages bring these scan statistics to R. There are also other packages that use different algorithms to detect disease clusters, such as AMOEBA, that is based in the Getis-Ord algorithm. As we will see, we are spoiled for choice in this regard. Here we will only provide a glimpse into how to perform a few of these tests with R.

In crime analysis, the study of **repeat victimisation** is also relevant. This refers to the observed pattern of repeated criminal victimisation against the same person or target. The idea that victimisation presages further victimisation was first observed in studies of burglary, where the single best predictor of new victimisation was past victimisation. This opened up an avenue of research led by British scholars Ken Pease and Graham Farrell, who noticed that the risk of re-victimisation is much higher in the days immediate to the first offence and that repeat victimisation accounts for a non-trivial proportion of crime (Pease 1998). There continues to be some debate as to whether this is to do with underlying stable vulnerabilities (*flag* hypothesis) or to increased risk resulting form the first victimisation (*boost* hypothesis), although likely both play a role. Subsequent research has extended the foci of application well beyond the study of burglary and the UK (for a review see Pease and Farrell (2017) and Pease, Ignatans, and Batty (2018)). From the very beginning this literature emphasised the importance for crime prevention policy, for this helps to identify potential targets for it. The advocates of this concept argue that using insights from repeat victimisation is a successful approach to crime reduction (Grove et al. 2014; Farrell and Pease 2017).

A related phenomenon is that of **near repeat victimisation**. Townsley, Homel, and Chaseling (2003), building in the epidemiological notion of contagion, proposed and tested the idea that "proximity to a burgled dwelling increases burglary risk for those areas that have a high degree of housing homogeneity and that this risk is similar in nature to the temporarily heightened risk of becoming a repeat victim after an initial victimisation" (p. 615). Or as, in a subsequent paper, Bowers and Johnson (2004) put it, "following a burglary at one home the risk of burglary at nearby homes is amplified" (p. 12). So if we combine the ideas of repeat victimisation and near repeats we would observe: (1) a heightened but short-lived risk of victimisation for the original victim following a first attack and (2) a heightened but also short-lived risk for those "targets" located nearby. Although most studies on near repeats have focused on studying burglary, there are some that have also successfully applied this concept to other forms of criminal victimisation. In 2006 and with funds from the US National Institute of Justice, Jerry Ratcliffe developed a free point-and-click interface to perform near repeat analysis (Ratcliffe 2020). This software can be accessed from his personal home-page. But our focus in this chapter will be in a package (`NearRepeat`) developed by Wouter Steenbeek in 2019 to bring this functionality into R Steenbeek (2021). At the time of writing, this package is not yet available from CRAN but needs to be installed from its GitHub repository. This can be done by using the `install_github()` function from the `remotes` package, with the code below:

```
remotes::install_github("wsteenbeek/NearRepeat")
```

Additionally, for this chapter we will need the following packages:

```
# Packages for handling data and for geospatial carpentry
library(dplyr)
library(readr)
library(sf)
library(raster)
# Packages for detecting clusters
library(spdep)
library(rgeoda)
library(DCluster)
library(spatstat)
# Packages for detecting near repeat victimisation
library(NearRepeat) # Available at https://github.com/wsteenbeek/NearRepeat
# Packages for visualisation and mapping
library(tmap)
library(mapview)
library(RColorBrewer)
# Packages with relevant data for the chapter
library(crimedata)
```

10.1.2 Burglaries in Manchester

We return to burglaries in Manchester for most of this chapter. The files we are loading include the geomasked locations of burglary occurrences in 2017, for the whole of Manchester city and separately for its city centre. Later in this chapter, we will look at the whole of Manchester and we will use counts of burglary per Lower Super Output Area (LSOA); so we will also load this aggregated data now.

```
# Create sf object with burglaries for Manchester city
manc_burglary <- st_read("data/manc_burglary_bng.geojson",
                         quiet=TRUE)
manc_burglary <- st_transform(manc_burglary , 4326)
# Create sf object with burglaries in city centre
cc_burglary <- st_read("data/manch_cc_burglary_bng.geojson",
                       quiet=TRUE)
cc_burglary <- st_transform(cc_burglary , 4326)
# Read the boundary for the Manchester city centre ward
city_centre <- st_read("data/manchester_wards.geojson",
                       quiet = TRUE) %>%
  filter(wd16nm == "City Centre")
city_centre <- st_transform(city_centre, 4326)
# Read the boundary for Manchester city
manchester <- st_read("data/manchester_city_boundary_bng.geojson",
                      quiet = TRUE)
manchester <- st_transform(manchester , 4326)
# Read into R the count of burglaries per LSOA
burglary_lsoa <- st_read("data/burglary_manchester_lsoa.geojson",
                         quiet = TRUE)
# Place last file in projected British National Grid
burglary_lsoa <- st_transform(burglary_lsoa, 27700)
```

Now that we have all the data we need in our environment, let's get started.

10.2 Using `raster` micro-grid cells to spot hot spots

Analysis which produces "hot spots" generally focuses on understanding small places or micro-geographical units where crime (or other topic of interest) may concentrate. There are different ways in which one could define this "micro-unit". A number of studies, recognising the nature of police data, look at this from the perspective of street segments. Another way of approaching the problem would be to define micro-grid cells. Chainey (2021) argues that if street segments are available, this should be the preferred level of analysis "because they act as a behavioral setting along which social activities are organized" (p. 42). Yet, as we will see throughout this chapter, most of the more traditional techniques that have been developed for detecting local clusters, and that are available in standard software for hot spot detection, were designed for points and polygons rather than lines. The focus of this chapter is on these more traditional techniques, and therefore we will use micro-grid cells as the vector object to work with throughout this chapter.

In Chapter 4 we discussed alternatives to choropleth maps, and one of these was to use a grid map. This is a process whereby a tessellating grid of a shape such as squares or hexagons is overlaid on top of our area of interest, and our points are aggregated to such a grid. We explored this using `ggplot2` package. We also used this approach when discussing quadrat counting with `ppp` objects and using the functionality provided by the `spatstat` package in Chapter 7.

Here we can revisit this with another approach, using the `raster` package. First we need to create the grid by using the `raster()` function. The first argument defines the extent for the grid. By using our `manchester sf` object, we ensure it includes all of our study area. Then the `resolution` argument defines how large we want the cells in this grid to be. You can play around with this argument and see how the results will vary.

```
# Create empty raster grid
manchester_r <- raster(manchester, resolution = 0.004)
```

We can then use the `rasterize()` function to count the number of points that fall within each cell. This is just a form of quadrat counting as the one we introduced in Chapter 7. The `rasterize()` function takes as the first argument the coordinates of the points. The second argument provides the grid we just created (`manchester_r`). The `field =` argument in this case is assigned a value of `1` so that each point is counted a unit. Then we use the `fun =` argument to specify what we want to do with the points. In this case, we want to "count" the points within each cell. The `rasterize()` function by default will set at `NA` all the values where there are no points. We don't want this to be the case, as no points is meaningful in our case (means 0 crimes in that grid). So that the raster includes those cells as zero, we use the `background =` argument and set it to zero:

```
# Fill the grid with the count of incidents within each cell
m_burglary_raster <- rasterize(manc_burglary,
                               manchester_r,
                               field = 1,
                               fun = "count",
                               background = 0)
```

We can now plot our results:

```
# Use base R functionality to plot the raster
plot(m_burglary_raster)
plot(st_geometry(manchester), border='#00000040', add=T)
```

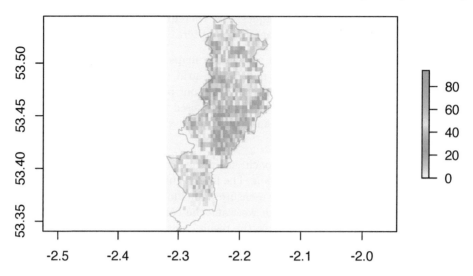

FIGURE 10.1: Raster map of burglary across Manchester

This kind of map is helpful if you want to get a sense of the distribution of crime over the whole city. However, from a crime analyst's operational point of view, the interest may be at a more granular scale. We want to dive in and observe the micro-places that may be areas of high crime within certain parts of the city. We can do this by focusing on particular neighbourhoods. In Manchester, City Centre ward has one of the highest counts of burglaries - so let's zoom in on here.

When we read in our files at the start of the chapter, we created the `city_centre` object, which contains the LSOAs in this ward, and also `cc_burglary` which contains only those burglaries which occurred in the city centre.

Let's repeat the steps of creating a raster, and then the grid which we demonstrated above for all Manchester Local Authority, but now for the City Centre only.

```
# Create empty grid for the City Centre
city_centre_r <- raster(city_centre, resolution = 0.0009)
# Produce raster object with the number of incidents within each cell
burglary_raster_cc <- rasterize(cc_burglary,
                                city_centre_r,
                                field = 1,
                                fun = "count",
                                background = 0)
# Plot results with base R
plot(burglary_raster_cc)
plot(st_geometry(city_centre), border='#00000040', add=T)
```

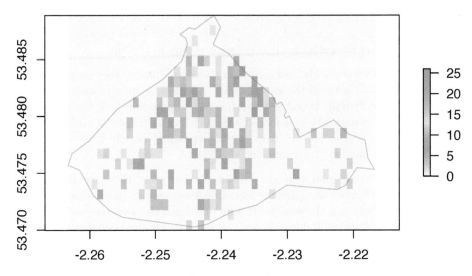

FIGURE 10.2: Raster map of burglary in City Centre only

You can see in the above map where hot spots might occur. However, you also see that we have created cells of 0 burglaries outside of the boundary of our City Centre ward. This is not quite true, as we didn't include burglaries which occurred outside of here. Therefore, we should not have in this raster object incidents outside the boundaries of the neighbourhoud. Instead, we want to declare those cells as NAs. We can do that with the mask() function in the raster package.

```
burglary_raster_cc <- mask(burglary_raster_cc, city_centre)
plot(burglary_raster_cc)
```

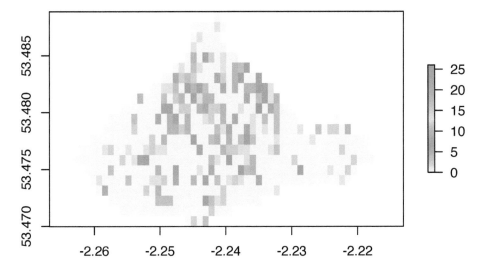

FIGURE 10.3: Raster plot of burglary

Now you can see we indicate more clearly where we have cells which contain no burglaries at all in our data, versus where we do not have data to be able to talk about burglaries.

To better understand what might be going on in our grids with higher burglary count, let's add some context. We will use mapview for this. This is another great package if you want to give your audience some interactivity. It does what is says in the tin:

> it provides functions to very quickly and conveniently create interactive visualisations of spatial data. Its main goal is to fill the gap of quick (not presentation grade) interactive plotting to examine and visually investigate both aspects of spatial data, the geometries and their attributes. It can also be considered a data-driven API for the leaflet package as it will automatically render correct map types, depending on the type of the data (points, lines, polygons, raster) (Appelhans et al. 2021).

With very few lines of code, you can get an interactive plot. Here we first will use brewer.pal() from the RColorBrewer package to define a convenient palette (see Chapter 5 for details on this) and then use the mapview() function from the mapview package to plot our raster layes we created above. Below you see how we increase the transparency with the alpha.region argument, and how we use col.regions to change from the default palette to the one we created.

```
my.palette <- brewer.pal(n = 9, name = "OrRd")
mapview(burglary_raster_cc,
        alpha.regions = 0.7,
        col.regions = my.palette)
```

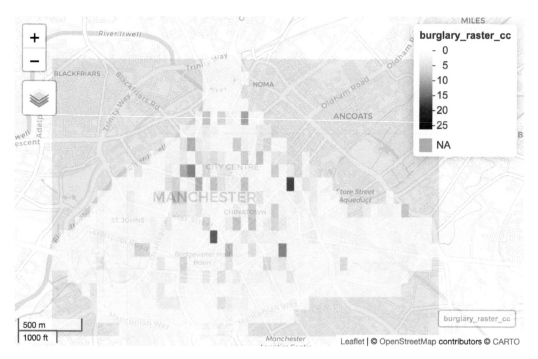

FIGURE 10.4: Interactive plot of burglaries.

We can now explore the underlying environmental backcloth behind the areas with high burglary count and build hypotheses and research questions about what might be going on there; for example whether these are crime attractors, crime generators, or other types of crime places. However, keep in mind that these are illustrations of the counts of observed burglaries, and there is nothing here which declares something a "hot spot". For that, keep reading!

10.3   Local Getis-Ord

The maps above and the ones we introduced in Chapter 7 (produced through kernel density estimation) are helpful for visualising the varying intensity of crime across the study region of interest. However, as we noted above, it may be hard to discern random variation from clustering of crime incidents simply using these techniques. There are a number of statistical tools that have been developed to extract the signal from the noise, to determine what is "hot" from random variation. Many of these techniques were originally designed to detect clusters of disease.

In crime analysis it is common to use a technique developed by Arthur Getis and JK Ord in the early 1990s (see Getis and Ord (1992) and Ord and Getis (1995)), partly because it was implemented in the widely successful ArcGIS software tool as part of their hot spot analysis tool. This statistic provides a measure of spatial dependence at the local level. They are not testing homogeneity, they are testing dependence of a given attribute around neighbouring areas (which could be operationalised as micro-cells in a grid, as above, or could be some form of administrative areas, like wards, or census boundaries like LSOAs or census blocks).

Whereas the tests we covered in the previous chapter measure dependence at the global level (whereas areas are surrounded by alike areas), these measures of local clustering try to identify pockets of dependence. They are also called **local indicators of spatial autocorrelation**. It is indeed useful for us to be able to assess quantitatively whether crime events cluster in a non-random manner. In the words of Jerry Ratcliffe (2010), however, "while a global Moran's I test can show that crime events cluster in a non-random manner, this simply explains what most criminal justice students learn in their earliest classes." (p. 13). For a crime analyst and practitioner, while clustering is of interest, learning about the existence and location of local clusters is of paramount importance.

There are two variants of the local Getis-Ord statistic. The local G is computed as a ratio of the weighted average of the values of the attribute of interest in the neighbouring locations, not including the value at the location. It is generally used for spread and diffusion studies. The local G^* on the other hand, includes the value at the location in both numerator and denominator. It is more generally used for clustering studies (for formula details, see Getis and Ord (1992) and Ord and Getis (1995)). High positive values indicate the possibility of a local cluster of high values of the variable being analysed ("hot spot"); very low relative values indicate a similar cluster of low values ("cold spot").

> A value larger than the mean (or, a positive value for a standardized z-value) suggests a High-High cluster or hot spot, a value smaller than the mean (or, negative for a z-value) indicates a Low-Low cluster or cold spot (Anselin 2020).

For statistical inference, a Bonferroni-type test is suggested in the papers by Getis and Ord, where tables of critical values are included. The critical values for the 95th percentile under the assumptions discussed by Ord and Getis (1995) are for $n=30$: 2.93, $n=50$: 3.08, $n=100$: 3.29, $n=500$: 3.71, and for $n=1000$: 3.89. Anselin (2020) suggests that this analytical approximation may not be reliable in practice and that conditional permutation is advisable. Inference with these local measures are also complicated by problems of multiple comparisons, so it is typically advisable to use some form of adjustment to this problem.

The first step in computing the Getis and Ord statistics involves defining the neighbours of each area. And to do that, first we have to turn each cell in our grid into a polygon, so we basically move from the raster to the vector model. This way we can use the methods discussed in Chapter 9 to operationalise neighbouringness in our data. To transform our raster grid cells into polygons, we can use the `rasterToPolygons()` function from the `raster` package.

```
# Move to vector model
grid_cc <- rasterToPolygons(burglary_raster_cc)
```

What is a neighbour and the code related to defining neighbours is a topic we covered in Chapter 9. For brevity, we will only use the queen criteria for this illustration. Only to indicate that `include.self()` ensures we get the G^* variation of the local Getis-Ord test, the second variation we mentioned above.

```
nb_queen_cc <- poly2nb(grid_cc)
lw_queen_cc <- nb2listw(include.self(nb_queen_cc), zero.policy=T)
```

Now that we have the data ready we can, analytically, compute the statistic using the `localG()` function from the `spdep` package. We can then add the computed local Gi^* to each cell. And we can then see the produced z scores:

```
# Perform the local G analysis (Getis-Ord GI*)
grid_cc$G_hotspot_z <- as.vector(localG(grid_cc$layer, lw_queen_cc))
# Summarise results
summary(grid_cc$G_hotspot_z)
```

```
##    Min. 1st Qu.  Median    Mean 3rd Qu.    Max.
## -1.411  -1.011  -0.398   0.033   0.835   5.231
```

How do you interpret these? They are z scores. So you expect large absolute values to index the existence of cold spots (if they are negative) or hot spots (if they are positive). Our sample size is 475 cells, and there are only a very small handful of cells with values that reach the critical value in our dataset.

A known problem with this statistic is that of multiple comparisons. As noted by Pebesma and Bivand (2021): "although the apparent detection of hot spots from values of local indicators has been quite widely adopted, it remains fraught with difficulty because adjustment of the inferential basis to accommodate multiple comparisons is not often chosen." So we need to ensure the critical values we use to adjust for this. As recommended by Pebesma and Bivand (2021), we can use the `p.adjust()` function to count the number of observations that this statistic identifies as hot spots using different solutions for the multiple comparison problem, such as Bonferroni, False Discovery Rate (`fdr`), or the method developed by Benjamini and Yekutielli (`BY`).

```
p_values_lg <- 2 * pnorm(abs(c(grid_cc$G_hotspot_z)), lower.tail = FALSE)
p_values_lgmc <- cbind(p_values_lg,
            p.adjust(p_values_lg, "bonferroni"),
            p.adjust(p_values_lg, "fdr"),
            p.adjust(p_values_lg, "BY"))
colnames(p_values_lgmc) <- c("None", "Bonferroni",
```

```
                      "False Discovery Rate", "BY (2001)")
apply(p_values_lgmc, 2, function(x) sum(x < 0.05))
```

```
##                  None           Bonferroni
##                   53                   10
## False Discovery Rate        BY (2001)
##                   15                   12
```

We can see that we get fewer significant hot (or cold) spots once we adjust for the problem of multiple comparisons. In general, Bonferroni correction is the more conservative option. Here, the p-value is multiplied by the number of comparisons. This is useful, when let's say we're trying to avoid false positives. We will progress with those areas identified with this method.

To map this, we can create a new variable, let's call it hotspot, which can take the value of "hot spot" if the Bonferroni adjusted p-value is smaller than 0.05 and the z score is positive, and "cold spot" if the Bonferroni adjusted p-value is smaller than 0.05 and the z score is negative. Otherwise, it will be labelled "Not significant".

```
# convert to simple features object
grid_cc_sf <- st_as_sf(grid_cc)
# Creates p values for Bonferroni adjustment
grid_cc_sf$G_hotspot_bonf <- p.adjust(p_values_lg, "bonferroni")
# Create character vector based on Z score and p values
grid_cc_sf <- grid_cc_sf %>%
  mutate(hotspot = case_when(grid_cc_sf$G_hotspot_bonf < 0.05 &
                              grid_cc_sf$G_hotspot_z > 0 ~ "Hot spot",
                             grid_cc_sf$G_hotspot_bonf < 0.05 &
                              grid_cc_sf$G_hotspot_z < 0 ~ "Cold spot",
                             TRUE ~ "Not significant"))
```

Now let's see the result:

```
table(grid_cc_sf$hotspot)
```

```
##
##        Hot spot Not significant
##              10             512
```

Great, we have our 10 significant hot spots, and no significant cold spots apparently. We can now map these values.

```
#set to view mode and make sure colours are colourblind safe
tmap_mode("view")
tmap_style("col_blind")

# make map of the z-scores
map1 <- tm_shape(grid_cc_sf) +
  tm_fill(c("G_hotspot_z"), title="Z scores", alpha = 0.5)
```

```
# make map of significant hot spots
map2 <- tm_shape(grid_cc_sf) +
  tm_fill(c("hotspot"), title="Hot spots", alpha = 0.5)

# print both maps side by side
tmap_arrange(map1, map2, nrow=1)
```

We can see here that not all the micro-grid cells with an elevated count of burglaries exhibit significant local dependence. There are only two clusters of burglaries where we see positive spatial dependence: one is located north of Picadilly Gardens in the Northern Quarter of Manchester (an area of trendy bars and shops, and significant gentrification, in proximity to a methadone delivery clinic) and the Deansgate area of the city centre (one of the main commercial hubs of the city not far from the criminal courts).

10.4   Local Moran's I and Moran scatterplot

10.4.1   Computing the local Moran's I

The Local Getis and Ord statistic is a local indicator of spatial association. In Chapter 9 of this book, we looked at using Moran's I as a global measure of spatial autocorrelation. There is also a local version of Moran's I, which allows for the decomposition of global indicators of spatial autocorrelation such as Moran's I discussed in Chapter 9, into the contribution of each observation. This allows us to elaborate on the location and nature of the clustering within our study area, once we have found that autocorrelation occurs globally.

Let's first look at Moran's scatterplot for our data. Remember from last chapter that Moran's scatterplot plots the values of the variable of interest, in this case the burglary count on each micro-grid, against the spatial lagged value. For creating Moran's scatterplot we generally use row standardisation in the weight matrix so that the y and the x-axes are more comparable.

```
# create row standardised weights
lw_queen_rs <-  nb2listw(nb_queen_cc, style='W')

# build Moran's scatterplot
m_plot_burglary <- moran.plot(as.vector(scale(grid_cc$layer)),
                      lw_queen_rs, cex=1, pch=".",
                      xlab="burglary count",  ylab="lagged burglary count")
```

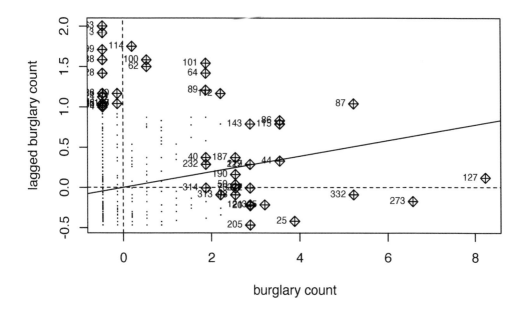

FIGURE 10.5: Moran's scatterplot for burglary in Manchester

Notice how the plot is split into four quadrants structured around the centered mean for the value of burglary and its spatial lag. The top right corner belongs to areas that have high level of burglary and are surrounded by other areas that have above the average level of burglary. These are **high-high** locations. The bottom left corner belongs to the **low-low areas**. These are areas with low level of burglary and surrounded by areas with below average levels of burglary. Both the high-high and low-low represent clusters. A high-high cluster is what you may refer to as a *hot spot*. And the low-low clusters represent *cold spots*.

In the opposite diagonal we have **spatial outliers**. They are not outliers in the standard statistical sense, as in extreme observations. Instead, they are outliers in that they are surrounded by areas that are very unlike them. So you could have *high-low spatial outliers*, areas with high levels of burglary and low levels of surrounding burglary, or *low-high spatial outliers*, areas that have themselves low levels of burglary (or whatever else we are mapping) and that are surrounded by areas with above average levels of burglary. This would suggest pockets of non-stationarity. Anselin (1996) warns these can sometimes surface as a consequence of problems with the specification of the weight matrix.

The slope of the line in the scatterplot gives you a measure of the global spatial autocorrelation. As we noted earlier, the local Moran's I helps you to identify the locations that weigh more heavily in the computation of the global Moran's I. The object generated by the `moran.plot()` function includes a measure of this leverage, a hat value, for each location in your study area and we can plot these values if we want to geographically visualise those locations:

```
# We extract the influence measure and place it in the sp object
grid_cc_sf$lm_hat <- m_plot_burglary$hat
# Then we can plot this measure of influence
tm_shape(grid_cc_sf) + tm_fill("lm_hat")
```

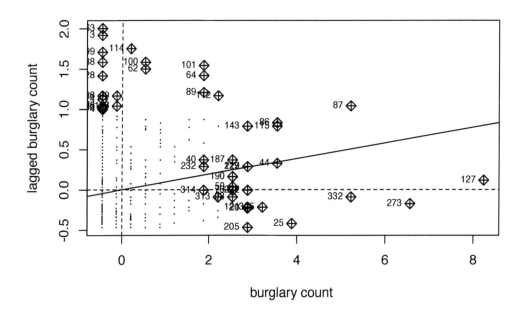

FIGURE 10.6: Map to visualise high leverage (hat value) areas

To compute the local Moran's I, we can use the `localmoran()` function from the `spdep` package.

```
locm_burglary <- localmoran(grid_cc_sf$layer, lw_queen_rs, alternative = "two.sided")
```

As with the local $G*$ it is necessary to adjust for multiple comparisons. The unadjusted p-value for the test is stored in the fifth column of the object of class `localmoran` that we generated with the `localmoran()` function. As before we can extract this element and pass it as an argument to the `p.adjust()` function in order to obtain p-values with some method of adjustment for the problem of multiple comparisons.

```
p_values_lm <- locm_burglary[,5]
sum(p_values_lm < 0.05)
```

```
## [1] 54
```

NOTE: It is important to note that by default, the localmoran() function does not adjust for the problem of multiple comparisons. This needs to be kept in mind, as it might lead us to erroneously conclude we have many hot or cold spots.

In our example, if we do not adjust for multiple comparisons, then, there would be 33 areas with a significant p value. If we apply again the Bonferroni correction, these get reduced to only 11.

```
bonferroni_lm <- p.adjust(p_values_lm, "bonferroni")
sum(bonferroni_lm < 0.05)
```

```
## [1] 10
```

Another approach is to use conditional permutation for inference purposes, instead of the analytical methods used above. In this case we would need to invoke the localmoran_perm() function also from the spdep package. This approach is based on simulations, so we also have to specify an additional argument nsim setting the number of simulations. Bivand and Wong (2018) suggest that "increasing the number of draws beyond 999 has no effect" (p. 741). The iseed argument ensures we use the same random seed and get reproducible results. As before adjusting for multiple comparisons notably reduces the number of significant clusters.

```
locm_burglary_perm <- localmoran_perm(grid_cc_sf$layer,
                                      lw_queen_rs,
                                      nsim=499,
                                      alternative="two.sided",
                                      iseed=1)
p_values_lmp <- locm_burglary_perm[,5]
sum(p_values_lmp < 0.05)
```

```
## [1] 54
```

```
bonferroni_lmp <- p.adjust(p_values_lmp, "bonferroni")
sum(bonferroni_lmp < 0.05)
```

```
## [1] 9
```

We go from 54 to 9 significant regions! This really illustrates the perils of not correcting for multiple comparisons!

10.4.2 Creating a LISA map

In order to map these Local Indicators of Spatial Association (LISA), we need to do some prep work.

First we must scale the variable of interest. As noted earlier, when we scale burglary, what we are doing is re-scaling the values so that the mean is zero. We use the scale() function, which is a generic function whose default method centers and/or scales the variable.

```
# Scale the count of burglary
grid_cc_sf$s_burglary <- scale(grid_cc_sf$layer)
# Creates p values for Bonferroni adjustment
grid_cc_sf$localmp_bonf <- p.adjust(p_values_lmp, "bonferroni")
```

To produce the LISA maps, we also need to generate a **spatial lag**: the average value of the burglary count in the areas that are considered neighbours of each LSOA. For this we need our `listw` object, which is the object created earlier, when we generated the list with weights using row standardisation. We then pass this `listw` object into the `lag.listw()` function, which computes the spatial lag of a numeric vector using a `listw` sparse representation of a spatial weights matrix.

```
#create a spatial lag variable and save it to a new column
grid_cc_sf$lag_s_burglary <- lag.listw(lw_queen_rs, grid_cc_sf$s_burglary)
```

Make sure to check the summaries to ensure the numbers are in line with our expectations, and nothing weird is going on.

```
summary(grid_cc_sf$s_burglary)
summary(grid_cc_sf$lag_s_burglary)
```

We are now going to create a new variable to identify the quadrant in which each observation falls within Moran's scatterplot, so that we can tell apart the high-high, low-low, high-low, and low-high areas. We will only identify those that are significant according to the p-value that was provided by the local Moran function adjusted for multiple comparisons with out Bonferroni adjustment. The data we need for each observation, in order to identify whether it belongs to the high-high, low-low, high-low, or low-high quadrants are the standardised burglary score (positive or negative), the spatial lag score, and the p-value.

Essentially all we'll be doing is assigning a variable values based on where in the plot it is. So for example, if it's in the upper right, it is high-high and has values larger than 0 for both the burglary and the spatial lag values. It it's in the upper left, it's low-high and has a value larger than 0 for the spatial lag value, but lower than 0 on the burglary value. And so on, and so on. Here's an image to illustrate:

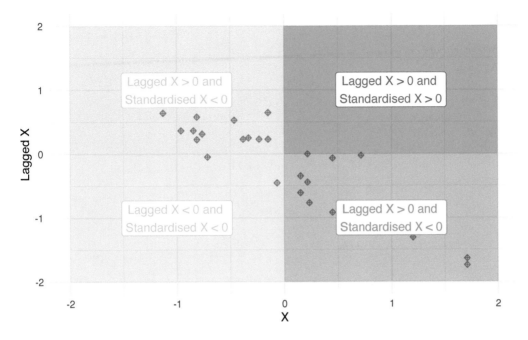

FIGURE 10.7: Quadrants of a Moran scatterplot

So let's first initialise this variable. In this instance we are creating a new column in the sp object and calling it "quad_sig".

```
grid_cc_sf <- grid_cc_sf %>%
  mutate(quad_sig = case_when(localmp_bonf >= .05 ~ "Not significant",
                       s_burglary > 0 &
                        lag_s_burglary > 0 &
                        localmp_bonf < 0.05 ~ "High-high",
                       s_burglary < 0 &
                        lag_s_burglary < 0 &
                        localmp_bonf < 0.05 ~ "Low-low",
                       s_burglary < 0 &
                        lag_s_burglary > 0 &
                        localmp_bonf < 0.05 ~ "Low-high",
                       s_burglary > 0 &
                        lag_s_burglary < 0 &
                        localmp_bonf < 0.05 ~ "High-low" ))
```

Now we can have a look at what this returns:

```
table(grid_cc_sf$quad_sig)
```

```
##
##       High-high        Low-high Not significant
##               5               4             513
```

So the 9 significant clusters we found split into 5 high-high and 4 low-high.

Let's put them on a map, using the standard colours used in this kind of maps:

```
tm_shape(grid_cc_sf) +
  tm_fill("quad_sig",
          palette= c("red","blue","white"),
          labels = c("High-High","Low-High", "non-significant"),
          alpha=0.5) +
  tm_borders(alpha=.5) +
  tm_layout(frame = FALSE,
            legend.position = c("right", "bottom"),
            legend.title.size = 0.8,
            legend.text.size = 0.5,
            main.title = "LISA with the p-values",
            main.title.position = "centre",
            main.title.size = 1.2)
```

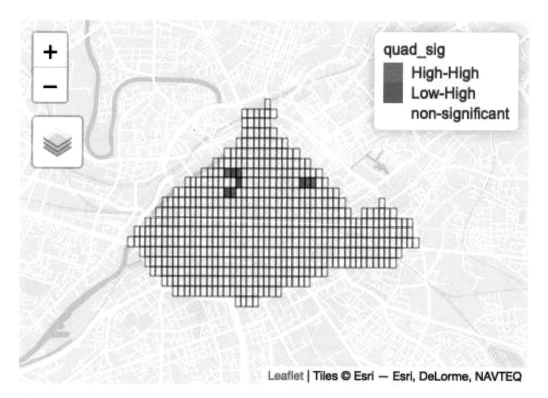

FIGURE 10.8: Significant clusters from local Moran's I

There we have it, our LISA map of a grid overlay over Manchester's City Centre ward, identifying local spatial autocorrelation in burglaries.

10.5 Scan satistics

10.5.1 The `DCluster` package

As we mentioned at the outset of the chapter, a number of alternative approaches using a different philosophy to those that are based on the idea of LISA have been developed within spatial epidemiology. Rather than focusing on local dependence, these tests focus on assessing homogeneity (Marshall 1991b):

> The null hypothesis when clustering is thought to be due to elevated local risk is usually that cases of disease follow a non-homogeneous Poisson process with intensity proportional to population density. Tests of the null Poisson model can be carried out by comparing observed and expected count (p. 423).

Here we will look at two well-known approaches with a focus on their implementation in the `DCluster` package: Openshaw's GAM and Kulldorf's scan statistic. The package `DCluster` (Gomez-Rubio, Ferrandiz, and Lopez 2005) was one of the first to introduce tools to assess spatial disease clusters to R. This package precedes `sf` and it likes the data to be stored in a data frame with at least four columns: observed number of cases, expected number of cases, and longitude and latitude. It is a bit opinionated in the way it is structured. `DCluster`, like many packages that are influenced by the epidemiological literature, takes as key inputs the *observed cases* (in criminology this will be criminal occurrences) and *expected cases*. In epidemiology, expected cases are often standardised to account for population heterogeneity. In crime analysis, this is not as common (although in some circumstances it should be!).

Straight away we see the problem we encounter here. The criminology of place emphasises small places, but for this level of aggregation (street segments or micro-grid cells) we won't have census variables indexing the population that we could easily use for purposes of standardisation. This means we need to (1) compromise and look for clusters at a level of aggregation for which we have some population measure that we could use for standardising or (2) get creative and try to find other ways of solving this problem. Increasingly, environmental criminologists are trying to find ways of taking this second route (see Andresen et al. (2021)). For this illustration and for convenience reasons, we will take the first one. Although it is important to warn we should try to stick whenever possible to the level of the spatial process we are interested in exploring.

For this example, the expected counts will be computed using the overall incidence ratio (i.e., total number of cases divided by the total population; in this case the number of residential dwellings in the area). We will use the unit of analysis of neighbourhoods operationalised as a commonly used census unit of geography in the UK, the Lower Super Output Area (LSOA).

Let's compute the expected count by dividing the total number of burglaries by the total number of dwellings (if burglaries were evenly distributed across all dwellings in our study area) and then multiplying this with the number of dwellings in each LSOA

```
# calculate rate per dwelling in whole study area
even_rate_per_dwelling <- sum(burglary_lsoa$burglary) / sum(burglary_lsoa$dwellings)
```

```
# expected count for each LSOA times rate per number of dwellings
burglary_lsoa <- burglary_lsoa %>%
  mutate(b_expected = dwellings * even_rate_per_dwelling)
```

This package requires as well the centroids for the polygons we are using. To extract this, we can use `st_centroid()` function form the `sf` package. If we want to extract the coordinates individually, we also need to use the `st_coordinates()` function. All the GEOS functions underlying `sf` need projected coordinates to work properly, so we need to ensure an adequate projection (which we did when we loaded the data at the outset of the chapter):

```
# Get coordinates for centroids
lsoa_centroid <- st_centroid(burglary_lsoa) %>%
  st_coordinates(lsoa_centroid)
# Place the coordinates as vectors in our sf data frame
burglary_lsoa$x <- lsoa_centroid[,1]
burglary_lsoa$y <- lsoa_centroid[,2]
```

For convenience, we will just place these four columns in a new object. Although this is not clear from the documentation, in our experience some of the code will only work if you initialise the data using these column names:

```
lsoas <- data.frame(Observed = burglary_lsoa$burglary,
                    Expected = burglary_lsoa$b_expected,
                    Population = burglary_lsoa$dwellings,
                    x = burglary_lsoa$x,
                    y = burglary_lsoa$y)
```

`DCluster` includes several tests for assessing the general heterogeneity of the relative risks. A chi-squared test can be run to assess global differences between observed and expected cases.

```
chtest <- achisq.test(Observed~offset(log(Expected)),
                      as(lsoas, "data.frame"),
                      "multinom",
                      999)
chtest
```

```
## Chi-square test for overdispersion
##
##   Type of boots.: parametric
##   Model used when sampling: Multinomial
##   Number of simulations: 999
##   Statistic:  4612
##   p-value :  0.001
```

It also includes the Pottoff and Withinghill test of homogeneity. The alternative hypothesis of this test is that the observed cases are distributed following a negative binomial distribution (Bivand, Pebesma, and Gómez-Rubio 2013).

```
pwtest <- pottwhitt.test(Observed~offset(log(Expected)),
                         as(lsoas, "data.frame"),
                         "multinom", 999)
oplus <- sum(lsoas$Observed)
1 - pnorm(pwtest$t0, oplus * (oplus - 1), sqrt(2 * 100 * oplus * (oplus - 1)))
```

```
## [1] 0
```

Detail in those two tests and the used code is available from Bivand, Pebesma, and Gómez-Rubio (2013). In what follows we introduce two of the approaches implemented here for the detection of clusters.

10.5.2 Openshaw's GAM

Stan Openshaw is a retired British geography professor with an interest in geocomputation. We already encountered him earlier, for he wrote one of the seminal papers on the modifiable areal unit problem. In 1987, together with several colleagues, he published his second most highly cited paper introducing **geographical analysis machine** (GAM). This was a technique developed to work with point pattern data with the aim of finding clusters (Openshaw et al. 1987). The machine is essentially an automated algorithm that tries to assess whether "there is an excess of observed points within x km of a specific location" (p. 338). A test statistics is computed for this circular search area, and some form of inference can then be made.

The technique requires a grid and will draw circles centred in the intersection of the cells of this grid. From each point of the grid backdrop, a radial distance is calculated assuming a Poisson distribution. The ratio of events to candidates is counted up and then the test is done. What we are doing is to observe case counts in overlapping circles and trying to identify potential clusters among these, by virtue of running separate significance test for each of the circles individually. The statistically trained reader may already note a problem with this. In a way, this is similar to tests for quadrat counting, only here we have overlapping "quadrats" and we have many more of them (Kulldorff and Nagarwalla 1995).

The `opgam()` function from `DCluster` does this for you. You can either specify the grid passing the name of the object containing this grid using the `thegrid` argument or specify the step size for computing this grid. For this you simply introduce a value for the `step` argument. You also need to specify the search `radius` and can specify an explicit significance level of the test performed with the `alpha` argument. The radius is typically 5 to 10 times the lattice spacing. It is worth mentioning that although one could derive this test using asymptotic theory, `DCluster` uses a bootstrap approach.

```
gam_burglary <- opgam(data = as(lsoas, "data.frame"),
                      radius = 50, step = 10, alpha = 0.002)
```

Once `opgam` has found a solution, we could plot either the circles or the centre of these circles. We can use the following code to do the map, which will be shown in the next section.

```
gam_burglary_sf <- st_as_sf(gam_burglary, coords = c("x", "y"), crs = 27700)
gam_burglary_sf <- st_transform(gam_burglary_sf, crs = 4326)
tmap_mode("plot")
```

```
## tmap mode set to plotting
```

```
map_gam <- tm_shape(burglary_lsoa) + tm_borders(alpha = 0.3) +
  tm_shape(gam_burglary_sf) + tm_dots(size=0.1, col = "red") +
  tm_layout(main.title = "Openshaw's GAM", main.title.size=1.2)
```

One of the problems with this method is that the tests are not independent; very similar clusters (i.e., most of their regions are the same) are tested (Bivand, Pebesma, and Gómez-Rubio 2013). Inference then is a bit compromised and the approach has received criticism over the years for this failure to solve the multiple testing problem: "any Bonferroni type of procedure to adjust for multiple testing is futile due to the extremely large number of dependent tests performed" (Kulldorff and Nagarwalla 1995). For these reasons and others (see Marshall (1991a)), it is considered hard to ask meaningful statistical questions from a GAM. Notwithstanding this, this technique, as noted by Bivand, Pebesma, and Gómez-Rubio (2013), can still be helpful as part of the exploration of your data, as long as you are aware of its limitations.

10.5.3 Kulldorf's Scan Statistic

More popular has been the approach developed by Kulldorff and Nagarwalla (1995). This test could be used for aggregated or point pattern data. It aims to overcome the limitations from previous attempts to assess clusters of disease. Unlike previous methods, the scan statistic constructs circles of varying radii (Kulldorff and Nagarwalla 1995):

> Each of the infinite number of circles thus constructed defines a zone. The zone defined by a circle consists of all individuals in those cells whose centroids lie inside the circle and each zone is uniquely identified by these individuals (p. 802).

The test statistic then compares the relative risk within the circle to the relative risk outside the circles (for more details, see Kulldorff and Nagarwalla (1995) and Kulldorff (1997)).

The scan statistic was implemented by Kulldorff in the SatScan^{TM} software, freely available from www.satscan.org. This software extends the basic idea to various problems, including spatio-temporal clustering. A number of R packages use this scan statistic (smerc, spatstat, SpatialEpi are some of the best known), although the implementation is different across these packages. There is also a package (rsatscan) that can allow us to interface with SatScan^{TM} via R. Here we will continue focusing on DCluster for parsimonious reasons. The vignettes for the other packages provide a useful platform for exploring those other solutions.

For this we return to the opgam() function. There are some key differences in the arguments. The icluster argument, which we didn't specify previously, needs to make clear now we are using the clustering function for Kulldorff and Nagarwalla scan statistic. R is the number of bootstrap permutations to compute. The model argument allows us to specify the model to generate the random observations. Bivand, Pebesma, and Gómez-Rubio (2013) suggest the use of a negative binomial model ("negbin") to account for over-dispersion which may result from unaccounted spatial autocorrelation. And `mle` specifies parameters that are needed for the bootstrap calculation. These can be computed with the `DCluster::calculate.mle()` function, which simply takes as arguments the data and the type of model. When illustrating `opgam` earlier, we used the `step` method to draw the grid; in this case we will explicitly provide a grid based on our data points.

```
mle <- calculate.mle(as(lsoas, "data.frame"), model = "negbin")
thegrid <- as(lsoas, "data.frame")[, c("x", "y")]
kns_results <- opgam(data = as(lsoas, "data.frame"),
                     thegrid = thegrid, alpha = 0.02, iscluster = kn.iscluster,
                     fractpop = 0.05, R = 99, model = "negbin",
                     mle = mle)
```

Once the computation is complete, we are ready to plot the data and to compare the result to those obtained with the GAM.

```
kulldorff_scan <- st_as_sf(kns_results,
                     coords = c("x", "y"),
                     crs = 27700)
kulldorff_scan <- st_transform(kulldorff_scan, crs = 4326)
map_scan <- tm_shape(burglary_lsoa) + tm_borders(alpha = 0.3) +
  tm_shape(kulldorff_scan) + tm_dots(size=0.1, col = "red") +
  tm_layout(main.title = "Kulldorff Scan", main.title.size=1.2)
tmap_arrange(map_gam, map_scan)
```

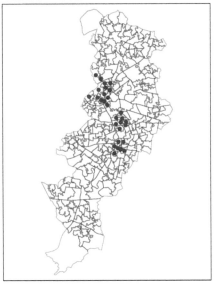

FIGURE 10.9: Difference in results between Openshaw's GAM and Kulldorf scan statistics

Whereas the GAM picks up a series of locations in East and North Manchester known by their high level of concentrated disadvantage, the Kulldorff scan only identifies the clusters around the city centred on areas of student accommodation south of the University of Manchester.

10.6   Exploring near repeat victimisation

Although a good deal of the near repeat victimisation literature focuses on burglary, our Manchester data is not good for illustrating the analysis of near repeat victimisation. For this we need data with more granular time information than the month in which it took place (the only one available from UK open data police portals). So we will return to the burglary data we explored in Chapter 7 when introducing spatial point pattern analysis. Remember, though, that there is a good deal of artificial "revictimisation" in those files for the way that geomasking is introduced. So although we can use that data for illustration purposes, beware the results we will report cannot be trusted as genuine.

```
# Large dataset, so it will take a while
nyc_burg <- get_crime_data(
  cities = "New York", #specifies the city for which you want the data
  years = 2010,        #reads the appropriate year of data
  type = "extended",   #select extended (a fuller set of fields)
  output = "sf") %>%   #Specify you want a sf object with WGS84
  dplyr::filter(offense_type == "residential burglary/breaking & entering" &
         nyc_boro_nm == "MANHATTAN")
manhattan <- st_read("data/manhattan.geojson", quiet=TRUE) %>%
  filter(BoroName == "Manhattan")
manhattan <- st_transform(manhattan, 32118)
nyc_burg <- st_transform(nyc_burg, 32118)
```

The `NearRepeat` package uses the Knox test for space time clustering, which has been around since 1964. This test whether there is a significant cluster within a defined distance and time period. The Knox algorithm counts the pair of points within a specified space and time interval and compares this to the expected number of points. "If many of the cases that are 'close' in time are also 'close' in space or vice versa, then there is space-time interaction" (Kulldorff and Hjalmars 1999). The idea underpinning the test, as highlighted by Mantel (1967), is that "if there is time-space clustering, cases in a cluster will be close both in time and space, while unrelated cases will tend to have a larger average separation in time and space." (p. 209).

The `NearRepeat` package is quite straightforward with just a couple of functions, as we will see. The key function is `NearRepeat()` that takes as inputs the data (a vector with the x coordinates, a vector with the y coordinates, and a vector with the time information, which can be an integer, numeric or date). You then need to specify the time and the spatial intervals. Ratcliffe (2020) indicates that:

> The number of spatial bands is dependent on how far you expect a pattern of near repeats to extend. For example, most environmental criminology research suggests that near

repeat patterns occur for only a few blocks or a few hundred metres at most. Any effects appear to peter out beyond this distance. Adding additional bands beyond any identified effects rarely adds much value; however the program is often most effective when you experiment with various settings. For fairly large data sets, ten spatial bands is often a good starting point, though experimentation is encouraged... Be advised that selecting too many bands or too narrow a bandwidth can reduce the number of observations in each category and potentially limit the findings by creating too many categories with low (or zero) values in the observed or expected matrices

And likewise for the temporal bands:

The number of temporal bands is dependent on how long you expect a pattern of near repeats to extend. For example, the research on repeat victimization suggests that a risk of repeat burglary increases rapidly after an initial burglary, but that this risk dissipates in the months after the incident and the risk returns to the background (normal level for the area) after a few months. Settings of a 30 day temporal bandwidth with 12 temporal bands, or 14 days with 13 temporal bands (to cover a six month period) are common, but experimentation is encouraged.

The Knox test basically uses these temporal and spatial bands to compute the expected number of pairs of events that are close in time and space and compares this with the observed count. Let's first prepare the data. We will extract the key vectors to a data frame.

```
nyc_burg_df <- matrix(unlist(nyc_burg$geometry), ncol = 2, byrow = T)
nyc_burg_df <- as.data.frame(nyc_burg_df)
nyc_burg_df$time <- as.Date(nyc_burg$date_single)
nyc_burg_df <- rename(nyc_burg_df, x = V1, y = V2)
```

We can then try the test. We look at three temporal intervals of seven days each (defined by tds) and four spatial bands (defined by sds), expressed in metres here. By default, this function relies on the Manhattan distances, but we can also set them to Euclidean if we prefer (method = "euclidean"). The p values are computed using a Monte Carlo simulation approach, so nrep sets the number of simulations we desire. The generic plot() function to the generated object will then generate a heat map with the results of the Knox test.

```
set.seed(123)
results_knox <- NearRepeat(x = nyc_burg_df$x,
                y = nyc_burg_df$y,
                time = nyc_burg_df$time,
                sds = c(0, 0.1, 100, 200, 400),
                tds = c(0, 7, 14, 21),
                nrep = 99)
plot(results_knox)
```

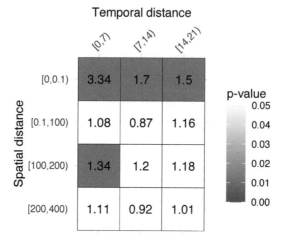

FIGURE 10.10: Results of the Knox test

By default, this will plot the cells with significant p values and the Knox ratios based on the mean across simulations that are higher than 1.2 (that is, the observed number of pairs is 20% greater than expected). This default can be changed to various options (as discussed in the vignette for the package).

This test was commonly used in epidemiology for preliminary analysis where an infectious origin is suspected before criminologists started to use it. It is worthwhile noticing that within the epidemiological literature there are known concerns with this test (see Kulldorff and Hjalmars (1999) and Townsley, Homel, and Chaseling (2003)). One is bias introduced by shifts in the population that, considering the temporal scale typically considered in criminological issues, is not the greatest concern. We also need to worry about arbitrary cut offs and the related problem of multiple comparisons. The near repeat calculator manual suggestion for experimentation around these cut-offs likely exacerbates this problem in the practice of crime analysis. Townsley, Homel, and Chaseling (2003) suggest using past findings reported in the literature to establish the time and space intervals. If there are seasonal effects, this test is not well suited to account for them. Finally, the literature also discusses edge effects. Within the epidemiological literature, several modifications to the Knox test have been proposed, and presumably we will see more of this used by criminologists in the future.

A recent article by Briz-Redón, Martínez-Ruiz, and Montes (2020a) proposes a modification of the traditional Knox test. The classic test assumes that crime risk is homogeneous in time and space, despite the fact that this assumption rarely holds. Their procedures relax this assumption, but have yet to be implemented in an R package.

10.7 Further reading

Hot spots of crime are simply a convenient perceptual construct. As Ned Levine (2013) (chapter 7, p. 1) highlights, *"Hot spots do not exist in reality, but are areas where there is sufficient clustering of certain activities (in this case, crime) such that they get labeled such. There is not a border around these incidents, but a gradient where people draw an imaginary line to indicate the location at which the hot spot starts."* Equally, there is not a unique solution to the identification of hot spots. Different techniques and algorithms will give you different answers. He further emphasises: *"It would be very naive to expect that a single technique can reveal the existence of hot spots in a jurisdiction that are unequivocally clear. In most cases, analysts are not sure why there are hot spots in the first place. Until that is solved, it would be unreasonable to expect a mathematical or statistical routine to solve that problem"* (chapter 7, p. 7).

So, as with most data analysis exercises, one has to try different approaches and use professional judgment to select a particular representation that may work best for a particular use. Equally, we should not reify what we produce and, instead, take the maps as a starting point for trying to understand the underlying patterns that are being revealed. Critically, you want to try several different methods. You will be more persuaded a location is a hot spot if several methods for hot spot analysis point to the same location. Keep in mind as well the points we raised about the problems with police data in earlier chapters. This data presents limitations. There is reporting bias, recording bias, and issues with geocoding quality. Briz-Redón, Martínez-Ruiz, and Montes (2020b), specifically, highlight that the common standard of a 85% match in geocoding is not good enough if the purpose is detecting clusters. What we see in our data may not be what it is, because of the quality of this data.

A more advanced approach at considering space time interactions and exploration of highly clustered even sequences than the one assumed by the Knox test relies on the idea of self-exciting points that are common in seismology. For details on this approach, see Moehler et al. (2011). Kulldorff et al. (2005) also developed a space time permutation scan statistic for detecting disease outbreaks that may be useful within the context of crime analysis (see, for example, Uittenbogaard and Ceccato (2012) use to study spatio-temporal clusters of crime in Stockholm or Cheng and Adepeju (2014) use of the prospective scan statistic in London). Gorr and Lee (2015) discuss an alternative approach to study and respond to chronic and temporary hot spot. Of interest as well is the work by Adepeju, Langton, and Bannister (2021), and the associated R package `Akmedoids`(Adepeju, Langton, and Bannister 2020), for classifying longitudinal trajectories of crime at micro-places.

Also, as noted in previous chapters, we need to acknowledge the underlying network structure constraining the spatial distribution of our crime data. Shiode et al. (2015) developed a hot spot detection method that uses a network-based space-time search window technique (see also Shiode and Shiode (2020) for more recent work). However, the software implementation of this approach has not yet seen the light as of this writing despite the authors mentioning a "GIS plug-in tool" to be in development in their 2015 paper. Also relevant is the work by Briz-Redón, Martínez-Ruiz, and Montes (2019b) that developed an R package (`DRHotNet`) for differential risk hot spots on a linear network. This is a procedure to detect whether specific type of event is over-represented in relation to the other types of events observed (say, for example, burglaries in relation to other crimes). There is a detailed tutorial for this package in Briz-Redón, Martínez-Ruiz, and Montes (2019a).

11

Spatial regression models

11.1 Introduction

In science one of our main concerns is to develop models of the world, models that help us to understand the world a bit better or to predict how things will develop better. Crime analysis also is interested in understanding the processes that may drive crime. This is a key element in the SARA process (*analysis*) that forms part of the problem oriented policing approach (Eck and Spelman 1997). Statistics provides a set of tools that help researchers build and test scientific models, and to develop predictions. Our models can be simple. We can think that unemployment is a factor that may help us to understand why cities differ in their level of violent crime. We could express such a model like this:

FIGURE 11.1: A simple model

Surely we know the world is complex and likely there are other things that may help us to understand why some cities have more crime than others. So, we may want to have tools that allow us to examine such models. Like, for example, the one below:

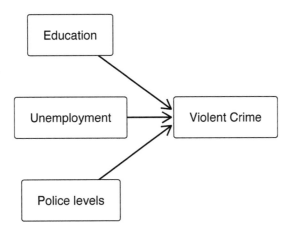

FIGURE 11.2: A slightly less simple model

In this chapter we cover regression analysis for spatial data, a key tool for testing models. Regression is a flexible technique that allows you to "explain" or "predict" a given outcome, variously called your outcome, response or dependent variable, as a function of a number of what is variously called inputs, features or independent, explanatory, or predictive variables. We assume that you have covered regression analysis in previous training. If that is not the case or you want a quick refresher, we suggest you consult *Appendix B: Regression Analysis (A Refresher)* in the website for this book. Our focus in this chapter will be on a series of regression models that are appropriate when your data has a spatial structure.

There is a large and developing field, spatial econometrics, dedicated to developing this kind of models. Traditionally, these models were fit using maximum likelihood, whereas in the last few years we have seen the development of new approaches. There has also been some parallel developments in different disciplines, some of which have emphasised **simultaneous autoregressive models** and others (such as disease mapping) that have emphasised **conditional autoregressive models**. If these terms are not clear, do not worry; we will come back to them later on. This is a practical R focused introductory text and, therefore, we cannot cover all the nuisance associated with some of these differences and the richness of spatial econometrics. Here we will just simply introduce some of the key concepts and more basic models for cross-sectional spatial data, illustrate them through a practical example, and later on provide guidance to more in-depth literature.

We need to load the libraries we will use in this chapter. Although there are various packages that can be used to fit spatial regression models, particularly if you are willing to embrace a Bayesian approach, our focus in this chapter will be on `spatialreg`. This package contains the estimation functions for spatial cross-sectional models that before were shipped as part of `spdep`, `sphet`, and `spse`. There are various ways of thinking about spatial regression, here we will focus on those that rely on frequentist methods and a spatial weight matrix (to articulate the spatial nature of the data).

```
# Packages for handling spatial data and for geospatial carpentry
library(sf)
library(sp)
library(spdep)
# Packages for regression and spatial regression
library(arm)
library(effects)
library(spatialreg)
# Packages for mapping and visualisation
library(tmap)
library(ggplot2)
# Packages with spatial datasets
library(geodaData)
```

Here we will return to the "ncovr" data we have used in previous chapters. This dataset was produced as part of the National Consortium on Violence Research (NCOVR) agenda, a National Science Foundation initiative that was led by Al Blumstein. A piece of work that resulted from this initiative was the paper led by Robert Baller in *Criminology*, for which this data was compiled (Baller et al. 2001). This paper, co-authored by Luc Anselin who is one of the key figures in the field of spatial econometrics, runs and describes spatial regression models of the kind we cover in this chapter with this dataset. It would be helpful, thus, to read that piece in parallel to this chapter.

```
data("ncovr")
```

11.2 Challenges for regression with spatial data

There are a number of issues when it comes to fitting regression with spatial data. We will highlight three: validity, spatial dependence, and spatial heterogeneity. One, validity, is common to regression in general, although with spatial data it acquires particular relevance. The other two (spatial dependence and spatial heterogeneity) are typical of spatial data.

The most important assumption of regression is that of **validity**. The data should be appropriate for the question that you are trying to answer. As noted by Gelman and Hill (2007):

> Optimally, this means that the outcome measure should accurately reflect the phenomenon of interest, the model should include all relevant predictors, and the model should generalize to all cases to which it will be applied... Data used in empirical research rarely meet all (if any) of these criteria precisely. However, keeping these goals in mind can help you be precise about the types of questions you can and cannot answer reliably.

Although validity is a key assumption for regression, whether spatial or not, it is the case that the features of spatial data make it particularly challenging (as discussed in detail by Haining (2003)).

The second challenge is **spatial dependence**. A key assumption of regression analysis is the independence between the observations. That is, a regression model is formally assuming that what happens in area X_i is not in any way related (it is independent of) what happens in area X_{ii}. But if those two areas are adjacent or proximal in geographical space, we know that there is a good chance that this assumption may be violated. Values that are close together in space are unlikely to be independent. Crime levels, for example, tend to be similar in areas that are close to each other. The problem for regression is not so much autocorrelation in the response or dependent variable, but the remaining autocorrelation in the residuals. If the spatial autocorrelation of our outcome y is fully accounted by the spatial structure of the observed predictor variables in our model, we don't really have a problem. On the other hand, when the residuals of your model display spatial autocorrelation, you are violating an assumption of the standard regression model, and you need to find a solution for this.

When positive spatial autocorrelation is observed, using a regression model that assumes independent errors will lead to an underestimation of the uncertainty around your parameters (or overestimation if the autocorrelation is negative). With positive spatial autocorrelation, you will be more likely to engage in a Type I error (concluding the coefficient for a given predictor is not zero when this decision is not justified). Essentially, we have less degrees of freedom, as Haining and Li (2020) note for positive spatial autocorrelation (the most common case in crime analysis and social science):

> Underlying the results... is the fact that 'less information' about the parameter of interest... is contained in a dataset where observations are positively autocorrelated... The

term 'effective'' sample size has been coined and used to measure the information content in a set of autocorrelated data. Compared with the case of N independent observations, if we have N autocorrelated data values then the effective sample size is less than N (how much less depends on how strongly autocorrelated values are). The effective sample size can be thought of as the *equivalent number of independent observations available for estimating the parameter...* It is this reduction in the information content of the data that increases the uncertainty of the parameter estimate... This arises because each observation contains what might call 'overlapping' or 'duplicate' information about other observations (p. 7).

In previous weeks we covered formal tests for spatial autocorrelation, which allow us to test whether this is a feature of our dependent (or predictor) variable. So before we fit a regression model with spatial data, we need to explore the issue of autocorrelation in the attributes of interest, but critically what we need to establish is whether there is spatial autocorrelation in the residuals. In this session, we will see that there are two basic ways of adjusting for spatial autocorrelation: through a spatial lag model or through a spatial error model. Although there are multiple types of spatial regression models (see Anselin and Rey (2014); Bivand, Millo, and Piras (2021); Chi and Zhu (2020); and Beale et al. (2010), specifically for a review of their relative performance), these are the two basic forms that you probably (particularly if your previous statistical training sits in the frequentist tradition) need to understand before you get into other types.

A third challenge for regression models with spatial data is **spatial heterogeneity** (also referred to as **non-stationarity**). We already discussed the concept of spatial homogeneity when introducing the study of point patterns. In that context we defined spatial homogeneity as the homogeneous intensity of the point pattern across the study surface. But spatial homogeneity is a more general concept that applies to other statistical properties. In the context of lattice or area level data, it may refer to the mean value of an attribute (such as crime rate) across areas, but it could also refer to how other variables are related to our outcome of interest across the study region. It could be the case, for example, that some predictors display spatial structure in how they affect the outcome. It could be, for example, that presence of licensed premises to sell alcohol have a different impact on the level of violent crime in different types of communities in our study region. If there is spatial heterogeneity in these relationships, we need to account for it. We could deal with this challenge through some form of data partition, spatial regime, or local regression model (where we allow the regression coefficient to be area specific) such as geographically weighted regression. We will discuss this third challenge to regression in greater detail in the next chapter.

Not so much a challenge but an important limitation is the presence of missing data. Unlike in standard data analysis where you have a sample of n observations and there are a number of approaches to impute missing data, in the context of spatial econometrics, you only have **one** realisation of the data generating mechanism. As such, if there is missing data for some of the areas, the models cannot be estimated; although some solutions have been proposed when the percentage of missing cases is very small (see Floch and LeSaout (2018)).

In general, the development of a spatial regression model requires taking into account these challenges. There are various aspects that building the model will require (Kopczewska 2021): selecting the right model for the data at hand, the estimation method, and the model specification (variables to include).

11.3 Fitting a non-spatial regression model with R

The standard approach traditionally was to start with a non-spatial linear regression model and then to test whether or not this baseline model needs to be extended with spatial effects (Elhorst 2014). Given that we need to check autocorrelation in residuals, part of the workflow for modelling through regression spatial data will involve fitting a non-spatial regression model. In order to fit the model we use the `stats::lm()` function using the formula specification (`Y ~ X`). Typically you want to store your regression model in a "variable", let's call it "fit1_90". Here we model homicide rate in the 1990s (`HR90`) using, to start with, just two variables of interest: `RD90` - Resource Deprivation/Affluence, and `SOUTH` - Counties in the southern region scored 1.

```r
ncovr$SOUTH_f <- as.factor(ncovr$SOUTH) # specify this is a categorical variable
fit1_90 <- lm(HR90 ~ RD90 + SOUTH_f, data=ncovr)
```

You will see in your R Studio global environment space that there is a new object called `fit1_90` with 13 elements on it. We can get a sense for what this object is and includes using the functions we introduced in previous weeks:

```r
class(fit1_90)
attributes(fit1_90)
```

R is telling us that this is an object of class `lm` and that it includes a number of attributes. One of the beauties of R is that you are producing all the results from running the model, putting them in an object, and then giving you the opportunity for using them later on. If you want to simply see the basic results from running the model, you can use the standard `summary()` function.

```r
summary(fit1_90)
```

```
##
## Call:
## lm(formula = HR90 ~ RD90 + SOUTH_f, data = ncovr)
##
## Residuals:
##    Min     1Q Median     3Q    Max
## -16.48  -3.00  -0.58   2.22  68.15
##
## Coefficients:
##              Estimate Std. Error t value Pr(>|t|)
## (Intercept)     4.727      0.139    33.9   <2e-16 ***
## RD90            2.965      0.111    26.8   <2e-16 ***
## SOUTH_f1        3.181      0.222    14.3   <2e-16 ***
## ---
## Signif. codes:
## 0 '***' 0.001 '**' 0.01 '*' 0.05 '.' 0.1 ' ' 1
##
```

```
## Residual standard error: 5.3 on 3082 degrees of freedom
## Multiple R-squared:  0.365,   Adjusted R-squared:  0.364
## F-statistic:  884 on 2 and 3082 DF,  p-value: <2e-16
```

The coefficients suggest that, according to our fitted model, when comparing two counties with the same level of resource deprivation, the homicide rate on average will be about 3 points higher in the Southern counties than in the Northern counties. And when comparing counties "adjusting" for their location, the homicide rate will be also around 3 points higher for every one unit increase in the measure of resource deprivation. Although it is tempting to talk of the "effect" of these variables in the outcome (homicide rate), this can mislead you and your reader about the capacity of your analysis to establish causality, so it is safer to interpret regression coefficients as comparisons (Gelman, Hill, and Vehtari 2020).

With more than one input, you need to ask yourself whether all of the regression coefficients are zero. This hypothesis is tested with an F test. Again we are assuming the residuals are normally distributed, though with large samples the F statistics approximates the F distribution. You see the F test printed at the bottom of the summary output and the associated p value, which in this case is way below the conventional .05 that we use to declare statistical significance and reject the null hypothesis. At least one of our inputs must be related to our response variable. Notice that the table printed also reports a t test for each of the predictors. These are testing whether each of these predictors is associated with the response variable when adjusting for the other variables in the model. They report the "partial effect of adding that variable to the model" (James et al. 2013, 77). In this case we can see that all inputs seem to be significantly associated with the output.

The interpretation or regression coefficients is sensitive to the scale of measurement of the predictors. This means one cannot compare the magnitude of the coefficients to compare the relevance of variables to predict the response variable. One way of dealing with this is by rescaling the input variables. A common method involves subtracting the mean and dividing by the standard deviation of each numerical input. The coefficient in these models is the expected difference in the response variable, comparing units that differ by one standard deviation in the predictor while adjusting for other predictors in the model. Instead, Gelman (2008) has proposed dividing each numeric variables *by two times its standard deviation*, so that the generic comparison is with inputs equal to plus/minus one standard deviation. As Gelman explains, the resulting coefficients are then comparable to untransformed binary predictors. The implementation of this approach in the arm package subtracts the mean of each binary input, while it subtracts the mean and divides by two standard deviations for every numeric input. The way we would obtain these rescaled inputs uses the standardize() function of the arm package, which takes as an argument the name of the stored fit model.

```
standardize(fit1_90)
```

```
##
## Call:
## lm(formula = HR90 ~ z.RD90 + c.SOUTH_f, data = ncovr)
##
## Coefficients:
## (Intercept)      z.RD90    c.SOUTH_f
##        6.18        5.93         3.18
```

Notice the main change affects the numerical predictors. The unstandardised coefficients are influenced by the degree of variability in your predictors, which means that typically they will be larger for your binary inputs. With unstandardised coefficients, you are comparing complete change in one variable (whether one is a Southern county or not) with one-unit changes in your numerical variable, which may not amount to much change. So, by putting in a comparable scale, you avoid this problem. Standardising in the way described here will help you to make fairer comparisons. These standardised coefficients are comparable in a way that the unstandardised coefficients are not. We can now see what inputs have a comparatively stronger effect. It is very important to realise, though, that one **should not** compare standardised coefficients *across different models*.

In the social sciences there is a great interest in what are called conditional hypotheses or interactions. Many of our theories do not assume simply **additive effects** but **multiplicative effects**. For example, you may think that a particular crime prevention programme may work in some environments but not in others. The interest in this kind of conditional hypothesis is growing. One of the assumptions of the regression model is that the relationship between the response variable and your predictors is additive. That is, if you have two predictors x_1 and x_2, regression assumes that the effect of $x1$ on y is the same at all levels of x_2. If that is not the case, you are then violating one of the assumptions of regression. This is in fact one of the most important assumptions of regression, even if researchers often overlook it.

One way of extending our model to accommodate for interaction effects is to add more terms to our model, a third predictor x_3, where x_3 is simply the product of multiplying x_1 by x_2. Notice we keep a term for each of the **main effects** (the original predictors) as well as a new term for the interaction effect. "Analysts should include all constitutive terms when specifying multiplicative interaction models except in very rare circumstances" (Brambor, Clark, and Golder 2006, 66).

How do we do this in R? One way is to use the following notation in the formula argument. Notice how we have added a third term RD90:SOUTH_f, which is asking R to test the conditional hypothesis that resource deprivation may have a different impact on homicide for southern and northern counties.

```
fit2_90 <- lm(HR90 ~ RD90 + SOUTH_f + RD90:SOUTH_f , data=ncovr)
# which is equivalent to:
# fit_2 <- lm(HR90 ~ RD90 * SOUTH_f , data=ncovr)
summary(fit2_90)


##
## Call:
## lm(formula = HR90 ~ RD90 + SOUTH_f + RD90:SOUTH_f, data = ncovr)
##
## Residuals:
##    Min     1Q Median     3Q    Max
## -17.05  -3.00  -0.57   2.23  68.14
##
## Coefficients:
##                Estimate Std. Error t value Pr(>|t|)
## (Intercept)       4.548      0.159   28.68   <2e-16 ***
## RD90              2.581      0.196   13.15   <2e-16 ***
## SOUTH_f1          3.261      0.225   14.51   <2e-16 ***
## RD90:SOUTH_f1     0.562      0.238    2.37    0.018 *
## ---
```

```
## Signif. codes:
## 0 '***' 0.001 '**' 0.01 '*' 0.05 '.' 0.1 ' ' 1
##
## Residual standard error: 5.29 on 3081 degrees of freedom
## Multiple R-squared:  0.366,  Adjusted R-squared:  0.365
## F-statistic:  592 on 3 and 3081 DF,  p-value: <2e-16
```

You see here that essentially you have only two inputs (resource deprivation and south) but several regression coefficients. Gelman and Hill (2007) suggest reserving the term input for the variables encoding the information and to use the term predictor to refer to each of the terms in the model. So here we have two inputs and three predictors (one for SOUTH, another for resource deprivation, and a final one for the interaction effect). In this case the test for the interaction effect is significant, which suggests there may be such an interaction. Let's visualise the results with the `effects` package:

```
plot(allEffects(fit2_90), ask=FALSE)
```

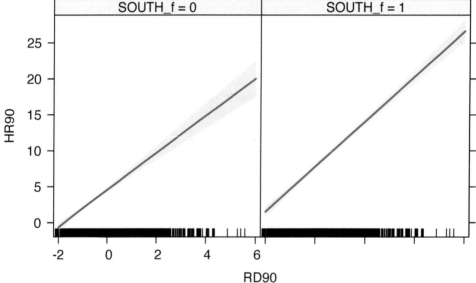

FIGURE 11.3: Effect plots showing different relationship between homicide rate and resource deprivation for counties in the South (South = 1) and the North (South =0)

Notice that essentially what we are doing is running two regression lines and testing whether the slope is different for the two groups. The intercept is different; we know that Southern counties are more violent, but what we are testing here is whether the level of homicide goes up in a steeper fashion (and in the same direction) for one or the other group as the level of resource deprivation goes up. We see that's the case here. The estimated lines are almost parallel, but the slope is a bit steeper in the Southern counties. In Southern counties, resource deprivation seems to have more of an impact on homicide than in Northern counties. This is related to one of the issues we raised earlier regarding spatial regression, that of hetereogeneity. Here what we see is that there is some evidence of non-stationarity in the relationship between our predictors and our outcome.

A word of warning about interactions: the moment you introduce an interaction effec,t the meaning of the coefficients for the other predictors changes (what is often referred to as the "main effects" as opposed to the interaction effect). You cannot retain the interpretation we introduced earlier. Now, for example, the coefficient for the SOUTH variable relates to the marginal effect of this variable when RD90 equals zero. The typical table of results helps you to understand whether the effects are significant but offers little of interest that will help you to meaningfully interpret what the effects are. For this, is better to use some of the graphical displays we have covered.

Essentially what happens is that the regression coefficients that get printed are interpretable only for certain groups. So now:

- The intercept still represents the predicted score of homicide for Southern counties and have a score of 0 in resource deprivation (as before).

- The coefficient of *SOUTH_f1* now can be thought of as the difference between the predicted score of homicide rate for Northern counties *that have a score of 0 in resource deprivation* and Southern counties *that have a score of 0 in resource deprivation*.

- The coefficient of *RD90* now becomes the comparison of mean homicide rate *for Southern* counties that differ by one point in resource deprivation.

- The coefficient for the interaction term represents the difference in the slope for *RD90* comparing Southern and Northern counties, the difference in the slope of the two lines that we visualised above.

Models with interaction terms are too often misinterpreted. We strongly recommend you read Brambor, Clark, and Golder (2006) to understand some of the issues involved.

11.4 Looking at the residuals and testing for spatial autocorrelation in regression

Residuals measure the distance between our observed Y values, Y, and the predicted Y values, $\hat{Y}$. So in essence they are deviations of observed reality from your model. Your regression line (or hyperplane in the case of multiple regression) is optimised to be the one that best represents your data if the assumptions of the model are met. Residuals are very helpful to diagnose, then, whether your model is a good representation of reality or not. Most diagnostics of the assumptions for regression rely on exploring the residuals.

Now that we have fitted the model, we can extract the residuals. If you look at the "fit_1" object in your RStudio environment or if you run the str() function to look inside this object, you will see that this object is a list with different elements, one of which is the residuals. An element of this object then includes the residual for each of your observations (the difference between the observed value and the value predicted by your model). We can extract the residuals using the residuals() function and add them to our spatial data set. Let's fit a model with all the key predictors available in the dataset and inspect the residuals.

```
# write model into object to save typing later
eq1_90 <- HR90 ~ RD90 + PS90 + MA90 + DV90 + UE90 + SOUTH_f
fit3_90 <- lm(eq1_90, data=ncovr)
summary(fit3_90)
```

```
##
## Call:
## lm(formula = eq1_90, data = ncovr)
##
## Residuals:
##     Min      1Q  Median      3Q     Max
## -17.64   -2.61   -0.70    1.65   68.51
##
## Coefficients:
##               Estimate Std. Error t value Pr(>|t|)
## (Intercept)     6.5165     1.0240    6.36  2.3e-10 ***
## RD90            3.8723     0.1427   27.13  < 2e-16 ***
## PS90            1.3529     0.1003   13.49  < 2e-16 ***
## MA90           -0.1011     0.0274   -3.69  0.00023 ***
## DV90            0.5829     0.0545   10.69  < 2e-16 ***
## UE90           -0.3059     0.0409   -7.47  1.0e-13 ***
## SOUTH_f1        2.1941     0.2205    9.95  < 2e-16 ***
## ---
## Signif. codes:
## 0 '***' 0.001 '**' 0.01 '*' 0.05 '.' 0.1 ' ' 1
##
## Residual standard error: 4.99 on 3078 degrees of freedom
## Multiple R-squared:  0.436,  Adjusted R-squared:  0.435
## F-statistic:  397 on 6 and 3078 DF,  p-value: <2e-16
```

```
ncovr$res_fit3 <- residuals(fit3_90)
```

If you now look at the dataset, you will see that there is a new variable with the residuals. In those cases where the residual is negative, this is telling us that the observed value is lower than the predicted (that is, our model is *overpredicting* the level of homicide for that observation); when the residual is positive, the observed value is higher than the predicted (that is, our model is *underpredicting* the level of homicide for that observation).

We could also extract the predicted values if we wanted. We would use the `fitted()` function.

```
ncovr$fitted_fit3 <- fitted(fit3_90)
```

Now look, for example, at the second county in the dataset: the county of Ferry in Washington. It had a homicide rate in 1990 of 15.89. This is the observed value. If we look at the new column we have created ("fitted_fit1"), our model predicts a homicide rate of 2.42. That is, knowing the level of resource deprivation and all the other explanatory variables (and assuming they are related in an additive manner to the outcome), we are predicting a homicide rate of 2.42. Now, this is lower than the observed value; so our model is underpredicting the level of homicide in this case. If you observed the residual, you will see that it has a value of 11.56, which is simply the difference between the observed (15.89) and the predicted value (2.42). This is normal. The world is too complex to be encapsulated by our models, and their predictions will depart from the observed world.

With spatial data one useful thing to do is to look at any spatial patterning in the distribution of the residuals. Notice that the residuals are the difference between the observed values for homicide and the predicted values for homicide, so you want your residual to not display any spatial patterning. If, on the other hand, your model displays a patterning in the areas of the study region where it predicts badly, then you may have a problem. This is telling you that your model is not a good representation of the social phenomena you are studying across the full study area: there is systematically more distortion in some areas than in others.

We are going to produce a choropleth map for the residuals, but we will use a common classification method we haven't covered yet: standard deviations. Standard deviation is a statistical technique type of map based on how much the data differs from the mean. You measure the mean and standard deviation for your data. Then, each standard deviation becomes a class in your choropleth maps. In order to do that, we will compute the mean and the standard deviation for the variable we want to plot and break the variable according to these values. The following code creates a new variable in which we will express the residuals in terms of standard deviations away from the mean. So, for each observation, we substract the mean and divide by the standard deviation. Remember, this is exactly what the scale function does:

```
# because scale is made for matrices, we just need to get the first column using [,1]
ncovr$sd_breaks <- scale(ncovr$res_fit3)[,1]
# this is equal to (ncovr$res_fit1 - mean(ncovr$res_fit1)) / sd(ncovr$res_fit1)
summary(ncovr$sd_breaks)
```

```
##    Min. 1st Qu.  Median    Mean 3rd Qu.    Max.
## -3.537  -0.524  -0.140   0.000   0.331  13.741
```

Next we use a new style, fixed, within the tm_fill function. When we break the variable into classes using the fixed argument we need to specify the boundaries of the classes. We do this using the breaks argument. In this case we are going to ask R to create 7 classes based on standard deviations away from the mean. Remember that a value of 1 would be 1 standard deviation higher than the mean, and -1 respectively lower. If we assume normal distribution, then 68% of all counties should lie within the middle band from -1 to +1.

```
my_breaks <- c(-14,-3,-2,-1,1,2,3,14)
tm_shape(ncovr) +
  tm_fill("sd_breaks", title = "Residuals", style = "fixed",
          breaks = my_breaks, palette = "-RdBu") +
  tm_borders(alpha = 0.1) +
  tm_layout(main.title = "Residuals", main.title.size = 0.7 ,
            legend.position = c("right", "bottom"), legend.title.size = 0.8)
```

Residuals

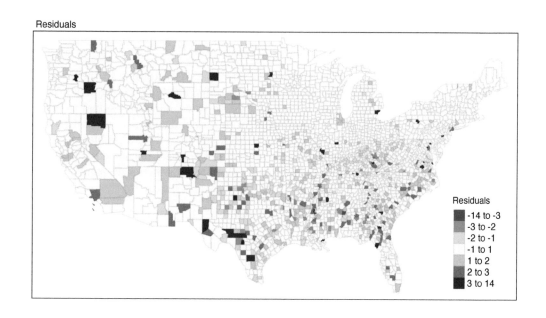

Residuals

- ■ -14 to -3
- ■ -3 to -2
- ☐ -2 to -1
- ☐ -1 to 1
- ☐ 1 to 2
- ■ 2 to 3
- ■ 3 to 14

FIGURE 11.4: Map of regression model residuals

Notice the spatial patterning of areas of over-prediction (negative residuals, or blue tones) and under-prediction (positive residuals, or reddish tones). This visual inspection of the residuals is suggesting that spatial autocorrelation in the residuals may be present here. This, however, would require a more formal test.

Remember from Chapter 9 that in order to do this first we need to turn our sf object into a sp class object and then create the spatial weight matrix. We cannot examine spatial dependence without making assumptions about the structure of the spatial relations among the units in our study. For this illustration we are going to follow Baller et al. (2001) and use 10 nearest neighbours to define the matrix to codify these spatial relations. If the code below and what it does is not clear to you, revise the notes from Chapter 9, when we first introduced it.

```
# We coerce the sf object into a new sp object
ncovr_sp <- as(ncovr, "Spatial")
# Then we create a list of neighbours using 10 nearest neighbours criteria
xy <- coordinates(ncovr_sp)
nb_k10 <- knn2nb(knearneigh(xy, k = 10))
rwm <- nb2mat(nb_k10, style='W')
rwm <- mat2listw(rwm, style='W')
```

We obtain Moran's test for regression residuals using the function lm.morantest() as below. It is important to realize that Moran's I test statistic for residual spatial autocorrelation takes into account the fact that the variable under consideration is a residual, computed from a regression. The usual Moran's I test statistic does not. It is therefore incorrect to simply apply a Moran's I test to the residuals from the regression without correcting for the fact that these are residuals.

```
lm.morantest(fit3_90, rwm, alternative="two.sided")
```

```
##
##   Global Moran I for regression residuals
##
## data:
## model: lm(formula = eq1_90, data = ncovr)
## weights: rwm
##
## Moran I statistic standard deviate = 12, p-value
## <2e-16
## alternative hypothesis: two.sided
## sample estimates:
## Observed Moran I       Expectation          Variance
##         9.544e-02        -1.382e-03        6.029e-05
```

You will notice that we obtain a statistically significant value for Moran's I. If we diagnose that spatial autocorrelation is an issue, that is, that the errors (residuals) are related systematically among themselves, then we likely have a problem, and we may need to use a more appropriate approach: a spatial regression model. Before going down that road, it is important to check this autocorrelation is not the result of an incorrectly specified model. It may appear when estimating a non-linear relationship with a linear model. In this scenario, Kopczewska (2021) suggests checking the remainders from the OLS estimated model on variable logarithms. It is also worth considering it may be the result of other forms of model mis-specification, such as omission of covariates or wrongful inclusion of predictors.

11.5 Choosing a spatial regression model

11.5.1 Basic spatial models

If the test is significant (as in this case), then we possibly need to think of a more suitable model to represent our data: a spatial regression model. Remember spatial dependence means that (more typically) there will be areas of spatial clustering for the residuals in our regression model. So our predicted line (or hyperplane) will systematically under-predict or over-predict in areas that are close to each other. That's not good. We want a better model that does not display any spatial clustering in the residuals.

There are two traditional general ways of incorporating spatial dependence in a regression model, through what we called a **spatial error model** or by means of a **spatially lagged model**. The Moran test doesn't provide information to help us choose, which is why Anselin and Bera (1998) introduced the **Lagrange multiplier tests**, which we will examine shortly.

The difference between these two models is both technical and conceptual. The **spatial error model** treats the spatial autocorrelation as a nuisance that needs to be dealt with. It includes a spatial lag of the error in the model (lambda in the R output). A spatial error model basically implies that the:

> spatial dependence observed in our data does not reflect a truly spatial process, but merely the geographical clustering of the sources of the behaviour of interest. For example, citizens in adjoining neighbohoods may favour the same (political) candidate not because they talk to their neighbours, but because citizens with similar incomes tend to cluster geographically, and income also predicts vote choice. Such spatial dependence can be termed attributional dependence (Darmofal 2015, 4).

In the example we are using, the spatial error model "evaluates the extent to which the clustering of homicide rates not explained by **measured independent variables** can be accounted for with reference to the clustering of error terms. In this sense, **it captures the spatial influence of unmeasured independent variables**" (Baller et al. 2001, 567, the emphasis is ours). As Kopczewska (2021) suggests, this spatial effect allows us to model "invisible or hardly measurable supra-regional characteristics, occurring more broadly than regional boundaries (e.g., cultural variables, soil fertility, weather, pollution, etc.)" (p. 218).

The **spatially lagged model** or **spatially autoregressive model**, on the other hand, incorporates spatial dependence by adding a "spatially lagged" variable y on the right-hand side of our regression equation (this will be denoted as rho in the R output). Its distinctive characteristic is that it includes a spatially lagged measure of our outcome as an endogenous explanatory factor (a variable that measures the value of our outcome of interest, say crime rate, in the neighbouring areas, as defined by our spatial weight matrix). It is basically explicitly saying that the value of y in the neighbouring areas of observation n_i is an important predictor of y_i on each individual area n_i. This is one way of saying that the spatial dependence may be produced by a spatial process such as the diffusion of behaviour between neighbouring units:

> If so the behaviour is likely to be highly social in nature, and understanding the interactions between interdependent units is critical to understanding the behaviour in question. For example, citizens may discuss politics across adjoining neighbours such that an increase in support for a candidate in one neighbourhood directly leads to an increase in support for the candidate in adjoining neighbourhoods (Darmofal 2015, 4).

So, in our example, what we are saying is that homicide rates in one county may be affecting the level of homicide in nearby counties (however you choose to define "nearby").

As noted by Baller et al. (2001):

> It is important to recognize that these models for spatial lag and spatial error processes are designed to yield indirect evidence of diffusion in cross-sectional data. However, any diffusion process ultimately requires "vectors of transmission," i.e., identifiable mechanisms through which events in a given place at a given time influence events in another place at a later time. The spatial lag model, as such, is not able to discover these mechanisms. Rather, it depicts a spatial imprint at a given instant of time that would be expected to emerge if the phenomenon under investigation were to be characterized by a diffusion process. The observation of spatial effects thus indicates that further inquiry into diffusion is warranted, whereas the failure to observe such effects implies that such inquiry is likely to be unfruitful (p.567).

A third basic model is the one that allows for autocorrelation in the predictors (**Durbin factor**). Here we allow for exogenous interaction effects, where the dependent variable of a given area is influenced by independent explanatory variables of other areas. In this case we include in the model a spatial lag of the predictors. These, however, have been less commonly used in practice.

These three models only have one spatial component, but this could be combined. In the last 15 years, interest in models with more than one spatial component has grown. You could have models with two spatial components and even consider together the three ways of spatial interaction (in the error, in the predictors, and in the dependent variable). Elhorst (2010) classified the different models according to the spatial interactions included.

TABLE 11.1: Models classified according to the spatial interactions included

Models	Features
Manski (GNS)	Includes spatial lag of dependent and explatory variables, and autocorrelated errors
Kelejian/Prucha (SAC or SARAR)	Includes spatial lag of dependent model and autocorrelated errors, often call a spatial autoregressive-autoregressive model
Spatial Durbin (SDM)	Allows for spatial lag of dependent variable and explanatory variables
Spatial Durbin Error (SDEM)	Allows for lag of explanatory variables and autocorrelated errors
Spatial Lag (SAR)	The traditional spatial lag of the dependent variable model, orspatial autoregressive model
Spatial Error (SEM)	The traditional model with autocorrelated errors

In practice, issues with overspecification mean you rarely find applications that allow for interaction at the three levels (the so-called **Manski model** or **general nesting model** (GNS)) simultaneously. But it still useful for testing the right specification.

11.5.2 Lagrange multipliers: the bottom-up approach to select a model

Moran's I test statistic has high power against a range of spatial alternatives. However, it does not provide much help in terms of which alternative model would be most appropriate. Until a few years ago, the main solution was to look at a series of tests that were developed by Anselin and Bera (1998) for this purpose: the Lagrange multipliers. The Lagrange multiplier test statistics do allow a distinction between spatial error models and spatial lag models, which was the key concern before interest developed in models with more than one type of spatial effect. The Lagrange multipliers also allow to evaluate if the Kelejian/Prucha (SAC) model is appropriate. This approach, starting with a baseline OLS model, tests residuals, runs Lagrange multipliers, and then selects the SAC or SEM model. It is referred to as the *bottom-up* or *specific to general* approach and was the dominant workflow until recently.

Both Lagrange multiplier tests (for the error and the lagged models, `LMerr` and `LMlag` respectively), as well as their robust forms (`RLMerr` and `RLMLag`, also respectively) are included in the `lm.LMtests()` function. Again, a regression object and a spatial `listw` object must be passed as arguments. In addition, the tests must be specified as a character vector (the default is only `LMerror`), using the `c( )` operator (concatenate), as illustrated below. Alternatively, one could ask to display all tests with `test="all"`. This would also run a test for the so-called SARMA model, but Anselin advises against its use. For sparser reporting, we may wrap this function within a summary() function.

```
summary(lm.LMtests(fit3_90, rwm, test = c("LMerr","LMlag","RLMerr","RLMlag")))
```

```
##   Lagrange multiplier diagnostics for spatial
##   dependence
## data:
## model: lm(formula = eq1_90, data = ncovr)
## weights: rwm
##
##          statistic parameter p.value
## LMerr    148.80            1 < 2e-16 ***
## LMlag    116.36            1 < 2e-16 ***
## RLMerr    35.41            1 2.7e-09 ***
## RLMlag     2.97            1   0.085 .
## ---
## Signif. codes:
## 0 '***' 0.001 '**' 0.01 '*' 0.05 '.' 0.1 ' ' 1
```

How do we interpret the Lagrange multipliers? The interpretation of these tests focuses on their significance level. First we look at the standard multipliers (LMerr and LMlag). If both are below the .05 level, like here, this means we need to have a look at the robust version of these tests (Robust LM).

If the non-robust version is not significant, the mathematical properties of the robust tests may not hold, so we don't look at them in those scenarios. It is fairly common to find that both the lag (LMlag) and the error (LMerr) non-robust LM are significant. If only one of them were, then it is problem solved. We would choose a spatial lag or a spatial error model according to this (i.e., if the lag LM was significant and the error LM was not, we would run a spatial lag model or vice versa). Here we see that both are significant, thus, we need to inspect their robust versions.

When we look at the robust Lagrange multipliers (RLMlag and RLMerr) here, we can see only the Lagrange for the error model is significant, which suggest we need to fit a spatial error model. But there may be occasions when both are as well significant. What do we do then? Luc Anselin (2008) proposes the following criteria:

> When both LM test statistics reject the null hypothesis, proceed to the bottom part of the graph and consider the Robust forms of the test statistics. Typically, only one of them will be significant, or one will be orders of magnitude more significant than the other (e.g., p < 0.00000 compared to p < 0.03). In that case the decision is simple: estimate the spatial regression model matching the (most) significant "robust" statistic. In the rare instance that both would be highly significant, go with the model with the largest value for the test statistic. However, in this situation, some caution is needed, since there may be other sources of misspecification. One obvious action to take is to consider the results for different spatial weight and/or change the basic (i.e., not the spatial part) specification of the model. There are also rare instances where neither of the Robust LM test statistics are significant. In those cases, more serious specification problems are likely present and those should be addressed first (p.199-200).

By other specification errors, Anselin refers to problems with some of the other assumptions of regression that you should be familiar with.

As noted, this way of selecting the "correct" model (the *bottom-up approach*), using the Lagrange multipliers, was the most popular until 15 years ago or so. It was attractive because it relied on observing the residuals in the non-spatial model, it was computationally very efficient, and when the correct model was the spatial error or the spatial lag model, simulation studies had shown it was the most effective approach (Floch and LeSaout 2018). Later we will explore other approaches. Also, since the spatial error and the spatial lag models are not nested, you could not compare them using likelihood ratio tests (Bivand, Millo, and Piras 2021).

11.6 Fitting and interpreting a spatial error model

We just saw that for the case of homicide in the 90s, the bottom-up approach suggested that the spatial error model was more appropriate when using our particular definition of the spatial weight matrix. In this case then, we can run a spatial error model. Maximum likelihood estimation of the spatial error model is implemented in the `spatialreg::errorsarlm()` function. The formula, data set and a `listw` spatial weights object must be specified, as illustrated below. We are still using the 10 nearest neighbours definition. When thinking about the spatial weight matrix to use in spatial regression analysis, it is worth considering T. Smith (2009) suggestions. His work showed that highly connected matrices may result in downward bias for the coefficients when using maximum likelihood.

```
fit3_90_sem <- errorsarlm(eq1_90, data=ncovr, rwm)
summary(fit3_90_sem)
```

```
##
## Call:
## errorsarlm(formula = eq1_90, data = ncovr, listw = rwm)
##
## Residuals:
##        Min         1Q     Median         3Q        Max
## -15.84281   -2.49313   -0.69615    1.60685   68.82483
##
## Type: error
## Coefficients: (asymptotic standard errors)
##                Estimate Std. Error z value  Pr(>|z|)
## (Intercept)    3.820725   1.147279  3.3302 0.0008677
## RD90           3.753935   0.159813 23.4895 < 2.2e-16
## PS90           1.245862   0.119091 10.4615 < 2.2e-16
## MA90          -0.044778   0.030298 -1.4779 0.1394354
## DV90           0.588880   0.062825  9.3734 < 2.2e-16
## UE90          -0.195440   0.046141 -4.2357 2.279e-05
## SOUTH_f1       2.126187   0.297569  7.1452 8.988e-13
##
## Lambda: 0.3661, LR test value: 113.9, p-value: < 2.22e-16
```

```
## Asymptotic standard error: 0.03254
##       z-value: 11.25, p-value: < 2.22e-16
## Wald statistic: 126.6, p-value: < 2.22e-16
##
## Log likelihood: -9276 for error model
## ML residual variance (sigma squared): 23.62, (sigma: 4.86)
## Number of observations: 3085
## Number of parameters estimated: 9
## AIC: 18571, (AIC for lm: 18683)
```

The spatial lag model is probably the most common specification and may be the most generally useful way to think about spatial dependence. But we can also introduce spatial dependence to the model through the error term in our regression equation. Whereas the spatial lag model sees the spatial dependence as substantively meaningful (in that y, say homicide, in county i is influenced by homicide in its neighbours), the spatial error model simply treats the spatial dependence as capturing the effect of unobserved variables. It is beyond the scope of this introductory course to cover the mathematical details and justification of this procedure, though you can use the suggested reading (particularly the highly accessible Ward and Gleditsch (2008), book or the more recent Darmofal (2015)) or some of the other materials we discussed at the end of the chapter.

How do you interpret these results? First, you need to look at the general measures of fit of the model. I know what you are thinking. Look at the R square, and compare them, right? Well, don't. This R square is not a real R square, but a pseudo-R square and is not comparable to the one we obtain from the OLS regression model. Instead we can look at the Akaike Information Criterion (AIC). We can see the AIC is better for the spatial model (18571) than for the non-spatial model (18683).

In this case, you can compare the regression coefficients with those from the OLS model, since we don't have a spatial lag capturing some of their effect. Notice how one of the most notable differences is the fact that median age is no longer significant in the new model.

We are using here the 10 nearest neighbours (following the original analysis of this data), but in real practice you would need to explore whether this is the best definition and one that makes theoretical sense. Various authors, such as Chi and Zhu (2020), suggest that you would want to run the models with several specifications of the spatial weight matrix and find out how robust the findings are to various specifications. They even propose a particular method to pick up the "best" specification of the weight matrix. Kopczewska (2021) provides R code and an example of checking for sensitivity of results to various specifications of the matrix.

11.7   Fitting and interpreting a spatially lagged model

We saw that in the previous case a spatial error model was perhaps more appropriate, but we also saw earlier in one of the simpler models that there was an interaction between the variable indicating whether the county is in the South and one of the explanatory variables. This is suggestive of the kind of spatial heterogeneity we were discussing a the outset of the chapter. When the relationship of the explanatory variables is not constant across the study area, we need to take this into account. Fitting the same model for the whole country may not be appropriate and there are so many interaction terms you can include before things get difficult to interpret. A fairly crude approach, instead of adding multiple interaction terms, is to partition the data. We could, in this example, partition the data into Southern and Northern counties and run separate models for this using two subsets of the data. For the next illustration, we are going to focus on the Southern countries and in a different decade, in which the lag model (as we will see) is more appropriate.

```
# Let's look at southern counties
ncovr_s_sf <- subset(ncovr, SOUTH == 1)
# We coerce the sf object into a new sp object
ncovr_s_sp <- as(ncovr_s_sf, "Spatial")
# Then we create a list of neighbours using 10 nearest neighbours criteria
xy_s <- coordinates(ncovr_s_sp)
nb_s_k10 <- knn2nb(knearneigh(xy_s, k = 10))
rwm_s <- nb2mat(nb_s_k10, style='W')
rwm_s <- mat2listw(rwm_s, style='W')
# We create new equation, this time excluding the southern dummy
eq2_70_S <- HR70 ~ RD70 + PS70 + MA70 + DV70 + UE70
fit4_70 <- lm(eq2_70_S, data=ncovr_s_sf)
summary(lm.LMtests(fit4_70, rwm_s, test = c("LMerr","LMlag","RLMerr","RLMlag")))
```

```
##  Lagrange multiplier diagnostics for spatial
##  dependence
## data:
## model: lm(formula = eq2_70_S, data = ncovr_s_sf)
## weights: rwm_s
##
##          statistic parameter p.value
## LMerr    1.47e+02          1  <2e-16 ***
## LMlag    1.75e+02          1  <2e-16 ***
## RLMerr   1.55e-03          1    0.97
## RLMlag   2.86e+01          1   9e-08 ***
## ---
## Signif. codes:
## 0 '***' 0.001 '**' 0.01 '*' 0.05 '.' 0.1 ' ' 1
```

Maximum Likelihood (ML) estimation of the spatial lag model is carried out with the spatreg::lagsarlm() function. The required arguments are again a regression "formula", a data set and a listw spatial weights object. The default method uses Ord's eigenvalue decomposition of the spatial weights matrix.

```
fit4_70_sar <- lagsarlm(eq2_70_S, data=ncovr_s_sf, rwm_s)
summary(fit4_70_sar)
```

```
##
## Call:
## lagsarlm(formula = eq2_70_S, data = ncovr_s_sf, listw = rwm_s)
##
## Residuals:
```

```
##         Min        1Q    Median        3Q       Max
## -17.13387  -4.09583  -0.95433   2.85297  65.93842
##
## Type: lag
## Coefficients: (asymptotic standard errors)
##               Estimate Std. Error z value  Pr(>|z|)
## (Intercept)   7.437000   1.413763  5.2604 1.437e-07
## RD70          2.247458   0.244104  9.2070 < 2.2e-16
## PS70          0.603290   0.256056  2.3561  0.018469
## MA70         -0.091496   0.041144 -2.2238  0.026160
## DV70          0.633473   0.211198  2.9994  0.002705
## UE70         -0.485272   0.099684 -4.8681 1.127e-06
##
## Rho: 0.4523, LR test value: 114.8, p-value: < 2.22e-16
## Asymptotic standard error: 0.04111
##     z-value: 11, p-value: < 2.22e-16
## Wald statistic: 121, p-value: < 2.22e-16
##
## Log likelihood: -4769 for lag model
## ML residual variance (sigma squared): 49.18, (sigma: 7.013)
## Number of observations: 1412
## Number of parameters estimated: 8
## AIC: 9555, (AIC for lm: 9668)
## LM test for residual autocorrelation
## test value: 7.662, p-value: 0.0056391
```

Remember what we said earlier in the spatial lag model that we are simply adding as an additional explanatory variable the values of y in the surrounding area. What we mean by "surrounding" will be defined by our spatial weight matrix. It's important to emphasise again that one has to think very carefully and explore appropriate definitions of "surrounding".

In our spatial lag model, you will notice that there is a new term *Rho*. What is this? This is our spatial lag. It is a variable that measures the homicide rate in the counties that are defined as surrounding each county in our spatial weight matrix. We are simply using this variable as an additional explanatory variable to our model, so that we can appropriately take into account the spatial clustering detected by our Moran's I test. You will notice that the estimated coefficient for this term is both positive and statistically significant. In other words, when the homicide rate in surrounding areas increases, so does the homicide rate in each country, even when we adjust for the other explanatory variables in our model.

You also see at the bottom further tests for spatial dependence, a likelihood ratio test. This is not a test for residual spatial autocorrelation after we introduce our spatial lag. What you want is for this test to be significant because in essence is further evidence that the spatial lag model is a good fit.

How about the coefficients? It may be tempting to look at the regression coefficients for the other explanatory variables for the original OLS model and compare them to those in the spatial lag model. But you should be careful when doing this. Their meaning now has changed:

> Interpreting the substantive effects of each predictor in a spatial lag model is much more complex than in a nonspatial model (or in a spatial error model) because of the presence of the spatial multiplier that links the independent variables to the dependent.

In the nonspatial model, it does not matter which unit is experiencing the change on the independent variable. The effect in the dependent variable of a change in the value of an independent variable is constant across all observations (Darmofal, 2015, p.107).

Remember we say, when interpreting a regression coefficient for variable x_i, that they indicate how much y goes up or down for every one unit increase in x_i when holding all other variables in the model constant. In our example, for the non-spatial model this "effect" is the same for every county in our dataset. But in the spatial lag model, things are not the same. We cannot interpret the regression coefficients for the substantive predictors in the same way because the "substantive effects of the independent variables vary by observation as a result of the different neighbours for each unit in the data" (Darmofal 2015, 107).

In the standard OLS regression model, the coefficients for any of the explanatory variables measure the absolute "impact" of these variables. It is a simpler scenario. We look at the "effect" of x in y within each county. So x in county A affects y in count A. In the spatial lag model there are two components to how x affect y. The predictor x affects y within each county directly, but remember we are also including the spatial lag, the measure of y in the surrounding counties (call them B, C, and D). By virtues of including the spatial lag (a measure of y in county B, C, and D) we are indirectly incorporating as well the "effects" that x has on y in counties B, C, and D. So the "effect" of a covariate (independent variable) is the sum of two particular "effects": a direct, local effect of the covariate in that unit, and an indirect, spillover "effect" due to the spatial lag.

In, other words, in the spatial lag model, the coefficients only focus on the "short-run impact" of x_i on y_i, rather than the net "effect". As Ward and Gleditsch (2008) explain, "Since the value of y_i will influence the level of homicide"in other" counties y_j and these y_j, in turn, feedback on to y_i, we need to take into account the additional effects that the short impact of x_i exerts on y_i through its impact on the level of" homicide "in other" counties. You can still read the coefficients in the same way but need to keep in mind that they are not measuring the net "effect". Part of their "effect" will be captured by the spatial lag. Yet, you may still want to have a look at whether things change dramatically, particularly in terms of their significance (which is not the case in this example).

In sum, this implies that a change in the *ith* region's predictor can affect the *jth* region's outcome. We have two situations: (a) the direct impact of an observation's predictor on its own outcome, and (b) the indirect impact of an observation's neighbour's predictor on its outcome. This leads to three quantities that we want to know:

- Average Direct Impact, which is similar to a traditional interpretation
- Average Total Impact, which would be the total of direct and indirect impacts of a predictor on one's outcome
- Average Indirect Impact, which would be the average impact of one's neighbours on one's outcome

These quantities can be found using the `impacts()` function in the `spdep` library. We follow the example that converts the spatial weight matrix into a "sparse" matrix and power it up using the `trW()` function. This follows the approximation methods described in LeSage and Pace (2009). Here, we use Monte Carlo simulation to obtain simulated distributions of the various impacts.

```
W <- as(rwm_s, "CsparseMatrix")
trMC <- trW(W, type="MC")
```

```
fit4_70_sar.im <- impacts(fit4_70_sar, tr=trMC, R=100)
summary(fit4_70_sar.im, zstats=TRUE, short=TRUE)
```

```
## Impact measures (lag, trace):
##          Direct Indirect    Total
## RD70   2.30369  1.79942   4.1031
## PS70   0.61838  0.48302   1.1014
## MA70  -0.09379 -0.07326  -0.1670
## DV70   0.64932  0.50719   1.1565
## UE70  -0.49741 -0.38853  -0.8859
## ========================================================
## Simulation results ( variance matrix):
## ========================================================
## Simulated standard errors
##          Direct Indirect    Total
## RD70 0.23328  0.27135 0.40515
## PS70 0.27102  0.22932 0.49148
## MA70 0.04071  0.03240 0.07195
## DV70 0.21484  0.19467 0.39866
## UE70 0.10801  0.09425 0.18968
##
## Simulated z-values:
##        Direct Indirect   Total
## RD70   9.776     6.664 10.092
## PS70   2.336     2.199  2.314
## MA70  -2.188    -2.165 -2.213
## DV70   3.147     2.775  3.051
## UE70  -4.541    -4.118 -4.632
##
## Simulated p-values:
##          Direct  Indirect Total
## RD70 < 2e-16 2.7e-11  < 2e-16
## PS70  0.0195  0.0279   0.0207
## MA70  0.0287  0.0304   0.0269
## DV70  0.0017  0.0055   0.0023
## UE70 5.6e-06 3.8e-05  3.6e-06
```

We see that all the variables have significant direct, indirect and total effects. You may want to have a look at how things differ when you just run a non-spatial model.

11.8   Beyond the SAR and SEM models

Given the "galaxy" of spatial models, newer approaches to select the one to run have been proposed. LeSage and Pace (2009) propose a *top-down* approach that involved starting with the spatial Durbin model (SDM) and based on likelihood ratio tests to select the best model. Elhorst (2010), on the other hand, proposes to start with the OLS model, run the

Lagrange multipliers, and then run a likelihood ratio test to see if the spatial Durbin model would be more appropriate. Why the Durbin model? Why not Manski or the SAR model? Elhorst (2010) corroborates that the Manski model can be estimated, but the parameters cannot be interpreted in a meaningful way (the endogenous and exogenous effects cannot be distinguished from each other) and the SAR or Kelejian/Prucha model "will suffer from omitted variables bias if the true data-generation process is a spatial Durbin or a spatial Durbin error model" and that similarly the same happens for the spatial Durbin error model" (p. 14).

If you paid attention, the summary of the lag model we ran in the previous section included at the bottom a test for remaining residual spatial autocorrelation. This means our model is still not filtering all the spatial dependency that is present here. Perhaps here, and following the workflow proposed by Elhorst (2010), we need to consider if the spatial Durbin model (SDM) would be more appropriate. We can run this model and then contrast it with the SAR model.

```
fit4_70_sdm <- lagsarlm(eq2_70_S, data=ncovr_s_sf, rwm_s, Durbin = TRUE,
                        method ="LU")
summary(fit4_70_sdm)
```

```
##
## Call:
## lagsarlm(formula = eq2_70_S, data = ncovr_s_sf, listw = rwm_s,
##     Durbin = TRUE, method = "LU")
##
## Residuals:
##       Min       1Q    Median       3Q      Max
## -18.09108  -4.13069  -0.88963  2.84179  67.03240
##
## Type: mixed
## Coefficients: (asymptotic standard errors)
##              Estimate Std. Error z value  Pr(>|z|)
## (Intercept)  9.996665   2.080609  4.8047 1.550e-06
## RD70         2.266521   0.350632  6.4641 1.019e-10
## PS70         0.297629   0.312080  0.9537 0.3402394
## MA70        -0.066899   0.055596 -1.2033 0.2288603
## DV70         0.819995   0.244796  3.3497 0.0008090
## UE70        -0.185133   0.129563 -1.4289 0.1530326
## lag.RD70     0.262121   0.515226  0.5087 0.6109276
## lag.PS70     1.071509   0.518284  2.0674 0.0386949
## lag.MA70    -0.023245   0.088356 -0.2631 0.7924840
## lag.DV70    -0.414450   0.429808 -0.9643 0.3349117
## lag.UE70    -0.737893   0.202560 -3.6428 0.0002697
##
## Rho: 0.415, LR test value: 79.91, p-value: < 2.22e-16
## Asymptotic standard error: 0.04542
##     z-value: 9.138, p-value: < 2.22e-16
## Wald statistic: 83.5, p-value: < 2.22e-16
##
## Log likelihood: -4759 for mixed model
## ML residual variance (sigma squared): 48.66, (sigma: 6.976)
```

```
## Number of observations: 1412
## Number of parameters estimated: 13
## AIC: 9544, (AIC for lm: 9622)
## LM test for residual autocorrelation
## test value: 13.98, p-value: 0.00018426
```

We can then compare the nested models using information criteria with `AIC()` and `BIC()`. Using this information criteria, we choose the model with the lowest AIC and BIC.

```
output_1 <- AIC(fit4_70, fit4_70_sar, fit4_70_sdm)
output_2 <- BIC(fit4_70, fit4_70_sar, fit4_70_sdm)
table_1 <- cbind(output_1, output_2)
table_1
```

```
##               df  AIC df  BIC
## fit4_70        7 9668  7 9704
## fit4_70_sar    8 9555  8 9597
## fit4_70_sdm   13 9544 13 9613
```

Our results indicate that AIC favours the SDM model, whereas the BIC favours the simpler SAR model. We can also run a likelihood ratio test to compare the two spatial nested models using `spatialreg::LR.Sarlm()` (that deprecates `spded::LR.sarlm()` you may still see mentioned in other textbooks). Here we look for the model with the highest log likelihood:

```
LR.Sarlm(fit4_70_sdm, fit4_70_sar)
```

```
##
##  Likelihood ratio for spatial linear models
##
## data:
## Likelihood ratio = 21, df = 5, p-value = 0.001
## sample estimates:
## Log likelihood of fit4_70_sdm
##                           -4759
## Log likelihood of fit4_70_sar
##                           -4769
```

The LR test is significant, suggesting we should favour the spatial Durbin model. If it had been insignificant, we would have had to choose the more parsimonious model.

As with the spatial lag model, here we need to be careful when interpreting the regression coefficients. We also need to estimate the "impacts" in a similar manner we illustrated above.

```
fit4_70_sdm.im <- impacts(fit4_70_sdm, tr=trMC, R=100)
summary(fit4_70_sdm.im, zstats=TRUE, short=TRUE)
```

```
## Impact measures (mixed, trace):
##        Direct Indirect   Total
## RD70   2.3255  1.99733  4.3228
## PS70   0.3562  1.98440  2.3406
```

```
## MA70 -0.0694 -0.08471 -0.1541
## DV70  0.8164 -0.12307  0.6933
## UE70 -0.2251 -1.35289 -1.5779
## ====--==========================================================
## Simulation results ( variance matrix):
## ================================================================
## Simulated standard errors
##        Direct Indirect  Total
## RD70 0.29855   0.6440 0.6312
## PS70 0.28645   0.7654 0.7383
## MA70 0.05711   0.1296 0.1137
## DV70 0.26395   0.6772 0.6877
## UE70 0.12024   0.2851 0.2694
##
## Simulated z-values:
##        Direct Indirect  Total
## RD70   7.681   3.1783  6.875
## PS70   1.243   2.6410  3.220
## MA70  -1.267  -0.6544 -1.382
## DV70   3.269  -0.2063  1.052
## UE70  -1.856  -4.8483 -5.961
##
## Simulated p-values:
##        Direct  Indirect Total
## RD70 1.6e-14 0.0015   6.2e-12
## PS70 0.2140  0.0083    0.0013
## MA70 0.2053  0.5129    0.1671
## DV70 0.0011  0.8366    0.2930
## UE70 0.0634  1.2e-06   2.5e-09
```

Notice how in this model only three inputs remain significant, resource deprivation, population structure, and unemployment. The sign of the coefficients remains the same for all the inputs. In most circumstances we see that when the direct effect is positive the indirect effect is also positive, and when the direct effect is negative the indirect effect is negative. Only for the insignificant effect of divorce we see a positive direct effect and a negative indirect effect.

As suggested by Kopczewska (2021), we can also examine the proportion of the direct effect in the total effect:

```
direct <- fit4_70_sdm.im$res$direct
indirect <- fit4_70_sdm.im$res$indirect
total <- fit4_70_sdm.im$res$total
direct/total
```

```
##   RD70   PS70   MA70   DV70   UE70
## 0.5380 0.1522 0.4503 1.1775 0.1426
```

The proportion of the (significant) direct effect ranges from 14% (for unemployment), fairly marginal, to 54% (for resource deprivation), about half of it. Another way of looking at it is by relation of the direct to the indirect effect by looking at its ratio in absolute values:

```
abs(direct)/abs(indirect)
```

```
##   RD70   PS70   MA70   DV70   UE70
## 1.1643 0.1795 0.8193 6.6332 0.1663
```

When the ratio is greater than 1, this suggests that the direct effect prevails, as we see for resource deprivation. With a score lower than 1, this indicates that the spillover effect is greater than that of internalisation as we see for population structure and unemployment.

11.9   Summary and further reading

This chapter has introduced some of the issues associated with spatial regression. We have deliberately avoided mathematical detail and focused instead on providing a general and more superficial introduction of some of the issues associated with fitting, choosing, and interpreting basic spatial regression models in R. We have also favoured the kind of models that have been previously discussed in the criminological literature, those relying on the frequentist tradition, use of maximum likelihood, simultaneous autoregressive models, and the incorporation of the spatial structure through a neighbourhood matrix. These are not the only ways of fitting spatial regression, though. The functions of `spatialreg`, for example, allow for newer generalised method of moments (GMM) estimators. On the other hand, other scientific fields have emphasised conditional autoregressive models (see Wall (2004) for the difference between CAR and SAR), and there is too a greater reliance, for example in spatial epidemiology, on Bayesian approaches. Equally, we have only focused on cross-sectional models, without any panel or spatio-temporal dimension to them. As Ralph Sockman was attributed to saying, "the larger the island of knowledge, the longer the shoreline of wonder". Learning more about spatial analysis only leads to discovering there is much more to learn. This chapter mostly aims to make you realise this about spatial regression and hopefully tempt you into wanting to continue this journey. In what follows we offer suggestions for this path.

There are a number of accessible, if somehow dated, introductory discussions of spatial regression for criminologists: Tita and Radil (2010) and Bernasco and Elffers (2010). The books by Ward and Gleditsch (2008), Darmofal (2015), and Chi and Zhu (2020) expand this treatment with the social student in mind. Pitched at a similar level is Anselin and Rey (2014). More mathematically detailed treatment and a wider diversity of models, including those appropriate for panel data, are provided in several spatial econometric books such as LeSage and Pace (2009), Elhorst (2014) or Kelejian and Piras (2017). In these econometric books you will also find more details about newer approaches to estimation, such as GMM. If you are interested in exploring the Bayesian approach to modelling spatial lattice data, we suggest you start with Haining and Li (2020).

There are also a number of resources that focus on R capabilities for spatial regression and that have inspired our writing. It is important to highlight that many of these, even the more recent ones, do not account properly for how `spatialreg` has deprecated the regression functionality of `spdep`, even if they still provide useful instruction (for the basic architecture of the functions that have migrated to `spatialreg` has not been started from scratch). The workbook of Anselin (2007) offers a detailed explanation of the spatial regression capabilities through `spdep`, most of which is still transferable to the equivalent functions in `spatialreg`. This workbook indeed uses the "ncovr" dataset and may be a useful continuation point for what we have introduced here. Kopczewska (2021) provides one of the most up-to-date discussions of spatial regression with R also pitched at a level the reader of this book should find accessible. Bivand, Millo, and Piras (2021) offer a more technical, up-to-date, and systematic review of the evolution of the software available for spatial econometrics in R. It is definitely a must read. For Bayesian approaches to spatial regression using R, which we have not had the space to cover in this volume, we recommend Lawson (2021a), Lawson (2021c), Gomez-Rubio, Ferrandiz, and Lopez (2005), and Blangiardo and Cameletii (2015), all with a focus on health data.

Aside from these reading resources, there are two series of lectures available on YouTube that provide good introductions to the field of spatial econometrics. Check out Prof. Mark Burkey's spatial econometric reboot[1] and Luc Anselin's lectures in the GeoDa Software channel.

Although the focus of this chapter has been on spatial regression, there are other approaches that are used within crime analysis for the purpose of predicting crime in particular areas such as risk terrain modelling or the use of other machine learning algorithms for prediction purposes. Those are well beyond the scope of this book. Risk terrain modeling (Caplan and Kennedy 2016) is generally actioned through the RTDMx sofware commercialised by its proponents. In a recent paper, Wheeler and Steenbeek (2021) illustrate how random forest, a popular machine learning algorithm, can also be used to generate crime predictions at micro-places. Dr. Gio Circo is currently developing a package (`quickGrid`) to apply the methods proposed by Wheeler and Steenbeek (2021)[2].

---

[1] https://tinyurl.com/mwnxpknb
[2] https://github.com/gmcirco/quickGrid

12

Spatial heterogeneity and regression

12.1 Introduction

One of the challenges of spatial data is that we may encounter exceptions to stationary processes. There may be, for example, parameter variability across the study area. When we have homogeneity, everything is the same everywhere; in terms of our regression equation, the parameters are constant. But as we illustrated via an interaction term in Chapter 11, that may not be the case. We can encounter situations where a particular input has a different effect in parts of our study area. In this chapter we explore this issue and a number of solutions that have been proposed to deal with it. It goes without saying that having extreme heterogeneity creates technical problems for estimation. If everything is different everywhere, we may have to estimate more parameters than we have data for.

One way of dealing with this is by imposing some form of structure, for example partitioning data for the United States to fit a separate model for Southern and Northern counties. Unlike data partition, **full spatial regimes** also imply using different coefficients for each subset of the data, but you fit everything in one go. This is essentially the same as running as many models as subsets you have. As Anselin (2007) highlights, this corrects for heterogeneity but does not explain it. We can then test whether this was necessary, using a *Chow test* comparing the simpler model with the one where we allow variability. Of course, we can also include spatial dependence in these models.

Another way of dealing with heterogeneity is by allowing continuous variation, rather than discrete (as when we subset the data), of the parameters. A popular method for doing this is **geographically weighted regression (GWR)**. This is a case of a local regression where you try to estimate a different set of parameters for each location and these parameters are obtained from a subset of observation using kernel regression. In traditional local regression we select the subset in the attribute space (observations with similar attributes); in GWR the local subset is defined in a geographical sense using a kernel (remember how we used kernels for density estimation to define a set of "neighbours"). We are estimating regression equations based on nearby locations as specified by our kernel, and this produces a different coefficient for each location. As with kernel density estimation, you can distinguish between fixed bandwidth and adaptive bandwidth.

For this chapter we will need to load the following packages:

```
library(tidyr)
library(dplyr)
# Packages for handling spatial data and for geospatial carpentry
library(sf)
library(sp)
```

```
library(spdep)
# Packages for regression and spatial regression
library(spatialreg)
# Packages for mapping and visualisation
library(tmap)
library(ggplot2)
# Packages with spatial datasets
library(geodaData)
# For geographicaly weighted regression
library(spgwr)
```

12.2   Spatial regimes

12.2.1   Fitting the constrained and the unconstrained models

To illustrate this solution, we will go back to the "ncovr" data we explored last week and use the same spatial weight matrix as in the previous chapter. If you have read the Baller et al. (2001) paper that we are in a way replicating here, you could see they decided that they needed to run separate models for the South and the North. Let's explore this here.

```
data("ncovr")
# We coerce the sf object into a new sp object
ncovr_sp <- as(ncovr, "Spatial")
# Then we create a list of neighbours using 10 nearest neighbours criteria
xy <- coordinates(ncovr_sp)
nb_k10 <- knn2nb(knearneigh(xy, k = 10))
rwm <- nb2mat(nb_k10, style='W')
rwm <- mat2listw(rwm, style='W')
# remove redundant objects
rm(list = c("ncovr_sp", "xy", "nb_k10"))
```

We can start by running a model for the whole country to use as a baseline model:

```
fit1_70 <- lm(HR70 ~ RD70 + PS70 + MA70 + DV70 + UE70, data = ncovr)
summary(fit1_70)
```

```
##
## Call:
## lm(formula = HR70 ~ RD70 + PS70 + MA70 + DV70 + UE70, data = ncovr)
##
## Residuals:
##    Min     1Q Median     3Q    Max
## -21.12  -3.14  -1.02   1.90  69.75
##
```

```
## Coefficients:
##               Estimate Std. Error t value Pr(>|t|)
## (Intercept)   11.1376     0.7602   14.65   <2e-16 ***
## RD70           4.1541     0.1147   36.?3   <2e-16 ***
## PS70           1.1449     0.1165    9.82   <2e-16 ***
## MA70          -0.2080     0.0233   -8.94   <2e-16 ***
## DV70           1.4066     0.1064   13.22   <2e-16 ***
## UE70          -0.4268     0.0497   -8.59   <2e-16 ***
## ---
## Signif. codes:
## 0 '***' 0.001 '**' 0.01 '*' 0.05 '.' 0.1 ' ' 1
##
## Residual standard error: 6.03 on 3079 degrees of freedom
## Multiple R-squared:  0.328,  Adjusted R-squared:  0.327
## F-statistic:  301 on 5 and 3079 DF,  p-value: <2e-16
```

A simple visualisation of the residuals suggests the model may not be optimal for the entire study region:

```
ncovr$res_fit1 <- fit1_70$residuals
ggplot(ncovr, aes(x = res_fit1, colour = as.factor(SOUTH))) +
  geom_density() +
  xlab("Residuals") +
  scale_colour_brewer(palette = "Set1", name = "Region (1 = South)") +
  theme_bw()
```

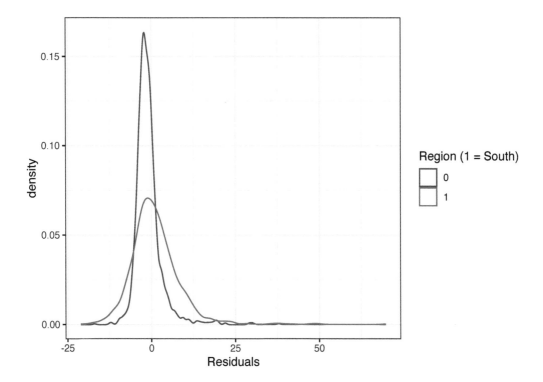

FIGURE 12.1: Distribution of residuals by study region (South versus North)

To be able to fit these spatial regimes, that will allow for separate coefficients for counties in the North and those in the South, first we need to define a dummy for the Northern counties, and we create a vector of 1.

```
ncovr$NORTH <- recode(ncovr$SOUTH, `1` = 0, `0` = 1)
ncovr$ONE <- 1
```

Having created these new columns, we can fit a standard OLS model interacting the inputs with our dummies and setting the intercept to 0, so that each regime is allowed its own intercept:

```
fit2_70_sr <- lm(HR70 ~ 0 + (ONE + RD70 + PS70 + MA70 + DV70 + UE70):(SOUTH + NORTH),
                 data = ncovr)
summary(fit2_70_sr)
```

```
##
## Call:
## lm(formula = HR70 ~ 0 + (ONE + RD70 + PS70 + MA70 + DV70 + UE70):(SOUTH +
##       NORTH), data = ncovr)
##
## Residuals:
##     Min      1Q Median     3Q    Max
## -17.47   -2.79  -0.76   1.71  65.60
##
## Coefficients:
##               Estimate Std. Error t value Pr(>|t|)
## ONE:SOUTH     15.0305     1.0714   14.03  < 2e-16 ***
## ONE:NORTH      6.6485     1.1308    5.88  4.6e-09 ***
## RD70:SOUTH     3.2364     0.1884   17.18  < 2e-16 ***
## RD70:NORTH     2.9523     0.2745   10.75  < 2e-16 ***
## PS70:SOUTH     0.9953     0.2114    4.71  2.6e-06 ***
## PS70:NORTH     0.8428     0.1402    6.01  2.0e-09 ***
## MA70:SOUTH    -0.1656     0.0338   -4.90  1.0e-06 ***
## MA70:NORTH    -0.1749     0.0325   -5.37  8.2e-08 ***
## DV70:SOUTH     0.6271     0.1750    3.58  0.00034 ***
## DV70:NORTH     1.4263     0.1313   10.86  < 2e-16 ***
## UE70:SOUTH    -0.7916     0.0817   -9.69  < 2e-16 ***
## UE70:NORTH    -0.0293     0.0631   -0.46  0.64265
## ---
## Signif. codes:
## 0 '***' 0.001 '**' 0.01 '*' 0.05 '.' 0.1 ' ' 1
##
## Residual standard error: 5.81 on 3073 degrees of freedom
## Multiple R-squared:  0.645,  Adjusted R-squared:  0.644
## F-statistic:  465 on 12 and 3073 DF,  p-value: <2e-16
```

We can see a higher intercept for the South, reflecting the higher level of homicide in these counties. Aside from this, the two more notable differences are the insignificant effect of unemployment in the North and the higher impact of the divorce rate in the North.

12.2.2 Running the Chow test

The Chow test is often used in econometrics to determine whether the independent variables have different impacts on different sub-groups of the population. Anselin (2007) provides an implementation for this test that extracts the residuals and the degrees of freedom from the constrained (the simpler OLS model) and the unconstrained models (the less parsimonious spatial regime), and then computes the test based on the F distribution.

```
chow.test <- function(rest, unrest) {
  er <- residuals(rest)
  eu <- residuals(unrest)
  er2 <- sum(er ^ 2)
  eu2 <- sum(eu ^ 2)
  k <- rest$rank
  n2k <- rest$df.residual - k
  c <- ((er2 - eu2) / k) / (eu2 / n2k)
  pc <- pf(c, k, n2k, lower.tail = FALSE)
  list(c, pc, k, n2k)
}
```

With the function in our environment we can then run the test, which will yield the test statistic, the p value, and two degrees of freedom.

```
 chow.test(fit1_70,fit2_70_sr)
```

```
## [[1]]
## [1] 40.16
##
## [[2]]
## [1] 2.674e-47
##
## [[3]]
## [1] 6
##
## [[4]]
## [1] 3073
```

How we interpret this test? When the Chow test is significant, as here, this provides evidence of spatial heterogeneity and in favour of fitting a spatial regime model, allowing for non-constant coefficients across the discrete subsets of data that we have specified.

12.2.3 Spatial dependence with spatial heterogeneity

So far we have only adjusted for a form of spatial heterogeneity in our model, but we may still have to deal with spatial dependence. Here what we covered in Chapter 11, still applies. We can run the various specification tests and then select a model that adjusts for spatial dependence.

```
summary(lm.LMtests(fit2_70_sr, rwm, test = c("LMerr","LMlag","RLMerr","RLMlag")))
```

```
##   Lagrange multiplier diagnostics for spatial
##   dependence
## data:
## model: lm(formula = HR70 ~ 0 + (ONE + RD70 +
## PS70 + MA70 + DV70 + UE70):(SOUTH + NORTH), data
## = ncovr)
## weights: rwm
##
##         statistic parameter p.value
## LMerr     227.20          1 < 2e-16 ***
## LMlag     252.00          1 < 2e-16 ***
## RLMerr      5.25          1   0.022 *
## RLMlag     30.05          1 4.2e-08 ***
## ---
## Signif. codes:
## 0 '***' 0.001 '**' 0.01 '*' 0.05 '.' 0.1 ' ' 1
```

The Lagrange multiplier tests, according to Anselin's suggestions for how to interpret these, point to a spatial lag model (since RLMlag is several orders more significant than the error alternative).

We could then fit this model using sp:

```
fit3_70_sr_sar <- lagsarlm(HR70 ~ 0 +
                           (ONE +
                               RD70 +
                               PS70 +
                               MA70 +
                               DV70 +
                               UE70):(SOUTH + NORTH),
                    data = ncovr,
                    rwm)
summary(fit3_70_sr_sar)
```

```
##
## Call:
## lagsarlm(formula = HR70 ~ 0 + (ONE + RD70 + PS70 + MA70 + DV70 +
##     UE70):(SOUTH + NORTH), data = ncovr, listw = rwm)
##
## Residuals:
##       Min        1Q    Median        3Q       Max
## -17.15273  -2.66726  -0.67209   1.49219  65.86918
##
## Type: lag
## Coefficients: (asymptotic standard errors)
##            Estimate Std. Error z value  Pr(>|z|)
## ONE:SOUTH  8.391260   1.103507  7.6042 2.864e-14
## ONE:NORTH  4.903963   1.091437  4.4931 7.019e-06
```

```
## RD70:SOUTH   2.355243    0.191445 12.3025  < 2.2e-16
## RD70:NORTH   2.669956    0.266521 10.0178  < 2.2e-16
## PS70:SOUTH   0.653640    0.203791  3.2074 0.0013394
## PS70:NORTH   0.778913    0.135054  5.7674 8.050e-09
## MA70:SOUTH  -0.100568    0.032699 -3.0756 0.0021009
## MA70:NORTH  -0.138235    0.031328 -4.4126 1.022e-05
## DV70:SOUTH   0.636867    0.168184  3.7867 0.0001527
## DV70:NORTH   1.157985    0.128559  9.0074  < 2.2e-16
## UE70:SOUTH  -0.522353    0.079210 -6.5945 4.267e-11
## UE70:NORTH  -0.039129    0.060706 -0.6446 0.5192148
##
## Rho: 0.3957, LR test value: 181.8, p-value: < 2.22e-16
## Asymptotic standard error: 0.02846
##     z-value: 13.9, p-value: < 2.22e-16
## Wald statistic: 193.3, p-value: < 2.22e-16
##
## Log likelihood: -9710 for lag model
## ML residual variance (sigma squared): 31.21, (sigma: 5.587)
## Number of observations: 3085
## Number of parameters estimated: 14
## AIC: 19448, (AIC for lm: 19628)
## LM test for residual autocorrelation
## test value: 0.7147, p-value: 0.39789
```

We see that the AIC is better than for the non-spatial model and that there is no more residual autocorrelation. We could explore other specifications (i.e., spatial Durbin model) and the same caveats about interpreting coefficients in a spatial lag model we raised in Chapter 11 would apply here. If the Lagrange multiplier tests had suggested a spatial error model, we could also have fitted the model in the way we have also already covered simply adjusting the regression formula as in this section.

We can now also run a Chow test; only in this case, we need to adjust the previous one to account for spatial dependence. Anselin (2007) also provides for an implementation of this test in R with the following ad-hoc function:

```
spatialchow.test <- function(rest, unrest) {
  lrest <- rest$LL
  lunrest <- unrest$LL
  k <- rest$parameters - 2
  spchow <- -2.0 * (lrest - lunrest)
  pchow <- pchisq(spchow, k, lower.tail = FALSE)
  list(spchow, pchow, k)
}
```

This function will test a spatial lag model for the country as a whole with the spatial lag model with our data partitions. So first we need to estimate the appropriate lag model to be our constrained model:

```
fit4_70_sar <-
  lagsarlm(HR70 ~ RD70 + PS70 + MA70 + DV70 + UE70,
```

```
    data = ncovr,
  rwm)
```

Which we can then use in the newly created function:

```
spatialchow.test(fit4_70_sar, fit3_70_sr_sar)
```

```
## [[1]]
##          [,1]
## [1,] 58.59
##
## [[2]]
##               [,1]
## [1,] 8.712e-11
##
## [[3]]
## [1] 6
```

The test is again highly significant providing evidence for the spatial lag model with the spatial regimes.

12.3   Geographically Weighted Regression

The basic idea behind Geographically Weighted Regression (GWR) is to explore how the relationship between a dependent variable (Y) and one or more independent variables (Xs) might vary geographically. Instead of assuming that a single model can be fitted to the entire study region, it looks for geographical differences in the relationship. This is achieved by fitting a regression equation to every feature in the dataset. We construct these separate equations by incorporating the dependent and explanatory variables of the features falling within the neighbourhood of each target feature.

GWR can be used for a variety of applications, including the following:

- Is the relationship between homicide rate and resource deprivation consistent across the study area?
- What are the key variables that explain high homicide rates?
- Where are the counties in which we see high homicide rates? What characteristics seem to be associated with these? Where is each characteristic most important?
- Are the factors influencing higher homicide rates consistent across the study area?

GWR provides three types of regression models: Continuous, Binary, and Count. These types of regression are known in statistical literature as Gaussian, Logistic, and Poisson, respectively. The Model Type for your analysis should be chosen on the level of measurement of the Dependent Variable Y.

So to illustrate, let us first build our linear model once again, regressing homicide rate in the 1970s for each county against resource deprivation (RD70), population structure (PS70), median age (MA70), divorce rate (DV70), and unemployment rate (UE70).

```
# make linear model
model <- lm(HR70 ~ RD70 + PS70 + MA70 + DV70 + UE70, data = ncovr)
summary(model)
```

```
##
## Call:
## lm(formula = HR70 ~ RD70 + PS70 + MA70 + DV70 + UE70, data = ncovr)
##
## Residuals:
##     Min     1Q Median     3Q    Max
## -21.12   -3.14  -1.02   1.90  69.75
##
## Coefficients:
##             Estimate Std. Error t value Pr(>|t|)
## (Intercept)  11.1376     0.7602   14.65   <2e-16 ***
## RD70          4.1541     0.1147   36.23   <2e-16 ***
## PS70          1.1449     0.1165    9.82   <2e-16 ***
## MA70         -0.2080     0.0233   -8.94   <2e-16 ***
## DV70          1.4066     0.1064   13.22   <2e-16 ***
## UE70         -0.4268     0.0497   -8.59   <2e-16 ***
## ---
## Signif. codes:
## 0 '***' 0.001 '**' 0.01 '*' 0.05 '.' 0.1 ' ' 1
##
## Residual standard error: 6.03 on 3079 degrees of freedom
## Multiple R-squared:  0.328,  Adjusted R-squared:  0.327
## F-statistic:  301 on 5 and 3079 DF,  p-value: <2e-16
```

And now, to get an idea of how our model might perform differently in different areas, we can map the residuals across each county:

```
# map residuals
ncovr$resids <- residuals(model)
qtm(ncovr, fill = "resids")
```

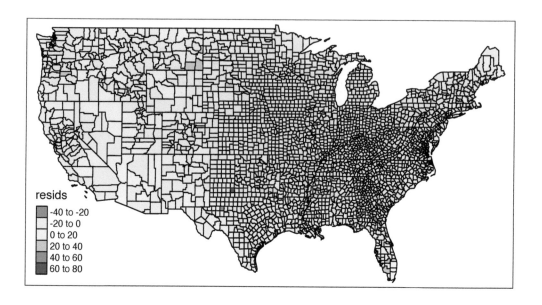

FIGURE 12.2: Map of model residuals

If we see that there is a geographic pattern in the residuals, it is possible that an unobserved variable may be influencing the dependent variable, and there might be some sort of spatial variation in our model's performance. This is what GWR allows us to explore, whether the relationship between our variables is stable over space.

12.3.1   Calculate kernel bandwidth

The big assumption on which the GWR model is based is the kernel bandwitdh. As mentioned earlier, in order to calculate local coefficients for our study regions, GWR takes into account the data points within our specified bandwith, weighting them appropriately. The bandwidth we choose will determine our results. If we choose something too large, we might mask variation and get results similar to the global results. If we choose something too small, we might get spikes, and lots of small scale variation, but also possibly large standard errors and unreliable estimates. Therefore, the process of choosing an appropriate bandwidth is paramount to GWR.

It is possible that we choose bandwidth manually, based on some prior knowledge, or a theoretically based argument. However, if we do not have such ideas to start, we can use data-driven processes to inform our selection. The most frequently used approach is to use cross-validation. Other approaches include Akaike Information Criteria, and maximum likelihood.

Here we use `gwr.sel()` which uses a cross-validation method for selecting an optimal method for GWR. The function finds a bandwidth for a given geographically weighted regression by optimizing a selected function. For cross-validation, this scores the root mean square prediction error for the geographically weighted regressions, choosing the bandwidth minimizing this quantity.

With the `adapt` parameter you can choose whether you want the bandwidth to adapt (i.e., find the proportion between 0 and 1 of observations to include in weighting scheme (k-nearest neighbours)), or whether you want to find a global bandwidth. A global bandwidth might make sense where the areas of interest are of equal size, and equally spaced; however, if we have variation, for example like in the case of the counties of the United States, it makes sense to adjust the bandwidth for each observation. Here we set to adapt.

Another important consideration is the function for the geographical weighting. We mentioned this above. In the `gwr.sel()` function you can set this with the `gweight=` parameter. By default, this is set to a Gaussian function. We will keep it that way here.

With an sf object (which our data are in), one of the parameters required is a coordinate point for each one of our observations. In this case, what we can do is grab the centroid of each one of our polygons. To do this, we can use the `st_coordinates()` function:

```
ncovr <- ncovr %>%
  mutate(cent_coords = st_coordinates(st_centroid(.)))
```

Now that we have our centroids, we can calculate our optimal bandwith, and save this into an object called `gwr_bandwidth`.

```
gwr_bandwidth <- gwr.sel(HR70 ~ RD70 + PS70 + MA70 + DV70 + UE70,
                  data = ncovr,
                  coords = ncovr$cent_coords,
                  adapt = T)
```

```
## Adaptive q: 0.382 CV score: 109039
## Adaptive q: 0.618 CV score: 110549
## Adaptive q: 0.2361 CV score: 107560
## Adaptive q: 0.1459 CV score: 105925
## Adaptive q: 0.09017 CV score: 104176
## Adaptive q: 0.05573 CV score: 102283
## Adaptive q: 0.03444 CV score: 100199
## Adaptive q: 0.02129 CV score: 98707
## Adaptive q: 0.01316 CV score: 98388
## Adaptive q: 0.01161 CV score: 98447
## Adaptive q: 0.01475 CV score: 98373
## Adaptive q: 0.0145 CV score: 98390
## Adaptive q: 0.01725 CV score: 98566
## Adaptive q: 0.01571 CV score: 98411
## Adaptive q: 0.01512 CV score: 98356
## Adaptive q: 0.01534 CV score: 98364
## Adaptive q: 0.01508 CV score: 98358
## Adaptive q: 0.01517 CV score: 98354
## Adaptive q: 0.01524 CV score: 98351
## Adaptive q: 0.01528 CV score: 98356
## Adaptive q: 0.01524 CV score: 98351
```

So we now have our bandwidth. It is:

```
gwr_bandwidth
```

```
## [1] 0.01524
```

As our data are in WGS84, this value is in degrees.

12.3.2 Building the geographically weighted model

Now we can build our model. The `gwr()` function implements the basic geographically weighted regression approach to exploring spatial non-stationarity for given bandwidth and chosen weighting scheme. We specify the formula and our data source, as with the linear and spatial regressions, and the coordinates of the centroids, as we did in the bandwidth calculation above; but now we also specify our bandwith, which we created earlier, but we do this with the `adapt=` parameter. This is because, by default, the function takes a global bandwidth with the `bandwidth=` parameter, and we have created an adaptive bandwidth, so we shall use this `adapt=` parameter to include our GWRbandwidth object. We also set the `hatmatrix =` parameter to TRUE, to return the hatmatrix as a component of the result, and the `se.fit=` parameter to TRUE to return local coefficient standard errors - as we set hatmatrix to TRUE, two effective degrees of freedom sigmas will be used to generate alternative coefficient standard errors.

```
gwr_model = gwr(
  HR70 ~ RD70 + PS70 + MA70 + DV70 + UE70,
  data = ncovr,
  coords = ncovr$cent_coords,
  adapt = gwr_bandwidth,
  hatmatrix = TRUE,
  se.fit = TRUE
)
```

```
gwr_model
```

```
## Call:
## gwr(formula = HR70 ~ RD70 + PS70 + MA70 + DV70 + UE70, data = ncovr,
##     coords = ncovr$cent_coords, adapt = gwr_bandwidth, hatmatrix = TRUE,
##     se.fit = TRUE)
## Kernel function: gwr.Gauss
## Adaptive quantile: 0.01524 (about 47 of 3085 data points)
## Summary of GWR coefficient estimates at data points:
##                  Min.   1st Qu.  Median  3rd Qu.    Max.
## X.Intercept. -4.2897    4.4361   7.4395  12.3212  30.1335
## RD70         -0.3625    2.0168   2.8788   3.8206   6.6405
## PS70         -4.0630    0.2914   1.0987   1.9019   3.4998
## MA70         -0.6831   -0.2072  -0.1028  -0.0186   0.2971
## DV70         -0.7513    0.5368   0.9652   1.4351   3.1027
## UE70         -1.4895   -0.4286  -0.1787  -0.0220   1.8486
##                  Global
```

```
## X.Intercept.   11.14
## RD70            4.15
## PS70            1.14
## MA70           -0.21
## DV70            1.41
## UE70           -0.43
## Number of data points: 3085
## Effective number of parameters (residual: 2traceS - traceS'S): 244
## Effective degrees of freedom (residual: 2traceS - traceS'S): 2841
## Sigma (residual: 2traceS - traceS'S): 5.405
## Effective number of parameters (model: traceS): 171.2
## Effective degrees of freedom (model: traceS): 2914
## Sigma (model: traceS): 5.337
## Sigma (ML): 5.186
## AICc (GWR p. 61, eq 2.33; p. 96, eq. 4.21): 19276
## AIC (GWR p. 96, eq. 4.22): 19082
## Residual sum of squares: 82986
## Quasi-global R2: 0.5022
```

This output tells us the 5-number summary for the coefficients (Y) for each predictor variable X. We can see for all predictor variables, some counties have negative values, while other counties have positive values. Evidently there is some variation locally in the relationships between these independent variables and our dependent variable.

For this to be informative, however, we should map this variation - this is where we can gain insight into how the relationships may change over our study area. In our `gwr_model` object, we have a *Spatial Data Frame*. This is a *SpatialPointsDataFrame* or *SpatialPolygonsDataFrame* object which contains fit.points, weights, GWR coefficient estimates, R square, and coefficient standard errors in its "data" slot. Let's extract this into another object called `gwr_results`:

```
gwr_results <- as.data.frame(gwr_model$SDF)
```

To map this, we can join these results back to our original data frame.

```
gwr_results <- cbind(ncovr, gwr_results)
```

Now we can begin to map our results. First, let's see how well the model performs across the different counties of the United States. To do this, we can map the `localR2` value which contains the local R^2 value for each observation.

```
qtm(gwr_results, fill = "localR2" )
```

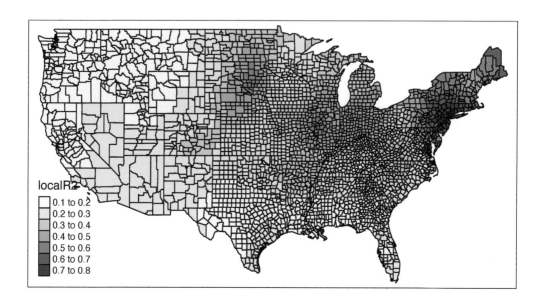

FIGURE 12.3: Map of local R square value for each county

We can see some really neat spatial patterns in how well our model performs in various regions of the United States. Our model performs really well in the North East for example, but not so well in the South.

We can also map the coefficients for specific variables. Let's say we're interested in how the relationship between resource deprivation (RD70) and homicide rate.

> **NOTE:** In our `gwr_results` data frame we have a variable called `RD70`; this contains the observed resource deprivation value for the county. We then have `RD70.1` which is the coefficient value for each county. In the results contained in the SDF element of our model (`gwr_model$SDF`) it was named `RD70` but when we joined the two objects, the `.1` was appended to allow us to differentiate between the two.

```
qtm(gwr_results, fill = "RD70.1" )
```

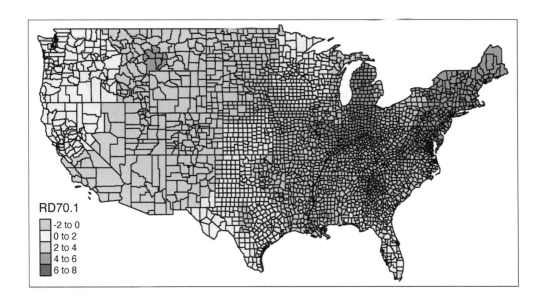

FIGURE 12.4: Map of coefficient for resource deprivation value for each county

We can see in this map again some regional patterns in the coefficient of resource deprivation. The relationship is stronger in the counties in the Northeast, while in other areas it is weak or even 0. And in some counties in Texas, we can see that the direction of the relationship flips. In these areas the coefficients are negative. This means that resource deprivation is associated with *lower* homicide rates in these counties.

This may be interesting and may raise questions; however, it is important to consider not only the coefficients, but also the *standard errors* associated with them, so we can get some ideas behind the reliability of these estimates. So let us have a look by mapping the standard errors for these coefficients, which are contained in the RD70_se variable.

```
qtm(gwr_results, fill =  "RD70_se" )
```

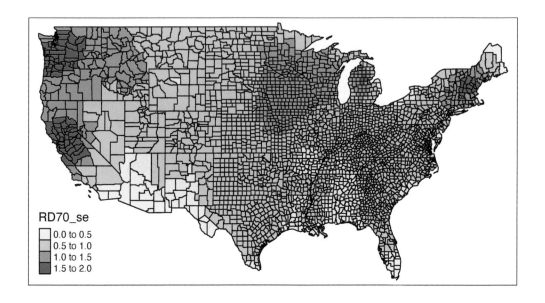

FIGURE 12.5: Map of standard errors for resource deprivation value for each county

Higher standard errors mean that our estimates may be less reliable. So it is important to keep in mind both maps: the one with the coefficients, and the one showing the standard errors, when drawing conclusions from our GWR results.

12.3.3   Using GWR

Overall, GWR is a good **exploratory** tool which can help illustrate how a particular model might apply differently in different regions of your study area, or how relationships between variables may vary spatially. This technique can be extended, for example to count models, or logit models, but (following the lead of Luc Anselin) we stop here, and pause to highlight some issues. Specifically, our *results are very sensitive to our choice of bandwidth*. The results we achieve with this approach might not show up when we use other bandwiths. Additionally, GWR does not really *explain* anything. What we do is demonstrate that the relationship between our variables is not stable over space. But we cannot with this technique explain *why* it is not stable over space.

Nevertheless, it is a great technique to identify spatial variation, which, unlike spatial regimes, does not rely on our a priory segmentation of our study region, but instead it allows for the variation to emerge from the data themselves.

12.4 Summary and further reading

This chapter has tackled some of the issues of spatial heterogeneity in regression results. When applying regression in a spatial context, it is possible to encounter situations where we might observe a different effect on our variables in different parts of our study area. Specifically, we covered two approaches to exploring this. The first one was to impose some a priori segmentation to our data, based on contextual knowledge and possible patterns in our regression results. We illustrated this by imposing spatial regimes in our NCOVR data set, splitting into separate North and South USA, as was done by the original authors of the paper by Baller et al. (2001). The second approach was to explore how the coefficients may vary across our study space by applying Geographically Weighted Regression to our study area. We provided a high-level overview of this process and recommended it as an illustrative, exploratory technique to raise questions about possible spatial heterogeneity in the processes we are trying to model in our study region.

To delve into greater detail on the topic of spatial heterogeneity, chapters 8 and 9 in Anselin (2007) discuss specification of spatial heterogeneity. This is particularly helpful as an introduction to the issues and solutions we have discussed in this chapter. For more details and applications of Geographically Weighted Regression, we recommend Fotheringham, Brunsdon, and Charlton (2003). For applications and an illustration of the importance to consider spatial variation within the context of criminology and crime analysis, read Cahill and Mulligan (2007) and Andresen and Ha (2020) as good examples.

And for those wishing to explore additional ways to model spatial interactions, such as CAR and Bayesian models, we suggest to read Banerjee, Carlin, and Gelfand (2014) as well as chapters 9 and 10 in Bivand, Pebesma, and Gómez-Rubio (2013).

Appendix A: A quick intro to R and RStudio

In this appendix we will cover all you need to know about R in order to benefit from the contents of this book. You will be introduced to the programming language, R and the RStudio interface. We assume you have successfully managed to install these two applications. Here we will acquaint you with them and will teach you three R basics: operators, objects, and packages.

Exploring RStudio

You can use R without using RStudio, but RStudio is an app that makes it easier to work with R. RStudio is what we call an IDE, an **integrated development environment**. It is a fancy way of saying that it is a cool interface designed to write programming code. Every time you open up RStudio, you are in fact starting an R session. RStudio automatically runs R in the background.

FIGURE A.1: Screenshot of RStudio IDE

When you first open RStudio, you will see (as in the image above) that there are three main windows. The bigger one to your left is the console. If you read the text in the console you will see that RStudio is indeed opening R, and you can see what version of R you are running. Since R is constantly being updated, the version you installed is likely more recent than the one we used at time of writing.

Opening up the script pane

The view in RStudio is structured so that you have four open windows in a regular session. However, when you first open, you might be starting with only three. To open the script pane (the one missing), click in the *File* drop-down Menu, select *New File*, then *R Script*.

FIGURE A.2: Navigate to File > New File> R Script to open script pane

You will now see the four window areas in display. In each of these areas you can shift between different views and panels. You can also use your mouse to re-size the different windows if that is convenient.

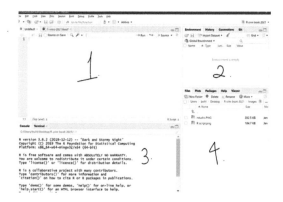

FIGURE A.3: The 4 panes of RStudio

The four panes of RStudio

The purposes of the four panes in the figure above are the following:

1. **Script and data view**: where you type your code that tells R what you want to do. These are essentially instructions that you type and save as a **script**, so that you can return to it later to remember what you did and to share it with others so that they can reproduce what you did.

2. **Environment and history view**:

i) *Environment* tab - gives you the names of all the (data) objects that you have defined during your current R session, including number of observations and rows in those objects. We learn more about objects later.

ii) *History* tab - shows you a history of all the code you have previously evaluated in the main console. One of the key advantages of doing data analysis this way, with code versus with a point and click interface like Excel or ArcGIS, is that you are producing a written record of every step you take in the analysis. First time around, it will take you time to write these instructions; it may be slower than pointing and clicking. And unlike with pointing and clicking you need to know the "words" and "grammar" of this language.

3. **Main console**: this is considered R's heart, and it is where R evaluates the codes that you run. You can type your code directly in the console, but for the sake of good habits, type them in the script and data view so you can save a record of them. Only type and run code from here if you want to debug or do some quick analysis.

4. **File Directory, Plots, Packages, Help**:

i) *Files* tab- allows you to see the files in the folder that is currently set as your working directory.

ii) *Plots* tab- you will see any data visualizations that you produce here. You have not produced any yet, so it is empty. now.

iii) *Packages* tab- you will see the packages that are currently available to install. We will explain what these are soon, but know that they are an essential feature when working with R.

iv) *Help* tab- you can access further information on the various packages.

Interacting with the four panes

In the previous section, you opened up the 'script' pane. We now write some code in it, and see what happens.

To do this, go to your open version of RStudio, and type in the script pane the following:

```
print("Hello world!")
```

When you have typed this, you will have typed your first bit of code. Yet nothing is happening? That is because you also have to **run** the code.

You can do this by highlighting the code you wish to run, and clicking on 'run' in the top right-hand corner. When you 'run' the code, it will print the text 'Hello World!' in the bottom pane, which is the **console**. That means you have written and executed your first line of code.

In the rest of the appendix, we will be unpacking how this all works, and getting more familiar and comfortable with using RStudio.

To recap: the **script** is where you write your programming code. A script is nothing but a text file with some code on it. Unlike other programs for data analysis you may have used in the past (Excel, SPSS), you need to interact with R by means of writing down instructions and asking R to evaluate those instructions. R is an *interpreted* programming language: you write instructions (code) that the R engine has to interpret in order to do something. And all the instructions we write can and should be saved in a script, so that you can return later to what you did.

As mentioned earlier, one of the key advantages of doing spatial data analysis with code versus with a point-and-click interface like ArcGIS or MapInfo (or even QGIS) is that you are producing a written record of every step you take in the analysis. First time around it may be slower than pointing and clicking, however over time you can re-run and re-use your code and will save lots of time (as well as build transparency and reproducibility into your analysis). Once you have written your instructions and saved them in a file, you will be able to share this file with others and run your every time you want in a matter of seconds. This creates a *reproducible* record of your analysis: something that your collaborators or someone else anywhere (including your future self, the one that will have forgotten how to do the stuff) could run and get the same results than you did at some point earlier. This makes science more transparent, and transparency brings with it many advantages. For example, it makes your research more trustworthy. Don't underestimate how critical this is. **Reproducibility** is becoming a key criteria to assess good quality research. You can read up on reproducibility and its importance in Leyser, Kingsley, and Grange (2015) or Ritchie (2020) for more detail.

Customising the RStudio look

RStudio allows you to customise the way it looks. Working with white backgrounds is not generally a good idea if you care about your eyesight. If you don't want to end up with dry eyes, not only it is good that you follow the 20-20-20 rule (every 20 minutes, look for 20 seconds to an object located 20 feet away from you), but it may also be a good idea to use more eye-friendly screen displays.

Click in the *Tools* menu and select *Global Options*. This will open up a pop-up window with various options. Select *Appearance*. In this section you can change the font type and size, but also the kind of theme background that R will use in the various windows.

FIGURE A.4: Find Tools > Global options

You can make any changes you'd like to here, including the theme background that R will use as the interface. For example, you may choose a darker theme like 'tomorrow night bright'.

FIGURE A.5: Select tomorrow night bright theme

Elements of R grammar

R is a programming language, and as a language it has a number of elements that are basic to its understanding. In this section we discuss functions (the "verbs" in the R grammar) and objects (the "nouns" in the R grammar).

Functions

Functions *do* things. They are called by a certain name, usually a name which represents what they do, and they are followed by brackets (). Within the brackets, you can put whatever it is that you want the function to work with. For example, the code we wrote in previously was the print() function. This function told R to print into the console whatever we put in the brackets ("Hello World!").

It is the same idea with a personalised greeting: if you want to print 'Hello Reka', you will need to have "Hello Reka" inside the brackets:

```
print("Hello Reka")
```

```
## [1] "Hello Reka"
```

There are so many functions in R. We will be learning many of them throughout the book. Print is fun, but most of the time, we will be using functions to help us with our data analysis. For example, getting the minimum, maximum, or mean of a list of numbers. R does this using functions in a very similar way.

For example, if we have a bunch of numbers, we just find the appropriate function to get the summary we want:

```
mean(c(10, 34, 5, 3, 77))
```

```
## [1] 25.8
```

```
min(c(10, 34, 5, 3, 77))
```

```
## [1] 3
```

```
max(c(10, 34, 5, 3, 77))
```

```
## [1] 77
```

How can you find the function you need? Throughout this book, you will learn a list that appears at the top of each lesson. A recommendation when you are starting with R is to also create a 'function cookbook', where you write down a list of functions, what the functions do, and some examples. Here is an example:

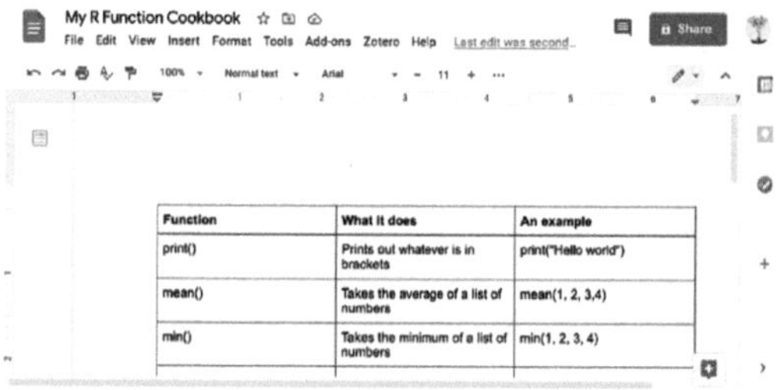

FIGURE A.6: An example of an R function cookbook

You can use Google to make your cookbook; and the website stackoverflow[1], in particular, can help you find the function you need. But be wary, especially in the beginning, that you understand what the function does. There can be several different functions for the same action.

NOTE: R is case-sensitive! For example:

```
# Calculating the logarithm
Log(100)
```

```
# ERROR!
```

```
# Instead, it should be:
log(100)
```

```
## [1] 4.605
```

---

[1]https://stackoverflow.com/

So make sure that you are specific about upper and lower case letters in your collection of relevant functions. In general, almost all functions will start with lower case letter. One exception is the `View()` function.

Objects

You can think of **objects** as boxes where you put things. Imagine a big, empty cardboard box. We can create this big empty box in R by simply giving it a name. Usually, you want your object/box to have a good descriptive name, which will tell people what is in it. Imagine moving house. If you have a cardboard box full of places, you might want to label it "plates". That way, when carrying, you know to be careful, and when unpacking, you know its contents will go in the kitchen. On the other hand, if you named it "box1", then this is a lot less helpful when it comes to unpacking.

Creating an object

Let us create an object called 'plates'. To do this, you go to your script, and type 'plates'.

But if you run this code, you will get an error. Try it and you will see the error 'Error! Object plates not found'. This is because you have not yet put anything inside the plates 'box'. Remember objects are like boxes, so there must be something inside our object 'plates'. In order for this object to exist, you have to put something inside it, or in R-speak *assign it some value.*

Therefore, we make an object by using an *assignment operator* (`<-`). In other words, we assign something to an object (i.e., put something in the box). For example:

```
plates <- "yellow plate"
```

Now if we run this, we will see no error message, but instead, we will see the `plates` object appear in our *environment pane.*

Here are some more examples to illustrate:

```
# Putting '10' in the 'a' box
a <- 10

# Putting 'Hello!' in the 'abc123' box
abc123 <- "Hello!"
```

In these examples, we are putting the value of `10` into the object `a`, and the value of 'Hello!' into the object `abc123`.

Earlier, we introduced you to the Environment and History pane. We mentioned that it lists objects you defined. After making the 'a' and 'abc123' objects, they should appear in that pane under the `Environment` tab.

Types of objects

Why are objects important? We will be storing everything in our data analysis process in these objects. Depending on what is inside them, they can become a different type of object. Here are some examples:

Data structures are important objects that store your data, and there are five main types but we focus on three for this course:

1. *(atomic) vector*: an ordered set of elements that are of the same *class*. Vectors are a basic data structure in R. Below are five different classes of vectors:

```
# 1. numeric vector with three elements
my_1st_vector <- c(0.5, 8.9, 0.6)

# 2. integer vector with addition of L at the end of the value
my_2nd_vector <- c(1L, 2L, 3L)

# 3. logical vector
my_3rd_vector <- c(TRUE, FALSE, FALSE)
# 'my_4th_vector' creates a logical vector using abbreviations of True and False,
#   but you should use the full words instead
my_4th_vector <- c(T, F)

# 4. character vector
my_5th_vector <- c("a", "b", "c")

# 5. complex vector (we will not use this for our class)
my_6th_vector <- c(1+0i, 2+4i)
```

2. *lists*: technically they, too, are vectors but they are more complex because they are not restricted on the length, structure, or class of the included elements. For example, to create a list containing strings, numbers, vectors and a logical, use the `list()` function, and inside the brackets, put everything you want to combine into a list:

```
list_data <- list("teal", "sky blue", c(10, 5, 10), TRUE, 68.26, 95.46, 99.7)
```

Above, we created `list_data`, an object that contains all those things that we put inside the `list()` function. This function serves to create a list from combining everything that is put inside its brackets.

Use the `class()` function to confirm that the objects have been defined as a list

```
class(list_data)
```

```
## [1] "list"
```

3. *data frames*: also store elements but differ from lists because they are defined by their number of columns and rows; the vectors (columns) must be of the same length. Data frames can contain different classes, but each column must be of the same class. For example, if you want to combine some related vectors to make a data frame on violent American cities, use the function `data.frame()`:

```
# Making some relevant vectors
TopVioCities <- c("St. Louis", "Detroit", "Baltimore") # some violent US cities
VioRatePer1k = c(20.8, 20.6, 20.3) # their violence rates per 1,000 persons
State <- c("Missouri", "Michigan", "Maryland") # in what states are these cities found

#Join them to make a data frame called 'df'
df<-data.frame(TopVioCities, VioRatePer1k, State)
```

We can then view the data frame, 'df', with the `View()` function:

```
View(df)
```

Doing things to objects

We have learned what functions are (i.e., things that do things) and what are objects (i.e., the boxes that hold things). We also saw some functions which helped us create objects. Functions can also do things to objects. For example, we saw the function `class()` that told us about what kind of object list_data was, and `View()` which allowed us to have a look at our dataframe we called `df`.

Let us look back at our `plates` object. Remember, it was the object that held our kitchen items. We added 'yellow plate' to it. Now let us add some more items and let us use the concatenate `c()` function for this again:

```
plates <- c("yellow plate", "purple plate", "silver plate", "orange bowl")
```

Let us say that we suddenly forgot what was in our object called 'plates'. Like what we learned earlier, we use the function `print()` to see what is inside this object:

```
print(plates)
```

```
## [1] "yellow plate" "purple plate" "silver plate"
## [4] "orange bowl"
```

This can apply to obtaining the mean, minimum, and maximum. You could assign those statistics to an object this time:

```
nums <- c(10, 34, 5, 3, 77)
```

Now if we want to know the mean, we can take the mean of the object nums, which we just created:

```
mean(nums)
```

```
## [1] 25.8
```

The object we will use most frequently though is data frames. These hold your data in a format whereby each column represents a variable, and each row an observation.

Just earlier, we had created a dataframe called df previously. If you have not yet copied this over into your own RStudio, do this now. You should have the object df in your environment. When you run View(df), you should see this dataset:

FIGURE A.7: View your dataframes

To do something to an entire dataframe, we would use the name of the object (df) to refer to it. In the case of the View() function, we want to see the whole thing, so we will call View(df). On the other hand, if we want to refer to only one variable in the data, (remember back to term 1 - each variable is held in each column) there is a special notation to do this.

To refer to a variable (column) inside a dataframe, you use:

dataframe name + $ + variable name

For example, to refer to the variable VioRatePer1k, we use the notation df$VioRatePer1k.

And if we wanted to View only that column, we use:

```
View(df$VioRatePer1k)
```

You should see one column of the dataframe appear in the top left pane.

Say, we wanted to know the mean violence rate across our units of analysis, the cities, for example, we would take the numeric column to calculate this:

```
mean(df$VioRatePer1k)
```

```
## [1] 20.57
```

Packages

Packages are a very important element of R. Packages are elements that add the functionality of R. What most packages do is they introduce new functions that allow you to ask R to do new different things. Anybody can write a package, so consequently R packages vary on quality and complexity. You can find packages in different places, as well, from official repositories (which means they have passed a minimum of quality control), something called GitHub (a webpage where software developers post work in progress), to personal webpages (danger danger!).

Throughout the book, and hopefully afterwards, you will find yourself installing numerous open-source software packages that allow R to do new and different things. There are loads of packages out there. In early 2020, there were over 150,000 packages available. Anyone can write one, so you will need to be careful on which ones you use, as the quality can vary. Official repositories, like CRAN[2], are your best bet for packages, as they will have passed some quality controls.

You can see what packages are available in your local install by looking at the *packages* tab in the *File Directory, Plots, Packages* pane.

A number of the packages we will use belong to a set of packages called **tidyverse**. These packages help make your data tidy. According to Statistician and Chief Scientist at RStudio, Hadley Wickham, transforming your data into *tidy data* is one of the most important steps of the data analysis process. It will ensure your data are in the format you need to conduct your analyses. We will also be using the simple features package **sf** and many more associated with spatial data analysis.

Packages can be installed using the `install.packages()` function. Remember that while you only need to install packages once, they need to be loaded with the `library()`function each time you open up RStudio. Let us install the package `dplyr` from `tidyverse` and load it:

```
library(dplyr)
```

A lot of code and activity appears in the console. Warnings may manifest. Most of the time, the warnings explain what is being loaded and confirm that the package is successfully loaded. If there is an error, you will have to figure out what the warnings are telling you to successfully load the package. This happens and is normal.

To double-check that you have actually installed `dplyr`, go to that *File Directory, Plots, Packages* pane and click on the *Packages* tab. The list of packages is in alphabetical order and `dplyr` should be there. If there is a tick in its box, it means that this package is currently loaded and you can use it; if there is no tick, it means that it is inactive, and you will have to bring it up with `library()`, or just tick its box.

On *masking*: sometimes packages introduce functions that have the same name as those that are already loaded into your session. When that happens, the newly loaded ones will override the previous ones. You can still use them, but you will have to refer to them explicitly by bringing them up and specifying to which package they belong with `library()`.

---

[2]https://cran.r-project.org/

How do you find out what a package does? You look at the relevant documentation. In the Packages window, scroll down until you find listed the new package we installed. Here you will see the name of the package (dplyr), a brief description of what the program is about, and the version you have installed (an indication that a package is a good package is that it has gone through several versions, that means that someone is making sure the package gets regular updates and improvements).

Click on the name dplyr. You will see that RStudio has now brought you to the Help tab. Here is where you find the help files for this package, including all the available documentation.

Every beginner in R will find these Help files a bit confusing. But after a while, their format and structure will begin to make sense to you. Click where it says *User Guides, Package Vignettes, and Other Documentation*. Documentation in R has become much better since people started to write **vignettes** for their packages. They are little tutorials that explain with examples what each package does.

Exploring data with R

Now that we know the basic component, let's play around with using R as we will throughout the book, for some data analysis.

We will get some data by installing a package which has data in it as well as functions, and then go on to produce some basic summaries. This should give some practice!

Playing around with data

We are going to look at some data that are part of the *fivethirtyeight* package. This package contains datasets and code behind the stories in this particular online magazine (http://fivethirtyeight.com/). This package is not part of the base installation of R, so you will need to install it first.

Remember, first we have to load the package if we want to use it:

```
library("fivethirtyeight")
data(package="fivethirtyeight") #Show all data frames available in named package
```

Notice that this package has some datasets that relate to stories covered in this newspaper that had a criminological angle. Let's look for example at the hate_crimes dataset. How do you that? First, we have to load the data frame into our global environment. To do so use the following code:

```
data("hate_crimes")
```

This function will search among all the *loaded* packages and locate the hate_crimes dataset.

Notice that it now appears in the global environment, although it also says "promise" next to it. To see the data in full, you need to do something to it first. So let's do that.

Every object in R can have **attributes**. These are names, dimensions (for matrices and arrays: number of rows and columns) and dimension names, class of object (numeric, character, etc.), length (for a vector this will be the number of elements in the vector), and other user-defined ones. You can access the attributes of an object using the `attributes()` function. Let's query R for the attributes of this data frame.

```
attributes(hate_crimes)
```

```
## $row.names
##  [1]  1  2  3  4  5  6  7  8  9 10 11 12 13 14 15 16 17
## [18] 18 19 20 21 22 23 24 25 26 27 28 29 30 31 32 33 34
## [35] 35 36 37 38 39 40 41 42 43 44 45 46 47 48 49 50 51
##
## $class
## [1] "tbl_df"      "tbl"           "data.frame"
##
## $names
##  [1] "state"
##  [2] "state_abbrev"
##  [3] "median_house_inc"
##  [4] "share_unemp_seas"
##  [5] "share_pop_metro"
##  [6] "share_pop_hs"
##  [7] "share_non_citizen"
##  [8] "share_white_poverty"
##  [9] "gini_index"
## [10] "share_non_white"
## [11] "share_vote_trump"
## [12] "hate_crimes_per_100k_splc"
## [13] "avg_hatecrimes_per_100k_fbi"
```

This prints out the row names (not very exciting here...) the class (see above when we used `class()` function) and the names, which are the column headers - or the *names of the variables within this dataset*. You can see there are things like state, and share who voted for Trump in the 2016 election.

Now use the `View()` function to glance at your data frame. What you get there is a spreadsheet with 12 variables and 51 observations. Each variable in this case is providing you with information (demographics, voting patterns, and hate crime) about each of the US states.

Ok, let's now have a quick look at the data. There are so many different ways of producing summary stats for data stored in R that is impossible to cover them all! We will just introduce a few functions that you may find useful for summarising data. Before we do any of that, it is important you get a sense for what is available in this dataset. Go to the help tab and in the search box input the name of the data frame, this will take you to the documentation for this data frame. Here you can see a list of the available variables.

Let's start with the *mean*. This function takes as an argument the numeric variable for which you want to obtain the mean. If you want to obtain the mean of the variable that gives us the proportion of people that voted for Donald Trump, you can use the following expression:

```
mean(hate_crimes$share_vote_trump)
```

```
## [1] 0.49
```

Another function you may want to use with numeric variables is `summary()`:

```
summary(hate_crimes$share_vote_trump)
```

```
##    Min. 1st Qu.  Median    Mean 3rd Qu.    Max.
##   0.040   0.415   0.490   0.490   0.575   0.700
```

This gives you the five number summary (minimum, first quartile, median, third quartile, and maximum, plus the mean and the count of missing values if there are any).

You don't have to specify a variable; you can ask for these summaries from the whole data frame:

```
summary(hate_crimes)
```

```
##     state            state_abbrev
##  Length:51          Length:51
##  Class :character   Class :character
##  Mode  :character   Mode  :character
##
##
##
##
##  median_house_inc share_unemp_seas share_pop_metro
##  Min.   :35521    Min.   :0.0280   Min.   :0.310
##  1st Qu.:48657    1st Qu.:0.0420   1st Qu.:0.630
##  Median :54916    Median :0.0510   Median :0.790
##  Mean   :55224    Mean   :0.0496   Mean   :0.750
##  3rd Qu.:60719    3rd Qu.:0.0575   3rd Qu.:0.895
##  Max.   :76165    Max.   :0.0730   Max.   :1.000
##
##   share_pop_hs     share_non_citizen share_white_poverty
##  Min.   :0.799    Min.   :0.0100    Min.   :0.0400
##  1st Qu.:0.841    1st Qu.:0.0300    1st Qu.:0.0750
##  Median :0.874    Median :0.0450    Median :0.0900
##  Mean   :0.869    Mean   :0.0546    Mean   :0.0918
##  3rd Qu.:0.898    3rd Qu.:0.0800    3rd Qu.:0.1000
##  Max.   :0.918    Max.   :0.1300    Max.   :0.1700
##                   NA's   :3
##    gini_index      share_non_white  share_vote_trump
##  Min.   :0.419    Min.   :0.060    Min.   :0.040
##  1st Qu.:0.440    1st Qu.:0.195    1st Qu.:0.415
##  Median :0.454    Median :0.280    Median :0.490
##  Mean   :0.454    Mean   :0.316    Mean   :0.490
```

```
##   3rd Qu.:0.467   3rd Qu.:0.420   3rd Qu.:0.575
##   Max.   :0.532   Max.   :0.810   Max.   :0.700
##
##   hate_crimes_per_100k_splc avg_hatecrimes_per_100k_fbi
##   Min.   :0.067             Min.   : 0.267
##   1st Qu.:0.143             1st Qu.: 1.293
##   Median :0.226             Median : 1.987
##   Mean   :0.304             Mean   : 2.368
##   3rd Qu.:0.357             3rd Qu.: 3.184
##   Max.   :1.522             Max.   :10.953
##   NA's   :4                 NA's   :1
```

There are multiple ways of getting results in R. Particularly for basic- and intermediate-level statistical analysis many core functions and packages can give you the answer that you are looking for. For example, there are a variety of packages that allow you to look at summary statistics using functions defined within those packages. You will need to install these packages before you can use them.

We are only going to introduce one of them here: skimr. You will need to install it before anything else.

Once you have loaded the *skimr* package you can use it. Its main function is *skim*. Like *summary* for data frames, skim() presents results for all the columns, and the statistics will depend on the class of the variable.

```
skim(hate_crimes)
```

Hopefully in your statistical modules you had taken previously, you have learned some things about how to graphically display variables. So you may have some memory about the amount of work involved with this. Hopefully R will offer some respite. Of course, there are many different ways of producing graphics in R. In this course we rely on a package called *ggplot2*, which is part of the tidyverse set of packages mentioned earlier.

```
library(ggplot2)
```

Then we will use one of its functions to create a scatterplot.

```
ggplot(hate_crimes, aes(x=share_vote_trump, y=avg_hatecrimes_per_100k_fbi)) +
    geom_point(shape=1) +
    geom_smooth(method=lm)
```

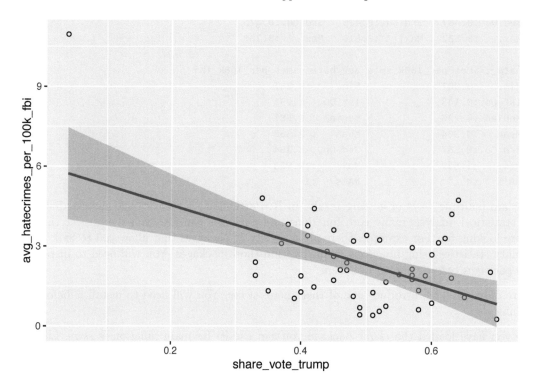

Graphing is very powerful in R, and much of the spatial visualisation we will produce throughout the book will build on this. If you are not already familiar with this, we recommend a read of the data visualisation chapter of Wickham and Grolemund (2017).

Getting organised: R Projects

One thing that can help you tremendously is keeping your code organised. RStudio helps with this by virtue of something called **R Projects**.

Technically, a RStudio project is just a directory with the name of the project, and a few files and folders created by RStudio for internal purposes. This is where you should hold your scripts, your data, and reports. You can manage this folder with your own operating system manager (discussed earlier, e.g., Windows) or through the RStudio file manager (that you access in the bottom-right corner set of windows in RStudio).

When a project is reopened, RStudio opens every file and data view that was open when the project was closed last time around. Trust me, this is a really helpful thing! If you create a project for this module, you can keep everything in one place, and refer back to your old code and your learnings throughout the module.

Saving your work and projects

First things first, hopefully you have already created a separate folder on your computer. Now save the script you've been working on into this folder. By clicking on "File" and "Save as...".

Then navigate to your folder for this module, and for your script make sure to give it some meaningful name like appendixlab.R or something like this. Then click 'save'.

Now, go back to "File" and select "New project..."

Then in the options that appear, choose "Existing Directory". This is because you already have a folder for this work; this is where you saved your script just before. For me this was my folder called 'crime_mapping'.

So select 'Existing Directory', and on the next page use the "Browse" button to select this folder (directory) where you saved the script earlier. Once you have done this, click on 'Create Project' on the bottom.

This will now open up a new RStudio window with your project. In the future, you can start right back up where you finished last time by navigating to the .Rproj file, and double clicking it. It helps you keep everything in one place, and lets R read everything from that folder.

With simple projects, a single script file and a data file is all you may have. But with more complex projects, things can rapidly become messy. So you may want to create sub-directories within this project folder. You could use the following structure to put all files of a certain type in the same sub-directory:

- *Scripts and code*: Here you put all the text files with the analytic code, including Rmarkdown files which is something we don't introduce here.

- *Source data*: Here you can put the original data. I would tend not to touch this once I have obtained the original data.

- *Documentation*: This is the sub-directory where you can place all the data documentation (e.g., codebooks, questionnaires, etc.)

- *Modified data*: All analysis involves doing transformations and changing things in the original data files. You don't want to mess up the original data files, so what you should do is create new data files as soon as you start changing your source data. I would go so far as to place them in a different sub-directory.

- *Literature*: Analysis is all about answering research questions. There is always a literature about these questions. I would place the relevant literature for the analytic project I am conducting in this sub-directory.

- *Reports and write-up*: Here is where I would file all the reports and data visualisations that are associated with my analysis.

You can read up on why projects are useful in Chan (2020).

Summary and further reading

This has been a very superficial overview of using R and RStudio to write code. If you were able to follow along, then you should be OK to follow along with the contents of this book as well. If you are not yet feeling confident, and are looking to further develop your R skills, we highly recommend Wickham and Grolemund (2017). Throughout the book we use Tidyverse packages and syntax wherever possible. For an overview of this, check out Wickham et al. (2019). And for those who want to hone their R skills even further, there is Wickham (2019) for generally more advanced R skills, and of course, Bivand, Pebesma, and Gómez-Rubio (2013) and Lovelace, Nowosad, and Muenchow (2019) are important reference texts.

Appendix B: Regression analysis (a refresher)

Introduction

In this appendix we present a very brief and high-level overview of regression analysis, in order to give some reference to when we introduce spatial regression in Chapter 11. We will follow a practical approach, working through one dataset to illustrate regression basics using R. For those who wish to have a more thorough grounding please refer to the further reading section at the end of this appendix.

Here we will return to the "ncovr" data set produced as part of the National Consortium on Violence Research (NCOVR) agenda (we use this also in chapter 11, for more information on this data see Baller et al. (2001)). The data set includes variables related to homicides in the US, as well as information in a number of sociodemographic variables that are often thought of as associated with the geographical distribution of homicides. This data set is available as part of the `geodaData` package.

```
library(geodaData)
data("ncovr")
```

The dataset contains information about 3085 counties in the United States and if you view it you will see it has information about several decades: the 60s, 70s, 80s, and 90s. The number at the end of the variable names denotes the relevant decade, and you will see that for each decade we have the same variables.

The purpose of regression analysis as we approach it here is to choose a model to represent the relationship between homicide and various predictors. You can think of a model as a map. A map aims to represent a given reality, but as you may have already discovered there are many ways of presenting the same information through a map. As an analyst you decide what the most appropriate representation for your needs is. Each representation you choose will involve an element of distortion. Maps (and models) are not exact representations of the real word, they are simply good approximations that may serve well a particular functional need. You may think about the map of the London Underground. It doesn't show you geographically where the Underground lines of London are; yet, if you want to use the Metro system, this map will be extremely helpful to you. It serves a need in a good way. The same happens with models. They may not be terribly good reflections of the world, but may give us approximations that allows us to develop useful insights.

Choosing a good model is like choosing a good way for displaying quantitative information in a map. Decisions, decisions, decisions. There are many parameters and options one can choose from. This can be overwhelming, particularly as you are learning how to model and map phenomena. How to make good decisions is something that you learn on earnest by practice, practice, practice. Nobody expects you to get the maps you are doing as you are learning, and the models you are developing as you are learning spot on. So please do not stress out about this. All we can do here is to learn some basic principles and start getting some practice, which you will be able to further develop in a professional context or in further training.

The first step in any analysis is to develop some familiarity with the data you are going to be working with. We have been here before. Read the codebook. Run summary statistics for your quantitative variables, frequency distributions for your categorical variables, and visualise your variables. This will help you to detect any anomalies and give you a sense for what you have. If, for example, you run a histogram for the homicide rate for 1990 (HR90), you will get a sense of the distribution form –which of course is skewed.

```
library(ggplot2)
qplot(x = HR90, data = ncovr)
```

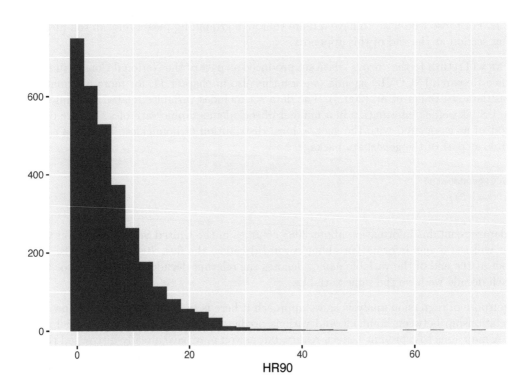

FIGURE B.1: Histogram of the HR90 variable

Once one has gone through the process of exploring the data in this way (or maybe using the skimr package - see *Appendix: A quick intro to R and RStudio*) for all the variables you want to work with, you can start exploring bivariate associations with your dependent, response or outcome variable. So, as an illustration, you could explore the association with resource deprivation (*RD90*), a measure of the level of concentrated disadvantage or social exclusion in an area, via a scatterplot:

```
ggplot(ncovr, aes(x = RD90, y = HR90)) +
  geom_point(alpha=.2)
```

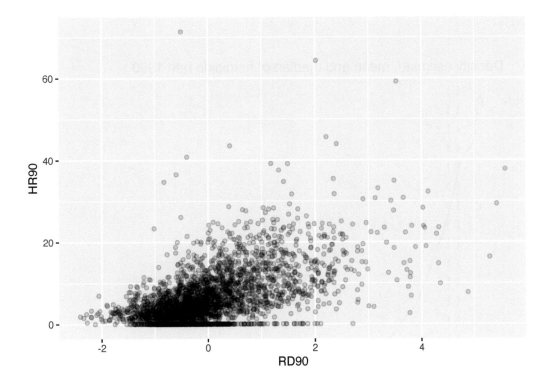

FIGURE B.2: Scatterplot showing relationship between resource deprivation (RD90) and homicide rate (HR90) in the 90s

What do you think when looking at this scatterplot? Is there a relationship between the variables? Does it look as if individuals that have a high score on the x axis also have a high score on the y axis? Or vice versa?

Motivating regression

Now, imagine that we play a game. Imagine I have all the respondents waiting in a room, and I randomly call one of them to the stage. You're sitting in the audience, and you have to guess the level of homicide (*HR90*) for that respondent. Imagine that I pay £150 to the student that gets the closest to the right value. What would you guess if you only have one guess and you knew (as we do) how homicide in the 90s is distributed?

```
ggplot(ncovr, aes(x = HR90)) +
  geom_density() +
```

```
geom_vline(xintercept = median(ncovr$HR90), linetype = "dashed", size = 1,
        color="red") + # median = 4.377
geom_vline(xintercept = mean(ncovr$HR90), linetype = "dashed", size = 1,
        color="blue") + # mean = 6.183
ggtitle("Density estimate, mean and median of homicide rate 1990")
```

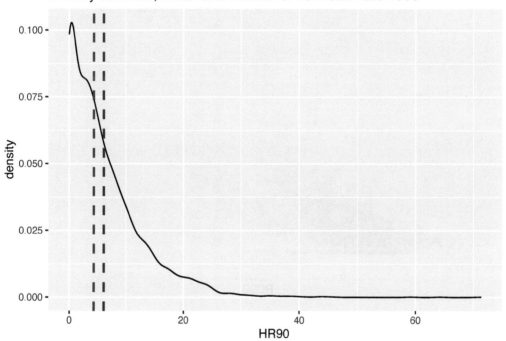

FIGURE B.3: Density estimate, mean, and median of homicide rate in 1990

```
summary(ncovr$HR90)
```

```
##    Min. 1st Qu.  Median    Mean 3rd Qu.    Max.
##    0.00    1.33    4.38    6.18    8.94   71.38
```

If I only had one shot, you could go for the median, in red, (given the skew) but the mean, in blue, perhaps would be your second best. Most of the areas here have values clustered around those values, which is another way of saying they are bound to be not too far from them.

Imagine, however, that now when someone is called to the stage, you are told the level of resource deprivation in the county - so the value of the *RD90* variable for the individual that has been selected (for example 4). Imagine as well that you have the scatterplot that we produced earlier in front of you. Would you still go for the value of "4.377" as your best guess for the value of the selected county?

I certainly would not go with the overall mean or median as my prediction anymore. If somebody said to me: the value *RD90* for the selected respondent is 4, I would be more inclined to guess the mean value for the level of homicide with that level of resource deprivation (the conditional mean), rather than the overall mean across all the counties. Wouldn't you?

If we plot the conditional means, we can see that the mean of homicide rate for counties that report a value of 4 in *RD90* is around 22. So you may be better off guessing that.

```
library(grid)
ggplot() +
  geom_point(data=ncovr, aes(x = RD90, y = HR90), alpha=.2) +
geom_line(data=ncovr, aes(x = round(RD90/0.12)*0.12, y = HR90),
          stat='summary', fun.y=mean, color="red", size=1) +
  annotate("segment", x=3, xend = 4, y = 25, yend= 22, color = "blue",
           size = 2, arrow = arrow()) +
  annotate("text", x = 3, y = 29, label = "Pick this one!",
           size =7, colour = "blue")
```

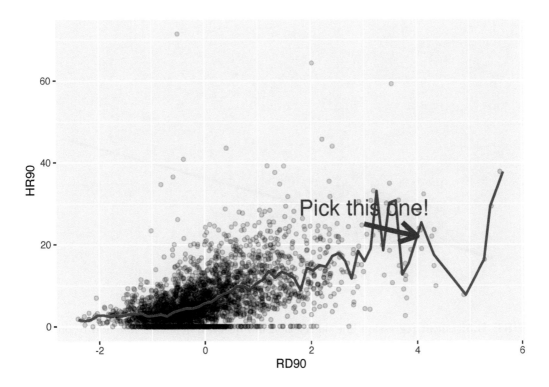

FIGURE B.4: Our best guess for homicide rate when resource deprivation score in the county is 4

Linear regression tackles this problem using a slightly different approach. Rather than focusing on the conditional mean (smoothed or not), it draws a straight line that tries to capture the trend in the data. If we focus on the regions of the scatterplot that are less sparse, we see that this is an upward trend, suggesting that as resource deprivation increases, so does the homicide rate.

Simple linear regression draws a single straight line of predicted values as the model for the data. This line would be a **model**, a *simplification* of the real world like any other model (e.g., toy pistol, architectural drawing, subway map), that assumes that there is approximately a linear relationship between X and Y. Let's draw the regression line using the function `geom_smooth()`. This asks for a `method=` parameter where we specify `lm` for the linear regression line. We also set `se=` to `FALSE`, so we ask for just the line to be printed but not the standard error around it. The other arguments specify the colour and thickness of the line.

```
ggplot(data = ncovr, aes(x = RD90, y = HR90)) +
  geom_point(alpha = .2) +
  geom_smooth(method = "lm", se = FALSE, color = "red", size = 1)
```

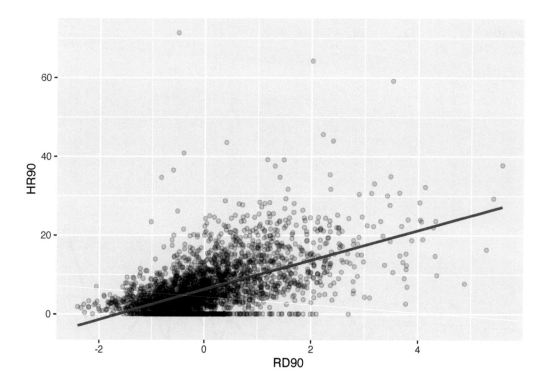

FIGURE B.5: Linear regression line on relationship between homicide rate and resource deprivation

What that line is doing is giving you guesses (predictions) for the values of homicide based on the information that we have about the level of resource deprivation. It gives you one possible guess for the value of homicide for every possible value of resource deprivation and links them all together in a straight line.

The linear model then is a model that takes the form of the equation of a straight line through the data. The line does not go through all the points.

Our regression line underpredicts at low levels of resource deprivation and does not seem to capture well the variability at higher levels of resource deprivation. But imperfect as a model as it might be it simplifies well the overall growing trend for homicide as resource deprivation increases.

As Veaux, Velleman, and Bock (2012) highlight: "like all models of the real world, the line will be wrong, wrong in the sense that it can't match reality exactly. But it can help us understand how the variables are associated" (p. 179). A map is never a perfect representation of the world, the same happens with statistical models. Yet, as with maps, models can be helpful.

Fitting a simple regression model

In order to draw a regression line (or in fact any line in a Cartesian coordinate system), we need to know two things:

- Where the line begins, what is the value of Y (our dependent variable) when X (our independent variable) is 0, so that we have a point from which to start drawing the line. The technical name for this point is the **intercept**.

- What is the **slope** of that line, that is, how inclined the line is, the angle of the line.

If you recall from elementary algebra (and you may not), the equation for any straight line is: $y = mx + b$. In statistics we use a slightly different notation, although the equation remains the same: $y = \beta_0 + \beta_1 x$.

We need the origin of the line (β_0) and the slope of the line (β_1). How does R get the intercept and the slope for the green line? How does R know where to draw this line? We need to estimate these **parameters** (or **coefficients**) from the data. How? We don't have the time to get into these more mathematical details now, but will suggest further reading at the end of this appendix for the curious. For now, suffice to say that for linear regression modes like the one we cover here, when drawing the line, R tries to minimise the distance from every point in the scatterplot to the regression line using a method called **least squares estimation**.

In order to fit the model, we use the `lm()` function using the formula specification $(Y\ X)$. Typically you want to store your regression model in a "variable", let's call it `fit_1`:

```
fit_1 <- lm(HR90 ~ RD90, data = ncovr)
```

You will see in your R Studio global environment space that there is a new object called `fit_1` with 12 elements on it. We can get a sense for what this object is and includes using the functions we introduced in previous weeks:

```
class(fit_1)
```

```
## [1] "lm"
```

```
attributes(fit_1)
```

```
## $names
##  [1] "coefficients"  "residuals"     "effects"
```

```
##   [4] "rank"          "fitted.values" "assign"
##   [7] "qr"            "df.residual"   "xlevels"
## [10] "call"           "terms"         "model"
##
## $class
## [1] "lm"
```

R is telling us that this is an object of class `lm` and that it includes a number of attributes. One of the beauties of R is that you are producing all the results from running the model, putting them in an object, and then giving you the opportunity for using them later on. If you want to simply see the basic results from running the model, you can use the `summary()` function.

```
summary(fit_1)
```

```
##
## Call:
## lm(formula = HR90 ~ RD90, data = ncovr)
##
## Residuals:
##    Min     1Q Median     3Q    Max
## -17.80  -3.42  -0.72   2.54  67.10
##
## Coefficients:
##              Estimate Std. Error t value Pr(>|t|)
## (Intercept)   6.1829     0.0984    62.8   <2e-16 ***
## RD90          3.7712     0.0985    38.3   <2e-16 ***
## ---
## Signif. codes:
## 0 '***' 0.001 '**' 0.01 '*' 0.05 '.' 0.1 ' ' 1
##
## Residual standard error: 5.47 on 3083 degrees of freedom
## Multiple R-squared:  0.322,  Adjusted R-squared:  0.322
## F-statistic: 1.47e+03 on 1 and 3083 DF,  p-value: <2e-16
```

Or if you prefer more parsimonious presentation, you could use the `display()` function of the arm package:

```
library(arm)
```

```
display(fit_1)
```

```
## lm(formula = HR90 ~ RD90, data = ncovr)
##             coef.est coef.se
## (Intercept) 6.18     0.10
## RD90        3.77     0.10
## ---
## n = 3085, k = 2
## residual sd = 5.47, R-Squared = 0.32
```

For now I just want you to focus on the numbers in the "Estimate" (or coef.est) column.

The value of 6.18 estimated for the **intercept** is the "predicted" value for Y when X equals zero. This is the predicted value of the fear of crime score *when resource deprivation has a value of zero.*

```
summary(ncovr$RD90)
```

```
##    Min. 1st Qu.  Median    Mean 3rd Qu.    Max.
## -2.410  -0.667  -0.202   0.000   0.439   5.583
```

RD90 is a variable that has been centered in 0. It has been created by the researchers in such a way that it has a mean value of 0. Since we only have one explanatory variable in the model, this corresponds to the mean of the homicide rate, 6.18. In many other contexts the intercept has less of a meaning.

We then need the b1 regression coefficient for our independent variable, the value that will shape the **slope** in this scenario. This value is 3.77. This estimated regression coefficient for our independent variable has a convenient interpretation. When the value is positive, it tells us that *for every one unit increase in X, there is a b1 increase on Y.* If the coefficient is negative, then it represents a decrease on Y. Here, we can read it as "for every one unit increase in the resource deprivation score, there is a 3.77 unit increase in the homicide rate."

Knowing these two parameters not only allows us to draw the line, but we can also solve for any given value of X. Let's go back to our guess-the-homicide-rate game. Imagine I tell you the level of resource deprivation is 1. What would be your best bet now? We can simply go back to our regression line equation and insert the estimated parameters:

$$y = b_0 + b_1 x$$
$$y = 6.18 + 3.77 \times 1$$

$$y = 9.95$$

Or if you don't want to do the calculation yourself, you can use the `predict()` function (differences are due to rounding error). First, you name your stored model and then you identify the new data (which has to be in a dataframe format and with a variable name matching the one in the original data set).

```
predict(fit_1, data.frame(RD90 = c(1)))
```

```
##     1
## 9.954
```

This is the expected value of Y, homicide rate, when X, resource deprivation is 1 **according to our model** (according to our simplification of the real world, our simplification of the whole cloud of points into just one straight line). Look back at the scatterplot we produced earlier. Does it look as if the red line when X is 1 corresponds to a value of Y of 9.95?

Residuals revisited: R squared

In the output above we saw there was something called the residuals. The residuals are the differences between the observed values of Y for each case minus the predicted or expected value of Y; in other words, the distances between each point in the dataset and the regression line (see the visual example below).

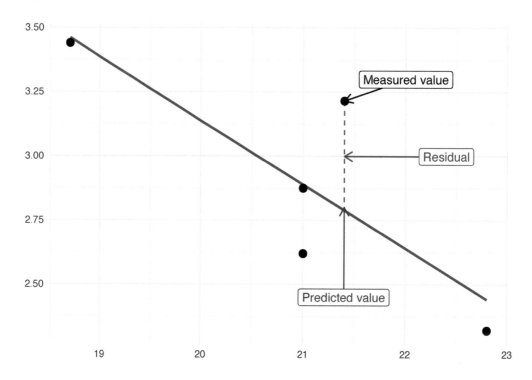

FIGURE B.6: Predicted values, measured value, and residual

You see that we have our line, which is our predicted values, and then we have the black dots which are our actually observed values. The distance between them is essentially the amount by which we were wrong, and all these distances between observed and predicted values are our residuals. Least square estimation essentially aims to reduce the average of the squares of all these distances: that's how it draws the line.

Why do we have residuals? Well, think about it. The fact that the line is not a perfect representation of the cloud of points makes sense, doesn't it? You cannot predict perfectly what the value of Y is for every observation just by looking ONLY at their level of resource deprivation! This line only uses information regarding resource deprivation. This means that there's bound to be some difference between our predicted level of homicide given our knowledge of deprivation (regression line) and the actual level of homicide (actual location of the points in the scatterplot). There are other things that matter that are not being taken into account by our model to predict the values of Y. There are other things that surely matter in terms of understanding homicide. And then, of course, we have measurement error and other forms of noise.

We can re-write our equation like this if we want to represent each value of Y (rather than the predicted value of Y) then: $y = \beta_0 + \beta_1 x +$ residuals.

The residuals capture how much variation is unexplained, how much we still have to learn if we want to understand variation in Y. A good model tries to maximise explained variation and reduce the magnitude of the residuals.

We can use information from the residuals to produce a measure of effect size, of how good our model is in predicting variation in our dependent variables. Remember our game where we try to guess homicide (Y)? If we did not have any information about X, our best bet for Y would be the mean of Y. The regression line aims to improve that prediction. By knowing the values of X, we can build a regression line that aims to get us closer to the actual values of Y (look at the figure below).

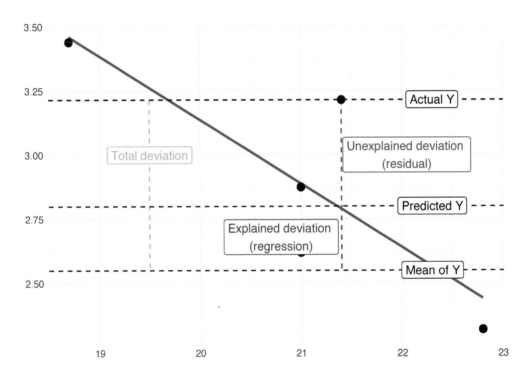

FIGURE B.7: Illustration of explained, unexplained, and total deviation

The distance between the mean (our best guess without any other piece of information) and the observed value of Y is what we call the **total variation**. The residual is the difference between our predicted value of Y and the observed value of Y. This is what we cannot explain (i.e, variation in Y that is *unexplained*). The difference between the mean value of Y and the expected value of Y (value given by our regression line) is how much better we are doing with our prediction by using information about X (i.e., in our previous example, it would be variation in Y that can be *explained* by knowing about resource deprivation), how much closer the regression line gets us to the observed values. We can then contrast these two different sources of variation (explained and unexplained) to produce a single measure of how good our model is. The formula is as follows:

$$R^2 = \frac{SSR}{SST} = \frac{\Sigma(\hat{y}_i - \bar{y})^2}{\Sigma(y_i - \bar{y})^2}$$

All this formula is doing is taking a ratio of the explained variation (squared differences between the regression line and the mean of Y for each observation) by the total variation (squared differences of the observed values of Y for each observation from the mean of Y). This gives us a measure of the **percentage of variation in Y that is "explained" by X**. If this sounds familiar, it is because it is a measure similar to eta squared (eta^2) in ANOVA.

As then we can take this value as a measure of the strength of our model. If you look at the R output, you will see that the R2 for our model was .32 (look at the multiple R square value in the output). We can say that our model explains 32% of the variance in the fear of homicide. When doing regression, you will often find that regression models with aggregate data such as county level data will give you better results than when dealing with individuals. It is much harder understanding individual variation than county level variation.

As an aside, and to continue emphasising your appreciation of the object oriented nature of R, when we run the `summary()` function, we are simply generating a list object of the class summary.lm.

```
attributes(summary(fit_1))
```

```
## $names
##  [1] "call"          "terms"          "residuals"
##  [4] "coefficients"  "aliased"        "sigma"
##  [7] "df"            "r.squared"      "adj.r.squared"
## [10] "fstatistic"    "cov.unscaled"
##
## $class
## [1] "summary.lm"
```

This means that we can access its elements if so we wish. So, for example, to obtain just the R Squared (R^2), we could ask for:

```
summary(fit_1)$r.squared
```

```
## [1] 0.3224
```

Knowing how to interpret this is important. R^2 ranges from 0 to 1. The greater it is, the more powerful our model is and the more explaining we are doing: the better we are able to account for variation in our outcome Y with our input. In other words, the stronger the relationship is between Y and X. As with all the other measures of effect size interpretation is a matter of judgment. You are advised to see what other researchers report in relation to the particular outcome that you may be exploring.

Weisburd and Britt (2014) suggest that in criminal justice you rarely see values for R^2 greater than .40. Thus, if your R^2 is larger than .40, you can assume you have a powerful model. When, on the other hand, R^2 is lower than .15 or .2 the model is likely to be viewed as relatively weak. Our observed R^2 here is rather poor. There is considerably room for improvement if we want to develop a better model to explain fear of violent crime. In any case, many people would argue that R^2 is a bit overrated. You need to be aware of what it measures and the context in which you are using it.

Inference with regression

In real applications, we have access to a set of observations from which we can compute the least squares line, but the population regression line is unobserved. So our regression line is one of many that could be estimated. A different sample would produce a different regression line. The same sort of ideas that we introduced when discussing the estimation of sample means or proportions also apply here. if we estimate b0 and b1 from a particular sample, then our estimates won't be exactly equal to b0 and b1 in the population. But if we could average the estimates obtained over a very large number of datasets, the average of these estimates would equal the coefficients of the regression line in the population.

We can compute standard errors for the regression coefficients to quantify our uncertainty about these estimates. These standard errors can in turn be used to produce confidence intervals. This would require us to assume that the residuals are normally distributed. For a simple regression model, you are assuming that the values of Y are approximately normally distributed for each level of X. In those circumstances we can trust the confidence intervals that we can draw around the regression line.

You can also then perform an standard hypothesis test on the coefficients. As we saw before when summarising the model, R will compute the standard errors and a t test for each of the coefficients.

```
summary(fit_1)$coefficients
```

```
##               Estimate Std. Error t value   Pr(>|t|)
## (Intercept)     6.183    0.09844    62.81   0.000e+00
## RD90            3.771    0.09846    38.30 6.536e-263
```

In our example, we can see that the coefficient for our predictor here is statistically significant.

We can also obtain confidence intervals for the estimated coefficients using the `confint()` function:

```
confint(fit_1)
```

```
##               2.5 % 97.5 %
## (Intercept)   5.990  6.376
## RD90          3.578  3.964
```

Fitting regression with categorical predictors

So far we have explained regression using a numeric input. It turns out we can also use regression with categorical explanatory variables. It is quite straightforward to run it.

There is only one categorical explanatory variable in this dataset, a binary indicator that indicates whether the county is in a Southern State or not. We can also explore this relationship using regression and a regression line. This is how you would express the model:

```
#We use the as.factor function to tell R that SOUTH is a categorical variable
fit_2 <- lm(HR90 ~ as.factor(SOUTH), data=ncovr)
```

Notice that there is nothing different in how we ask for the model.

Let's have a look at the results:

```
summary(fit_2)
```

```
##
## Call:
## lm(formula = HR90 ~ as.factor(SOUTH), data = ncovr)
##
## Residuals:
##     Min     1Q Median     3Q    Max
##   -9.55  -3.34  -1.17   1.93  68.04
##
## Coefficients:
##                     Estimate Std. Error t value Pr(>|t|)
## (Intercept)            3.342      0.144    23.2   <2e-16
## as.factor(SOUTH)1      6.208      0.212    29.2   <2e-16
##
## (Intercept)         ***
## as.factor(SOUTH)1   ***
## ---
## Signif. codes:
## 0 '***' 0.001 '**' 0.01 '*' 0.05 '.' 0.1 ' ' 1
##
## Residual standard error: 5.88 on 3083 degrees of freedom
## Multiple R-squared:  0.217,  Adjusted R-squared:  0.217
## F-statistic:  854 on 1 and 3083 DF,  p-value: <2e-16
```

As you will see, the output does not look too different. But notice in the print out how the row with the coefficient and other values for our input variable SOUTH we see a value of 1 after the name of the variable. What does this mean?

It turns out that a linear regression model with just one dichotomous categorical predictor is just the equivalent of a t test. When you only have one predictor, the value of the intercept is the mean value of the **reference category** and the coefficient for the slope tells you how much higher (if it is positive) or how much lower (if it is negative) is the mean value for the other category in your factor.

The reference category is the one for which R does not print the *level* next to the name of the variable for which it gives you the regression coefficient. Here we see that the named level is "1". That's telling you that the reference category here is "0". If you look at the codebook, you will see that 1 means the county is in a Southern state. Therefore, the Y intercept in this case is the mean value of fear of violent crime for the northern counties, whereas the coefficient for the slope is telling you how much higher (since it is a positive value) the mean value is for the southern counties. Don't believe me?

```
mean(ncovr$HR90[ncovr$SOUTH == 0], na.rm=TRUE)
```

```
## [1] 3.342
```

```
mean(ncovr$HR90[ncovr$SOUTH == 1], na.rm=TRUE) - mean(ncovr$HR90[ncovr$SOUTH == 0], na.rm=TRUE)
```

```
## [1] 6.208
```

So, to reiterate, for a binary predictor, the coefficient is nothing else than the difference between the mean of the two levels in your factor variable, between the averages in your two groups.

With categorical variables encoded as **factors**, you always have a situation like this: a reference category and then as many additional coefficients as there are additional levels in your categorical variable. Each of these additional categories is included into the model as **"dummy" variables**. Here our categorical variable has two levels, thus we have only one dummy variable. There will always be one fewer dummy variable than the number of levels. The level with no dummy variable, northern counties in this example, is known as the **reference category** or the **baseline**.

It turns out then that the regression table is printing out for us a t test of statistical significance for every input in the model. If we look at the table above, this t value is 29.22 and the p value associated with it is near 0. This is indeed considerably lower than the conventional significance level of 0.05. So we could conclude that the probability of obtaining this value if the null hypothesis is true is very low. The R^2 is not too bad either, although lower than we saw when using resource deprivation.

Motivating multiple regression

So we have seen that we can fit models with just one predictor. We can build better models by expanding the number of predictors (although keep in mind you should also aim to build models as parsimonious as possible).

Another reason why it is important to think about additional variables in your model is to control for spurious correlations (although here you may also want to use your common sense when selecting your variables!). You must have heard before that correlation does not equal causation. Just because two things are associated, we cannot assume that one is the cause for the other. Typically we see how the pilots switch the secure the belt button when there is turbulence. These two things are associated, they tend to come together. But the pilots are not causing the turbulences by pressing a switch! The world is full of **spurious correlations**, associations between two variables that should not be taken too seriously. You can explore a few here[3]. It's funny.

Looking only at covariation between a pair of variables can be misleading. It may lead you to conclude that a relationship is more important than it really is. This is no trivial matter, but one of the most important ones we confront in research and policy.

It's not an exaggeration to say that most quantitative explanatory research is about trying to control for the presence of **confounders**, variables that may explain away observed associations. Think about any criminology question: Does marriage reduce crime? Or is it that people which get married are different from those that don't (and are those pre-existing differences that are associated with less crime)? Do gangs lead to more crime? Or is it that young people that join gangs are more likely to be offenders to start with? Are the police being racist when they stop and search more members of ethnic minorities? Or is it that there are other factors (i.e., offending, area of residence, time spent in the street) that, once controlled, would mean there is no ethnic disproportionality in stop and searches? Does a particular program reduce crime? Or is the observed change due to something else?

These things also matter for policy. Wilson and Kelling (1982), for example, argued that signs of incivility (or antisocial behaviour) in a community lead to more serious forms of crime later on as people withdraw to the safety of their homes when they see those signs of incivilities, and this leads to a reduction in informal mechanisms of social control. Many policies to tackle antisocial behaviour are very much informed by this model and were heavily influenced by broken windows theory.

But is the model right? Sampson, Raudenbush, and Earls (1997) argue it is not entirely correct. They tried to show that there are other confounding factors (poverty, collective efficacy) that explain the association of signs of incivility and more serious crime. In other words, the reason why you see antisocial behaviour in the same communities that you see crime is because other structural factors explain both of those outcomes. They also argue that perceptions of antisocial behaviour are not just produced by observed antisocial behaviour, but also by stereotypes about social class and race. If you believe them, then the policy implications are that only tackling antisocial behaviour won't help you to reduce crime (as Wilson and Kelling (1982) have argued). So as you can see this stuff matters for policy not just for theory.

Multiple regression is one way of checking the relevance of competing explanations. You could set up a model where you try to predict crime levels with an indicator of broken windows and an indicator of structural disadvantage. If after controlling for structural disadvantage you see that the regression coefficient for broken windows is still significant, you may be onto something, particularly if the estimated effect is still large. If, on the other hand, the t test for the regression coefficient of your broken windows variable is no longer significant, then you may be tempted to think that perhaps Sampson, Raudenbush, and Earls (1997) were onto something.

---

[3]http://tylervigen.com/

Fitting and interpreting a multiple regression model

It could not be any easier to fit a multiple regression model. You simply modify the formula in the `lm()` function by adding terms for the additional inputs.

```
ncovr$SOUTH_f <- as.factor(ncovr$SOUTH)
fit_3 <- lm(HR90 ~ RD90 + SOUTH_f, data=ncovr)
summary(fit_3)
```

```
##
## Call:
## lm(formula = HR90 ~ RD90 + SOUTH_f, data = ncovr)
##
## Residuals:
##     Min     1Q Median     3Q    Max
## -16.48  -3.00  -0.58   2.22  68.15
##
## Coefficients:
##              Estimate Std. Error t value Pr(>|t|)
## (Intercept)     4.727      0.139    33.9   <2e-16 ***
## RD90            2.965      0.111    26.8   <2e-16 ***
## SOUTH_f1        3.181      0.222    14.3   <2e-16 ***
## ---
## Signif. codes:
## 0 '***' 0.001 '**' 0.01 '*' 0.05 '.' 0.1 ' ' 1
##
## Residual standard error: 5.3 on 3082 degrees of freedom
## Multiple R-squared:  0.365,  Adjusted R-squared:  0.364
## F-statistic:  884 on 2 and 3082 DF,  p-value: <2e-16
```

With more than one input, you need to ask yourself whether all of the regression coefficients are zero. This hypothesis is tested with the F test. Again we are assuming the residuals are normally distributed, though with large samples the F statistics approximates the F distribution. You see the F test printed at the bottom of the summary output and the associated p value, which in this case is way below the conventional .05 that we use to declare statistical significance and reject the null hypothesis. At least one of our inputs must be related to our response variable.

Notice that the table printed also reports a t test for each of the predictors. These are testing whether each of these predictors is associated with the response variable when adjusting for the other variables in the model. They report the "partial effect of adding that variable to the model" (James et al. 2013: 77). In this case we can see that both variables seem to be significantly associated with the response variable.

If we look at the R^2 we can now see that it is higher than before. R^2 will always increase as a consequence of adding new variables, even if the new variables added are weakly related to the response variable.

We see that the coefficients for the predictors change somehow. The coefficient goes down a bit for *RD90* and it halves for *SOUTH*. **But their interpretation now changes**. A common interpretation is that now the regression for each variable tells you about changes in Y related to that variable **when the other variables in the model are held constant**. So, for example, you could say the coefficient for *RD90* represents the increase in homicide for every one-unit increase in the measure of resource deprivation *when holding all other variables in the model constant* (in this case that refers to holding constant *SOUTH*). But this terminology can be a bit misleading.

Other interpretations are also possible and are more generalisable. Gelman and Hill (2007) emphasise what they call the *predictive interpretation* that considers how "the outcome variable differs, on average, when comparing two groups of units that differ by 1 in the relevant predictor while being identical in all the other predictors" (p. 34). So if you're regressing y on u and v, the coefficient of u is the average difference in y per difference in u, comparing pairs of items that differ in u but are identical in v.

For example, in this case we could say that comparing counties that have the same level of resource deprivation but that differed in whether they are South or North, the model predicts an expected difference of 3.18 in their homicide rate. And that respondents that do not vary in whether they are South or North, but that differ by one point in the level of resource deprivation, we would expect to see a difference of 2.96 in their homicide rate. So we are interpreting the regression slopes **as comparisons of cases that differ in one predictor while being at the same levels of the other predictors**.

As you can see, interpreting regression coefficients can be kind of tricky. The relationship between the response y and any one explanatory variable can change greatly depending on what other explanatory variables are present in the model.

For example, if you contrast this model with the one we run with only *SOUTH* as a predictor, you will notice the intercept has changed. You no longer read the intercept as the mean value of homicide rate for Northern counties. *Adding predictors to the model changes their meaning.* Now the intercept index the value of homicide for southern counties that score 0 in *RD90*. In this case you have cases that meet this condition (equal zero in all your predictors), but often you may not have any case that does meet the definition of the intercept. More often than not, then, there is not much value in bothering to interpret the intercept.

Something you need to be particularly careful about is to interpret the coefficients in a causal manner. At least your data come from an experiment, this is unlikely to be helpful. With observational data, regression coefficients should not be read as indexing causal relations. This sort of textbook warning is, however, often neglectfully ignored by professional researchers. Often authors carefully draw sharp distinctions between causal and correlational claims when discussing their data analysis, but then interpret the correlational patterns in a totally causal way in their conclusion section. This is what is called the causation[4] or causal[5] creep. Beware. Don't do this, as tempting as it may be.

Comparing the simple models with this more complex model, we could say that adjusting for *SOUTH* does not change much the impact of *RD90* in homicide, but that adjusting for resource deprivation halves the impact of the regional effect on homicide.

---

[4]http://junkcharts.typepad.com/numbersruleyourworld/2012/07/the-causation-creep.html
[5]http://www.carlislerainey.com/2012/12/05/another-example-of-causal-creep/

Presenting your regression results

Communicating your results in a clear manner is incredibly important. We have seen the tabular results produced by R. If you want to use them in a paper, you may need to do some tidying up of those results. There are a number of packages (`textreg`, `stargazer`) that automatise that process. They take your `lm` objects and produce tables that you can put straight away in your reports or papers. One popular trend in presenting results is the **coefficient plot** as an alternative to the table of regression coefficients. There are various ways of producing coefficient plots with R for a variety of models.

We are going to use instead the `plot_model()` function of the `sjPlot` package, that makes it easier to produce this sort of plots.

```
library(sjPlot)
```

Let's try with a more complex example:

```
fit_4 <- lm(HR90 ~ RD90 + SOUTH_f + DV90 + MA90 + PS90, data=ncovr)
plot_model(fit_4, breakLabelsAt = 30)
```

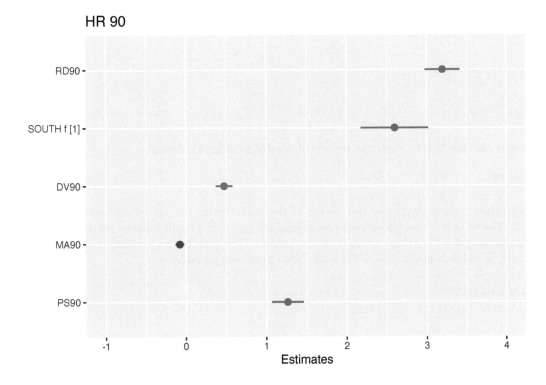

FIGURE B.8: Regression forest plot for the results of a regression

What you see plotted here is the point estimates (circles), the confidence intervals around those estimates (the longer the line the less precise the estimate), and the colours represent whether the effect is negative (red) or positive (blue). There are other packages that also provide similar functionality, like the `dotwhisker` package that you may want to explore.

The `sjPlot` package also allows you to produce html tables for more professional presentation of your regression tables. For this we use the `tab_model()` function.

```
tab_model(fit_4)
```

You can customise this table. For example, you can change the name that is displayed for the dependent variable with the `dv.labels` = parameter, and you could change the labels for the independent variables with the `pred.labels` = parameter.

```
tab_model(fit_4,
          dv.labels = "Homicide rate 1990",
          pred.labels = c("(Intercept)",
                          "Resource deprivation","South",
                          "Percent divorced males", "Median age",
                          "Population structure"))
```

The `tab_model()` function is great if you have an html output. In case you're making your tables for a PDF output, you might want to consider the `stargazer` package, from which the function `stargazer()` stargazer command produces LaTeX code, HTML code and ASCII text for well-formatted tables.

```
library(stargazer)
```

```
stargazer(fit_4, header = FALSE, title = "Model summary table using stargazer package")
```

You could also use this to show the results of several models side-by-side.

```
stargazer(fit_3, fit_4, header = FALSE, title = "Example table to display two models in one table")
```

Besides tables, visual display of the effects of the variables in the model are particularly helpful. The `effects` package allows us to produce plots to visualise these relationships (when adjusting for the other variables in the model). Here's an example going back to our model fit_3 which contained *SOUTH* and *RD90* predictor variables:

```
library(effects)
plot(allEffects(fit_3), ask=FALSE)
```

TABLE B.1: Model summary table using stargazer package

	Dependent variable:
	HR90
RD90	3.199***
	(0.112)
SOUTH_f1	2.600***
	(0.216)
DV90	0.476***
	(0.053)
MA90	−0.076***
	(0.027)
PS90	1.265***
	(0.100)
Constant	4.204***
	(0.985)
Observations	3,085
R^2	0.426
Adjusted R^2	0.425
Residual Std. Error	5.035 (df = 3079)
F Statistic	457.300*** (df = 5; 3079)
Note:	*p<0.1; **p<0.05; ***p<0.01

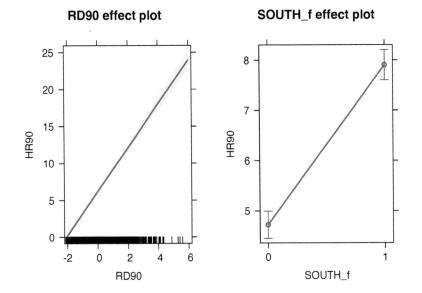

FIGURE B.9: Effect plots for a regression

TABLE B.2: Example table to display two models in one table

	Dependent variable:	
	HR90	
	(1)	(2)
RD90	2.965***	3.199***
	(0.111)	(0.112)
SOUTH_fl	3.181***	2.600***
	(0.222)	(0.216)
DV90		0.476***
		(0.053)
MA90		−0.076***
		(0.027)
PS90		1.265***
		(0.100)
Constant	4.727***	4.204***
	(0.139)	(0.985)
Observations	3,085	3,085
R^2	0.365	0.426
Adjusted R^2	0.364	0.425
Residual Std. Error	5.295 (df = 3082)	5.035 (df = 3079)
F Statistic	884.400*** (df = 2; 3082)	457.300*** (df = 5; 3079)
Note:		*p<0.1; **p<0.05; ***p<0.01

Notice that the line has a confidence interval drawn around it and that the predicted means for southern and northern counties (when controlling for *RD90*) also have a confidence interval.

Rescaling input variables to assist interpretation

The interpretation or regression coefficients is sensitive to the scale of measurement of the predictors. This means one cannot compare the magnitude of the coefficients to compare the relevance of variables. Let's look at the more recent model: how can we tell what predictors have a stronger effect?

```
summary(fit_4)
```

```
##
## Call:
## lm(formula = HR90 ~ RD90 + SOUTH_f + DV90 + MA90 + PS90, data = ncovr)
##
## Residuals:
##     Min     1Q Median     3Q    Max
## -15.74  -2.59  -0.68   1.71  69.18
##
## Coefficients:
##               Estimate Std. Error t value Pr(>|t|)
## (Intercept)     4.2035     0.9847    4.27   2e-05 ***
## RD90            3.1992     0.1117   28.65  <2e-16 ***
## SOUTH_f1        2.5998     0.2156   12.06  <2e-16 ***
## DV90            0.4759     0.0531    8.97  <2e-16 ***
## MA90           -0.0761     0.0274   -2.77  0.0056 **
## PS90            1.2645     0.1005   12.59  <2e-16 ***
## ---
## Signif. codes:
## 0 '***' 0.001 '**' 0.01 '*' 0.05 '.' 0.1 ' ' 1
##
## Residual standard error: 5.04 on 3079 degrees of freedom
## Multiple R-squared:  0.426,  Adjusted R-squared:  0.425
## F-statistic:  457 on 5 and 3079 DF,  p-value: <2e-16
```

We just cannot. One way of dealing with this is by rescaling the input variables. A common method involves subtracting the mean and dividing by the standard deviation of each numerical input. The coefficients in these models is the expected difference in the response variable, comparing units that differ by one standard deviation in the predictor while adjusting for other predictors in the model.

Instead, Gelman (2008) has proposed dividing each numeric variables *by two times its standard deviation*, so that the generic comparison is with inputs equal to plus/minus one standard deviation. As Gelman explains, the resulting coefficients are then comparable to untransformed binary predictors. The implementation of this approach in the arm package subtracts the mean of each binary input while it subtracts the mean and divides by two standard deviations for every numeric input.

The way we would obtain these rescaled inputs uses the standardize() function of the arm package, which takes as an argument the name of the stored fit model.

```
arm::standardize(fit_4)
```

```
##
## Call:
## lm(formula = HR90 ~ z.RD90 + c.SOUTH_f + z.DV90 + z.MA90 + z.PS90,
##     data = ncovr)
##
## Coefficients:
```

```
## (Intercept)        z.RD90      c.SOUTH_f       z.DV90
##        6.183         6.398          2.600        1.650
##        z.MA90        z.PS90
##       -0.548         2.529
```

Notice the main change affects the numerical predictors. The unstandardised coefficients are influenced by the degree of variability in your predictors, which means that typically they will be larger for your binary inputs. With unstandardised coefficients you are comparing complete change in one variable (whether one is a Southern county or not) with one-unit changes in your numerical variable, which may not amount to much change. So, by putting in a comparable scale, you avoid this problem.

Standardising in the way described here will help you to make fairer comparisons. This standardised coefficients are comparable in a way that the unstandardised coefficients are not. We can now see what inputs have a comparatively stronger effect. It is very important to realise, though, that one **should not** compare standardised coefficients *across different models*.

<hr>

Testing conditional hypothesis: interactions

In the social sciences there is a great interest in what are called conditional hypothesis or interactions. Many of our theories do not assume simply **additive effects** but **multiplicative effects**. For example, Wikström, Tseloni, and Karlis (2011) suggest that the threat of punishment only affects the probability of involvement in crime for those with a propensity to offend, but are largely irrelevant for people who do not have this propensity. Or you may think that a particular crime prevention programme may work in some environments but not in others. The interest in this kind of conditional hypothesis is growing.

One of the assumptions of the regression model is that the relationship between the response variable and your predictors is additive. That is, if you have two predictors x_1 and x_2. Regression assumes that the effect of x_1 on y is the same at all levels of x_2. If that is not the case, you are then violating one of the assumptions of regression. This is in fact one of the most important assumptions of regression, even if researchers often overlook it.

One way of extending our model to accommodate for interaction effects is to add terms to our model, a third predictor x_3, where x_3 is simply the product of multiplying x_1 by x_2. Notice we keep a term for each of the **main effects** (original predictors) as well as a new term for the interaction effect. "Analysts should include all constitutive terms when specifying multiplicative interaction models except in very rare circumstances" (Brambor, Clark, and Golder (2006) p.66).

How do we do this in R? One way is to use the following notation in the formula argument. Notice how we have added a third term `RD90:SOUTH_f`, which is asking R to test the conditional hypothesis that resource deprivation may have a different impact on homicide for southern and northern counties.

```
fit_5 <- lm(HR90 ~ RD90 + SOUTH_f + RD90:SOUTH_f , data=ncovr)
# which is equivalent to:
# fit_5 <- lm(HR90 ~ RD90 * SOUTH_f , data=ncovr)
summary(fit_5)
```

```
##
## Call:
## lm(formula = HR90 ~ RD90 + SOUTH_f + RD90:SOUTH_f, data = ncovr)
##
## Residuals:
##     Min      1Q Median      3Q     Max
## -17.05   -3.00  -0.57    2.23   68.14
##
## Coefficients:
##                Estimate Std. Error t value Pr(>|t|)
## (Intercept)       4.548      0.159   28.68   <2e-16 ***
## RD90              2.581      0.196   13.15   <2e-16 ***
## SOUTH_f1          3.261      0.225   14.51   <2e-16 ***
## RD90:SOUTH_f1     0.562      0.238    2.37    0.018 *
## ---
## Signif. codes:
## 0 '***' 0.001 '**' 0.01 '*' 0.05 '.' 0.1 ' ' 1
##
## Residual standard error: 5.29 on 3081 degrees of freedom
## Multiple R-squared:  0.366,  Adjusted R-squared:  0.365
## F-statistic:  592 on 3 and 3081 DF,  p-value: <2e-16
```

You see here that essentially you have only two inputs (resource deprivation and south) but several regression coefficients. Gelman and Hill (2007) suggest reserving the term input for the variables encoding the information and term predictor to refer to each of the terms in the model. So here we have two inputs and three predictors (one for SOUTH, another for resource deprivation, and a final one for the interaction effect).

In this case the test for the interaction effect is significant, which suggests there is such an interaction. Let's visualise the results with the `effects` package:

```
plot(allEffects(fit_5), ask=FALSE)
```

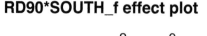

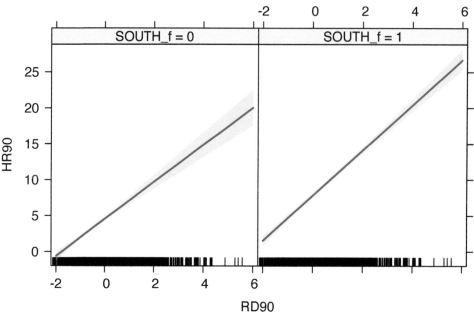

FIGURE B.10: Effect plots for regression testing conditional hypothesis

Notice that essentially what we are doing is running two regression lines and testing whether the slope is different for the two groups. The intercept is different: we know that Southern counties are more violent, but what we are testing here is whether the level of homicide goes up in a steeper fashion (and in the same direction) for one or the other group as the level of resource deprivation goes up. We see that's the case here. The estimated lines are almost parallel, but the slope is a bit more steep in the Southern counties. In Southern counties resource deprivation seems to have more of an impact on homicide than in northern counties.

A word of warning, the moment you introduce an interaction effect, the meaning of the coefficients for the other predictors changes (what it is often referred as the "main effect" as opposed to the interaction effect). You cannot retain the interpretation we introduced earlier. Now, for example, the coefficient for the SOUTH variable relates the marginal effect of this variable when RD90 equals zero. The typical table of results helps you to understand whether the effects are significant but offers little of interest that will help you to meaningfully interpret what the effects are. For this, is better you use some of the graphical displays we have covered.

Essentially what happens is that the regression coefficients that get printed are interpretable only for certain groups. So now:

- The intercept still represents the predicted score of homicide for southern counties and have a score of 0 in resource deprivation (as before).

- The coefficient of *SOUTH_f1* now can be thought of as the difference between the predicted score of homicide rate for northern counties *that have a score of 0 in resource deprivation* and northern counties *that have a score of 0 in resource deprivation.*

- The coefficient of *RD90* now becomes the comparison of mean homicide rate *for southern* counties who differ by one point in resource deprivation.

- The coefficient for the interaction term represents the difference in the slope for *RD90* comparing southern and northern counties, the difference in the slope of the two lines that we visualised above.

Model building and variable selection

How do you construct a good model? This partly depends on your goal, although there are commonalities. You do want to start with theory as a way to select your predictors and when specifying the nature of the relationship to your response variable (e.g., additive, multiplicative). Gelman and Hill (2007) provide a series of general principles. I would like to emphasise at this stage two of them:

- Include all input variables that, for substantive reasons, might be expected to be important in predicting the outcome.

- For inputs with large effects, consider including their interactions as well.

It is often the case that for any model, the response variable is only related to a subset of the predictors. There are some scenarios where you may be interested in understanding what is the best subset of predictors. Imagine that you want to develop a risk assessment tool to be used by police officers that respond to a domestic violence incident, so that you could use this tool for forecasting the future risk of violence. There is a cost to adding too many predictors. A police officer's time should not be wasted gathering information on predictors that are not associated with future risk. So you may want to identify the predictors that will help in this process.

Ideally, we would like to perform variable selection by trying out a lot of different models, each containing a different subset of the predictors. There are various statistics that help in making comparisons across models. Unfortunately, as the number of potentially relevant predictors increases, the number of potential models to compare increases exponentially. So you need methods that help you in this process. There are a number of tools that you can use for **variable selection** but this goes beyond the aims of this introduction. If you are interested, you may want to read this[6].

---

[6]http://link.springer.com/chapter/10.1007/978-1-4614-7138-7_6

Regression assumptions

Although so far we have discussed the practicalities of fitting and interpreting regression models, in practical applications you want to first check your model and proceed from there. There is not much point spending time interpreting your model until you know that the model reasonably fits your data.

In previous data analysis modules, you surely covered assumptions made by various statistical tests. The regression model also makes assumptions of its own. In fact, there are so many that we could spend an entire class discussing them. Gelman and Hill (2007) point out that the most important regression assumptions by decreasing order of importance are:

- **Validity**. The data should be appropriate for the question that you are trying to answer:

  Optimally, this means that the outcome measure should accurately reflect the phenomenon of interest, the model should include all relevant predictors, and the model should generalize to all cases to which it will be applied... Data used in empirical research rarely meet all (if any) of these criteria precisely. However, keeping these goals in mind can help you be precise about the types of questions you can and cannot answer reliably Gelman and Hill (2007: 46).

- **Additiviy and linearity**. These are the most important mathematical assumptions of the model. We already talked about additivity in previous sessions and discussed how you can include interaction effects in your models if the additivity assumption is violated. We will discuss problems with non-linearities today as well as ways to diagnose and solve this problem. If the relationship is non-linear (e.g., it is curvilinear), predicted values will be wrong in a biased manner, meaning that predicted values will systematically miss the true pattern of the mean of y (as related to the x-variables).

- **Independence of errors**. Regression assumes that the errors from the prediction line (or hyperplane) are independent. If there is dependency between the observations (you are assessing change across the same units, working with spatial units, or with units that are somehow grouped such as students from the same class), you may have to use models that are more appropriate (e.g., multi-level models, spatial regression, etc.).

- **Equal variances of errors**. When the variance of the residuals is unequal, you may need different estimation methods. This is, nonetheless, considered a minor issue. There is a small effect on the validity of t-test and F-test results, but generally regression inferences are robust with regard to the variance issue.

- **Normality of errors**. The residuals should be normally distributed. Gelman and Hill (2007: 46) discuss this as the least important of the assumptions and in fact "do *not* recommend diagnostics of the normality of the regression residuals". If the errors do not have a normal distribution, it usually is not particularly serious. Regression inferences tend to be robust with respect to normality (or non-normality of the errors). In practice, the residuals may appear to be non-normal when the wrong regression equation has been used. So, I will show you how to inspect normality of the residuals not because this is a problem in itself, but because it may be give you further evidence that there is some other problem with the model you are applying to your data.

Apart from this, it is convenient to diagnose multi-collinearity (this affects interpretation) and influential observations.

So these are the assumptions of linear regression, and today we will go through how to test for them, and also what are some options that you can consider if you find that your model violates them. While finding that some of the assumptions are violated does not necessarily mean that you have to scrap your model, it is important to use these diagnostics to illustrate that you have considered what the possible issues with your model are, and if you find any serious issues that you do address them.

Summary and further reading

Friedman, Hastie, and Tibshirani (2001) is a classic book for anyone looking to develop their skills modelling data with comprehensive examples and exercises to follow in R.

Models with interaction terms are too often misinterpreted. We strongly recommend a read of Brambor, Clark, and Golder (2006) to facilitate correct interpretation. Equally, Fox (2003) piece on the `effects` package goes to great detail to explain the logic and some of the options that are available when producing plots to show interactions with this package.

Appendix C: Sourcing geographical data for crime analysis

Throughout this book we provide example datasets and geometry files taken from various sources. Some, we get from data released with research papers, such as the NCOVR dataset on homicides in U.S. Counties. Some are administrative data, such as open crime data, which are linked with geometries from similarly available data provided on government or other institutional websites. In some cases we use APIs, for example to get data from Open Street Map, or from the Transport for London API, and in other cases we make use of resources made available for R such as the R package `rnaturalearth` which makes available boundaries from the natural earth project through installing the package.

As the kind of data, attribute and geometry, will vary between each reader of this book, and even for the same reader in different times, we cannot offer a definitive guide to sourcing the appropriate spatial data for your specific needs. However, we will instead show some examples of how we sourced our spatial data, in hopes that it might inspire and be useful to you in your mapping adventures.

Specifically, we will highlight the following useful resources: getting UK boundary data from:

- `rnaturalearth`
- Open Street Map API
- UK Data Service.

The `rnaturalearth` package

The package `rnaturalearth` is an R package created by Andy South to hold and facilitate interaction with Natural Earth map data (South 2017). Natural Earth is a public domain map dataset collected and maintained by the North American Cartographic Information Society. This dataset contains both vector and raster data, which are available to download and use in your crime-mapping endeavours. The package `rnaturalearth` allows us to access this database from within R, and to return the results as either `sp` or `sf` objects.

To do so, we need to install and load the package, and have a little understanding of the ways we can access and subset the data. We will give some small examples here but recommend the documentation and vignettes available with the package for further detail.

So step 1 is to install the package if you have not yet done so, and then load it with the `library()` function:

```
library(rnaturalearth)
```

Let's start by plotting the outline of all the countries in the world. We can retrieve all countries in the dataset with the `ne.countries()` function. Inside the function we specify the parameter `returnclass =` which allows us to choose whether we want a `sf` or `sp` object. Here let's choose `sf`:

```
all_countries <- ne_countries(returnclass = "sf")
```

If we wanted now to plot this, we can use any of the many methods covered in this book from `plot()` to `ggplot()` to `tmap` and so on.

```
library(ggplot2)

ggplot() +
  geom_sf(data = all_countries) +
  theme_void()
```

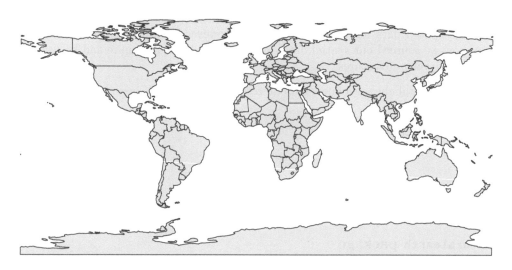

FIGURE C.1: Map of countries of the world

And here we see the boundaries for all the countries in the dataset. How exciting! We can also have a look at the attribute data included by viewing the `sf` object with the `View()` function.

```
View(all_countries)
```

You can see most of the information contained all relate to the ways to identify the different geometries, such as `geounit` or `name` or various codes. You can use these to filter, so that we only include certain countries we are interested in, or you can use this to join additional attribute data from elsewhere.

Let's say we wanted to select only Uganda. We could subset from the `all_countries` object we called above:

```
library(dplyr)

uganda <- all_countries %>% filter(name == "Uganda")
```

Or we could filter within the original call, which saves us having to download all the data in the first place:

```
uganda_2 <- ne_countries(country = 'Uganda', returnclass = "sf")
```

You should be able to see the two results are the same.

```
u1 <- ggplot() + geom_sf(data = uganda) + theme_void()

u2 <- ggplot() + geom_sf(data = uganda_2) + theme_void()

gridExtra::grid.arrange(u1, u2, nrow = 1)
```

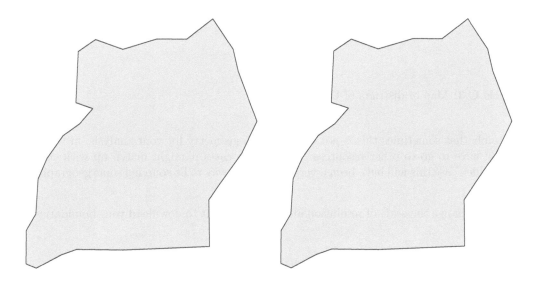

FIGURE C.2: Two identical maps showing border of Uganda

If we are focusing on one country, we might want to see some boundaries *within* said country. For this, we can use the function `ne_states()` when we request the file, and this will include boundaries.

```
uganda <- ne_states(country = "Uganda", returnclass = "sf")
```

Now, if we have a look, we can see the different districts within Uganda have been also downloaded.

```
ggplot() + geom_sf(data = uganda) + theme_void()
```

FIGURE C.3: Map of districts of Uganda

It is possible that sometimes this is not the ideal sub-geometry for your analysis, in which case you'll have to go to other resources; but in some cases it might match up with your needs, in which case this will have been a very convenient way to be sourcing some geography data!

You can also change the scale of resolution at which you want to download your boundaries.

```
# small
uganda_1 <- ne_countries(country = "Uganda", returnclass = "sf", scale = 110)

# medium
uganda_2 <- ne_countries(country = "Uganda", returnclass = "sf", scale = 50)

# large
uganda_3 <- ne_countries(country = "Uganda", returnclass = "sf", scale = 10)


# plot them to compare

u1 <- ggplot() + geom_sf(data = uganda_1) + theme_void()
u2 <- ggplot() + geom_sf(data = uganda_2) + theme_void()
```

```
u3 <- ggplot() + geom_sf(data = uganda_3) + theme_void()

gridExtra::grid.arrange(u1, u2, u3, nrow = 1)
```

FIGURE C.4: Three maps of Uganda with difference scale of resolution for the boundary

You can see as we move to lower numbers we get better resolution, which results in more detail around the border of the country (but also in larger file sizes!). It will be a decision for you to make about what is relevant and appropriate for the specific map you are making.

A final note on `rnaturalearth` before we move on. With the `ne_states()` function, we specified which country we wanted with the `country` argument. This was OK for Uganda, but what if we wanted to plot something in France instead?

```
france <- ne_states(country = "France", returnclass = "sf")

ggplot() + geom_sf(data = france) + theme_void()
```

FIGURE C.5: Map of France which shows also overseas territories of France

You see France appears, but so do the overseas territories of France, such as French Guiana, Guadeloupe, French Polynesia, and so on. It is possible you want to focus only on France. In this case, you can specify, instead of `country`, the parameter `geounit`. Like so:

```
france <- ne_states(geounit = "France", returnclass = "sf")

ggplot() + geom_sf(data = france) + theme_void()
```

FIGURE C.6: Map of France that shows only France without overseas territories

Now we see only France. But what if you do want to include it all? Well, in that case, one approach you might try is to add these territories in a way they appear closer to France, by inserting them by way of the *inset map*. We demonstrate how to add an inset map in Chapter 5.

Data from APIs

APIs stands for Application Programming Interface. An API can be understood as a tool which defines an interface for a programme to interact with a software component. For example, it defines the kind of requests or calls which can be made, and how these calls and requests can be carried out. Here, we are using the term 'API' to denote tools created by an open data provider to give access to different subsets of their content. Such APIs facilitate scripted and programmatic extraction of content, as permitted by the API provider (Olmedilla, Martínez-Torres, and Toral 2016). APIs can take many different forms and be of varying quality and usefulness (Foster et al. 2016). APIs may also have *wrappers* which refer to any interface which makes them easier to access. For example, Open Street Map has a graphical user interface called Overpass Turbo https://overpass-turbo.eu/, which allows users to build queries using any internet browser, and download and save the result.

Another wrapper comes in the form of an R package called `osmdata`. This package allows us to query the Open Street Map API and return the results of our query directly into R. We used this in Chapter 2 of this book, to query the location of pubs in Manchester. But we can also use this to source our boundary data as well.

In this section we want to illustrate the process of getting data from APIs by getting geographic data from Open Street Map. Open Street Map is a database of geospatial information built by a community of mappers, enthusiasts and members of the public, who contribute and maintain data about all sorts of environmental features, such as roads, green spaces, restaurants and railway stations, amongst many other things, all over the world (Open Street Map 2021a).

You can view the information contributed to Open Street Map using their online mapping platform (https://www.openstreetmap.org/). The result of people's contributions is a database of spatial information rich in local knowledge which provides invaluable information about places and their features, without being subject to strict terms on usage.

Open Street Map data can be accessed using its API.

Let's begin by loading the library:

```
library(osmdata)
```

Our first task is to create a bounding box, so that we return only those data that fall within this box. Let's continue to try and source data about Uganda, but let's narrow in to the capital of Kampala. To get a bounding box the shape of Kampala, we can use the `getbb()` function:

```
kampala_bb <- getbb("kampala", format_out = "sf_polygon")
```

We should check on this object to make sure it looks the way we would expect:

```
ggplot() + geom_sf(data = kampala_bb) + theme_void()
```

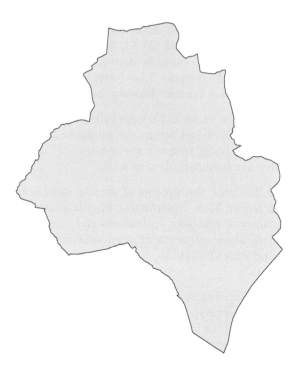

FIGURE C.7: Boundary which follows the shape of Kampala

That's looking Kampala-shaped to us, so we can continue. To query Open Street Map, we have to build an Overpass Query, Overpass being the Open Street Map API. We can create this with the `opq()` function (which stands for **O**verpass **Q**uery - we assume...!) where we specify our bounding box object we created above. We can then pipe (`%>%`) our query elements to this. In the next step, we specify what features we want with the `add_osm_feature()` function. The parameters should define what we want, following the Overpass query language. We can look for things (nodes, ways, or relations if we're being specific) using key value pairs. For example, to return all post boxes within an area, we need to use the *amenity* key and the *post box* value, or to find museum you would use the *tourism* key and the *museum* value. To learn more about this, visit the OSM wiki Open Street Map (2021a).

To get administrative boundaries, we can use the key of *admin_level*, and for the value, we pass a number which can be any number between 2 and 10, where bigger numbers mean more granular resolution. For example admin_level = 2 is almost always a de-facto independent country, and admin_level=4 is usually equivalent to a "province". However, numbers higher than 4 values vary in meaning between countries. You can look up the country specific levels on Open Street Map (2021b). In Uganda, we can see that level 2 represents the borders of the 112 districts of Uganda, 4 the boundary of counties, and level 8 the boundary of sub-counties. Let's use admin level 8 to map sub-counties in Kampala.

Finally, to make sure your query returns in `sf` format, append the `osmdata_sf()` to the end of the pipe.

```
kampala_boundaries <- opq(kampala_bb) %>%
add_osm_feature(key = "admin_level", value = "8") %>%
osmdata_sf()
```

```
## bb_poly has more than one polygon; the first will be selected.
```

Now that we have the results, we can extract our boundary polygons, which are stored in the `multiploygon` object inside the query results. Let's extract this now.

```
kampala_boundaries <- kampala_boundaries$osm_multipolygons
```

And now we can plot our results, labelling each region for clarity:

```
ggplot() + geom_sf(data = kampala_boundaries) +
  geom_sf_label(data = kampala_boundaries, aes(label = name)) + theme_void()
```

FIGURE C.8: Kampala with regions labelled

Data from data repositories

There are many places where you can access data relevant to your particular area of interest. For example, in the United Kingdom, one such service is the **UK Data Service**. This is a research council-funded comprehensive collection which includes major UK government-sponsored surveys, cross-national surveys, longitudinal studies, UK census data, international aggregate, business data, and qualitative data. A specific subset relevant to us is the *Census Boundaries* dataset, which is accessible through the Census Support Boundary Datasets.

To access this resource, we visit https://borders.ukdataservice.ac.uk/, where we can access all Census geography products since 1971. On this page, there is a link to the **UK Data Service Census Geography eLearning Modules** which are a set of resources designed to make users more familiar with the data and services available. There are also tutorials for **QGIS** and **Python** users.

To acquire boundary data from here, we can select the **Boundary Data Selector** option. When you get to the link, you will see on the top there is some notification to help you with the boundary data selector. If in the future you are looking for UK boundary data and you are feeling unsure at any point, feel free to click on that note, "**How to use Boundary Data Selector**", which will help to guide you.

For a quick example here, we can download some Local Authority boundaries for England. In this case that means, for *Country* select "England", for *Geography* select "Administrative", and for *Dates* let's go for "2011 and later". Then, hit the button which says "Find" and see some options appear in the *Boundaries* box below. From these, let's select "English Districts, UAs and London Boroughs, 2011".

Once you have the file, you could select sub-areas. For this, hit the "List Areas" button, and select those relevant to you. Otherwise, you can move to *Extract Boundary Data*, where you will be taken to the next page.

Here you will have a list of possible choices of what versions of the boundary files to download. At the time of writing, these options are:

- English Districts, UAs and London Boroughs, 2011
- English Districts, UAs and London Boroughs, 2011, Clipped
- English Districts, UAs and London Boroughs, 2011, Generalised (simplified polygon geometry)
- English Districts, UAs and London Boroughs, 2011, Clipped and Generalised (simplified polygon geometry)

The choice here is similar to that of the `scale` argument in the `ne_countries()` function above when we were getting boundary data from the `rnaturalearth` package. The first option is not changed, and so is the most geographically accurate file, but also therefore the largest in file size (which will later affect computation speed). Then there are increasing steps of simplification applied. For those interested in the Generalisation and Clipping process, see UK Data Service (2013). Generally, for crime-mapping purposes, we tend to go with the last option, but it will depend on the unit of analysis at which your data are collected, and the kinds of analysis you will do whether precision of the borders is important or not in your specific case.

Once you have chosen which file to download, you can also choose the format. This is between the options of: CSV, KML, MapInfo, or Shapefile. We make use of the shapefile format in this book; it is a very commonly used format for those originally trained on proprietary GIS software such as ESRI Arc Suite, as both the authors of this book were. This is probably why we use it... On the other hand, KML files are neat, simple, and can easily be imported into R using the `st_read()` function from the `sf` package. Again, it is up to you which format to choose, what you are familiar with, and whether other co-authors or collaborators might be using other GIS software which supports some formats better than others. R is flexible, so you can be too.

Here, let's download the KML format of the most simplified geometry. Once you select this download, it will save a `.zip` file to your computer. Make sure to put it in your working directory, and if you use sub-folders, then the relevant sub-folder. We have saved it to our "data" folder. We can extract (unzip) using R:

```
unzip('data/England_lad_2011_gen_clipped.zip',
      exdir = "data/England_lad_2011_gen_clipped")
```

```
## Warning in unzip("data/
## England_lad_2011_gen_clipped.zip", exdir = "data/
## England_lad_2011_gen_clipped"): error 1 in extracting
## from zip file
```

Now you see (if you are following the same structure as we are) there is a new sub-folder in the 'data' directory called 'England_lad_2011_gen_clipped' which contains the `.kml` file, and a `TermsAndConditions.html` file. This will contain information about how you can use this map. For example, all your maps will have to mention where you got all the data from. So since you got this boundary data from the UKDS, you will have to note the following:

"Contains National Statistics data © Crown copyright and database right [year] Contains OS data © Crown copyright [and database right] (year)"

You can read more about this in the terms and conditions document.

Now we can import our `.kml` file with `st_read()` from the `sf` package :

```
las <- st_read("data/England_lad_2011_gen_clipped/england_lad_2011_gen_clipped.kml")
```

And plot it to see:

```
ggplot() + geom_sf(data = las) + theme_void()
```

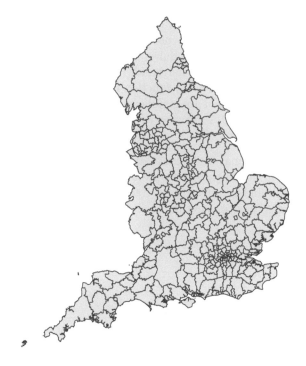

FIGURE C.9: Map of English Local Authorities boundaries

International resources

And here we have our English Local Authorities. Now we realise that this is useful mainly for our audiences based in the UK, but we include this illustration as other countries use similar data repositories where you can get the relevant census geographies. For example, the United States Census Bureau has census shapefiles for the United States (United States Census Bureau 2021), Census Canada for Canada (Statistics Canada 2021), the National Institute of Statistics and Geography (INEGI) for Mexico (INEGI 2021), and many more.

There are also resources which contain geography files for multiple countries. DIVA-GIS (Hijmans et al. 2021) is one such resource, which aims to collect free geographic data for any country in the world. The creators focus on studying the distribution of biodiversity; however, these maps can be useful for the crime mapper as well. Another collection comes from MIT (2021) who collect links to various GIS resources. Finally, international organisations or organisations which focus on international data can also provide a good starting point for sourcing your geographic data. For example, the Centers for Disease Control and Prevention (CDC) have a list of shapefiles for many countries on their website (Centers for Disease Control and Prevention 2021). There are probably many more, and sometimes it is just a case of a thorough internet search to find the right geography for your data.

Summary and further reading

Here we demonstrated a few ways in which you can source geodata for your crime-mapping projects and endeavors. The amazing resource that is `rnaturalearth` will no doubt come in handy for all, so do read up on the documentation in South (2017). Administrative boundaries will also no doubt be more and more available from official government statistical resources, or collective initiatives like Hijmans et al. (2021) and others, many collected on the MIT (2021) resource. And of course, Open Street Map is a bottomless well of georeferenced information. Not only for boundaries, but roads, transport paths, and any feature which your heart may desire is probably available here. Mooney and Minghini (2017) and Kounadi (2009) are good resources to read for those interested in making more use of Open Street Map data.

References

Abad, Lorena, Robin Lovelace, and Lucas van der Meer. 2019. "Spatial Networks in r with Sf and Tidygraph." https://r-spatial.org/r/2019/09/26/spatial-networks.html.

Adepeju, Monsuru, and Andy Evans. 2017. "Comparative Analysis of Two Variants of the Knox Test: Inferences from Space-Time Crime Pattern Analysis." In *International Conference on Computational Science and Its Applications*, 770–78. Springer.

Adepeju, Monsuru, Samuel Langton, and Jon Bannister. 2020. "Akmedoids r Package for Generating Directionally-Homogeneous Clusters of Longitudinal Data Sets." *The Journal of Open Software* 5 (56). https://doi.org/10.21105/joss.02379.

———. 2021. "Anchored k-Medoids: A Novel Adaptation of k-Medoids Further Refined to Measure Long-Term Instability in the Exposure to Crime." *Journal of Computational Social Science.* https://doi.org/10.1007/s42001-021-00103-1.

Aldstadt, Jared, and Arthur Getis. 2006. "Using AMOEBA to Create a Spatial Weights Matrix and Identify Spatial Clusters." *Geographical Analysis* 38 (4): 327–43.

Andresen, Martin. 2011. "The Ambient Population and Crime Analysis." *The Professional Geographer* 63 (2): 193–212.

Andresen, Martin, Andrea Curman, and Shannon Linning. 2017. "The Trajectories of Crime at Places: Understanding the Patterns of Disaggregated Crime Types." *Journal of Quantitative Criminology* 33: 427–49.

Andresen, Martin, and Olivia Ha. 2020. "Spatially Varying Relationships Between Immigration Measures and Property Crime Types in Vancouver Census Tracts, 2016." *The British Journal of Criminology* 60 (5): 1342–67.

Andresen, Martin, Cory Haberman, Shane Johnson, and Wouter Steenbeek. 2021. "Advances in Place-Based Methods: Editors' Introduction." *Journal of Quantitative Criminology*, doi.org/10.1007/s10940-021-09517-6.

Andresen, Martin, Shannon Linning, and Nick Malleson. 2017. "Crime at Places and Spatial Concentrations: Exploring the Spatial Stability of Property Crime in Vancouver BC, 2003–2013." *Journal of Quantitative Criminology* 33: 255–75.

Andresen, Martin, and Nick Malleson. 2011. "Testing the Stability of Crime Patterns: Implications for Theory and Policy." *Journal of Research in Crime and Delinquency* 48 (1): 58–82.

Andresen, Martin, Nick Malleson, Wouter Steenbek, Michael Townsley, and Christophe Vandeviver. 2020. "Minimum Geocoding Match Rates: An International Study of the Impact of Data and Areal Unit Sizes." *International Journal of Geographical Information Science* 34 (7): 1306–22.

Ang, Qi, Adrian Baddeley, and Gopalan Nair. 2012. "Geometrically Corrected Second Order Analysis of Events on a Linear Network, with Applications to Ecology and Criminology." *Scandinavian Journal of Statistics* 39: 591–617.

Anselin, Luc. 1995. "Local Indicators of Spatial Association: LISA." *Geographical Analysis* 27 (2): 93–115.

———. 1996. "The Moran Scatterplot as an ESDA Tool to Assess Local Instability in Spatial Association." In *Spatial Analytical Perspectives on GIS*, edited by Manfred Fischer, Henk Scholten, and David Unwin, 111–25. London: Routledge.

———. 2007. *Spatial Regression Analysis in r: A Workbook*. Center for Spatially Integrated Social Science.

———. 2020. "Local Spatial Autocorrelation (2): Other Local Spatial Autocorrelation Statistics." https://geodacenter.github.io/workbook/6b_local_adv/lab6b.html#getis-ord-statistics.

———. 2021. "GeoDa Center Playlists." https://www.youtube.com/user/GeoDaCenter/playlists.

Anselin, Luc, and Anil Bera. 1998. "Spatial Dependence in Linear Regression Models with an Introduction to Spatial Econometrics." In *Hanbook of Applied Economic Statistics*, edited by A Ullah and D Giles, 237–89. New York: Marcel Dekker.

Anselin, Luc, Grant Morrison, Angela Li, and Karina Acosta. 2021. "Tutorials for the Center for Spatial Data Science, the University of Chicago." https://spatialanalysis.github.io/tutorials/.

Anselin, Luc, and Sergio Rey. 2014. *Modern Spatial Econometrics in Practice*. Chicago: GeoDa Press.

Appelhans, Tim, Florian Detsch, Christoph Reudenbach, and Stefan Woellauer. 2021. *Mapview: Interactive Viewing of Spatial Data in r*. https://CRAN.R-project.org/package=mapview.

ArcGIS Pro. 2021. "Why Hexagons?" https://pro.arcgis.com/en/pro-app/latest/tool-reference/spatial-statistics/h-whyhexagons.htm.

Ashby, Matthew. 2018. "Studying Crime and Place with the Crime Open Database." SocArXiv pre-print. https://doi.org/10.31235/osf.io/9y7qz.

———. 2019. "Studying Crime and Place with the Crime Open Database: Social and Behavioural Scienes." *Research Data Journal for the Humanities and Social Sciences* 4 (1): 65–80.

———. 2020. "Why You Can't Identify Changes in Crime by Comparing This Month to Last Month." https://the-sra.org.uk/SRA/Blog/whyyoucantidentifychangesincrimeby comparingthismonthtolastmonth.aspx.

Baddeley, Adrian, Gopalan Nair, Suman Rakshit, Greg McSwiggan, and Tilman Davies. 2021. "Analysing Point Patterns on Networks — a Review." *Spatial Statistics* 42: doi.org/10.1016/j.spasta.2020.100435.

Baddeley, Adrian, Ege Rubak, and Rolf Turner. 2015. *Spatial Point Patterns: Methodology and Applications with r*. Boca Raton, FL: CRC Press.

Baddeley, Adrian, and Rolf Turner. 2005. "spatstat: An R Package for Analyzing Spatial Point Patterns." *Journal of Statistical Software* 12 (6): 1–42. https://www.jstatsoft.org/v12/i06/.

Baller, Robert, Luc Anselin, Steven Messner, Glenn Deane, and Darnell Hawkins. 2001. "Structural Covariates of US County Homicide Rates: Incorporating Spatial Effects." *Criminology* 39 (3): 561–88.

Banerjee, Sudipto, Bradley Carlin, and Alan Gelfand. 2014. *Hierarchical Modeling and Analysis for Spatial Data*. CRC Press.

Baumer, Eric, and Richard Wright. 1996. "Crime Seasonality and Serious Scholarship: A Comment on Farrell and Pease." *Brit. J. Criminology* 36: 579.

Beale, Colin, Jack Lennon, John Yearsley, Mark Brewer, and David Elston. 2010. "Regression Analysis of Spatial Data." *Ecology Letters* 13 (2): 246–64.

Bernasco, Wim, and Henk Elffers. 2010. "Statistical Analysis of Spatial Crime Data." In *Handbook of Quantitative Criminology*, edited by Alex Piquero and David Weisburd, 699–724. New York: Springer.

Bivand, Roger. 2019. "ECS530: (III) Coordinate Reference Systems." https://rsbivand.github.io/ECS530_h19/ECS530_III.html.

———. 2020. "Migration to PROJ6/GDAL3." http://rgdal.r-forge.r-project.org/articles/PROJ6_GDAL3.html.

Bivand, Roger, Giovanni Millo, and Gianfranco Piras. 2021. "A Review of Software for Spatial Econometrics in r." *Mathematics* 9 (11).

Bivand, Roger, Edzer Pebesma, and Virgilio Gómez-Rubio. 2013. *Applied Spatial Data Analysis with r.* 2nd ed. New York: Springer.

Bivand, Roger, and David Wong. 2018. "Comparing Implementations of Global and Local Indicators of Spatial Association." *TEST* 27: 716–48.

Blangiardo, Marta, and Michela Cameletii. 2015. *Spatial and Spatio-Temporal Bayesian Models with r-INLA.* Chichester: John Wiley & Sons.

Blumstein, Al, J. Cohen, J. Roth, and C. Visher. 1986. "Criminal Careers and Career Criminals." Research Report. National Science Foundation.

Boba, Rachel. 2013. *Crime Analysis with Crime Mapping.* 3rd ed. Thousand Oaks, CA: Sage.

Bolstad, Paul. 2019. *GIS Fundamentals: A First Text on Geographic Information Systems.* 6th ed. White Bear Lake, MN: Eider Press.

Bowers, Kate. 2021. "Risky Places: Crime Absorbers, Crime Radiators as Risky Places." https://play.kth.se/media/Risky+PlacesA+Crime+absorbers,+crime+radiators+as+risky+places,+Prof+Kate+Bowers,+UCL/0_zya4j354.

Bowers, Kate, and Shane Johnson. 2004. "Who Commits Near Repeats? A Test of the Boost Explanation." *Western Criminology Review* 5 (3).

Braga, Anthony, Brandon Turchan, Andrew Papachristos, and David Hureau. 2019. "Hot Spots Policing and Crime Reduction: An Update of an Ongoing Systematic Review and Meta-Analysis." *Journal of Experimental Criminology* 15: 289–311.

Braga, Anthony, and David Weisburd. 2010. *Policing Problem Places: Crime Hot Spots and Effective Prevention.* New York: Oxford University Press.

———. 2020. "Does Hot Spots Policing Have Meaningful Impacts on Crime? Findings from an Alternative Approach to Estimating Effect Sizes from Place-based Program Evaluations." *Journal of Quantitative Criminology,* doi.org/10.1007/s10940-020-09481-7.

Brambor, Thomas, William Clark, and Matt Golder. 2006. "Understanding Interaction Models: Improving Empirical Analyses." *Political Analysis* 14 (1): 63–82.

Brantingham, Patricia, and Paul Brantingham. 1982. "Mobility, Notoriety and Crime: A Study of Crime Patterns in Urban Nodal Points." *Journal of Environmental Systems* 11 (1): 89–99.

———. 1995. "Criminality of Place: Crime Generators and Crime Attractors." *European Journal on Criminal Policy and Research* 3: 5–26.

Brewer, Cynthia. 1994. "Color Use Guidelines for Mapping." *Visualization in Modern Cartography* 1994: 123–48.

———. 2006. "Basic Mapping Principles for Visualizing Cancer Data Using Geographic Information Systems (GIS)." *American Journal of Preventive Medicine* 30 (2S): S25–36.

Briggs, Daniel, and Rubén Monge-Gamero. 2017. *Dead-End Lives: Drugs and Violence in the City Shadows.* Bristol, UK: Policy Press.

Briz-Redón, Álvaro, Francisco Martínez-Ruiz, and Francisco Montes. 2019a. "DRHotNet: An r Package for Detecting Differential Risk Hotspots on a Linear Network." https://arxiv.org/abs/1911.07827.

———. 2019b. "Identification of Differential Risk Hotspots for Collision and Vehicle Type in a Directed Linear Network." *Accident Analysis and Prevention,* doi.org/10.1016/j.aap.2019.105278.

———. 2020a. "Adjusting the Knox Test by Accounting for Spatio-Temporal Crime Risk Heterogeneity to Analyse Near-Repeats." *European Journal of Criminology,* doi.org/10.1177/1477370820905106.

————. 2020b. "Reestimating a Minimum Acceptable Geocoding Hit Rate for Conducting a Spatial Analysis." *International Journal of Geographical Information Science* 34 (7): 1283–1305.

Bruinsma, Gerben, and Shane Johnson, eds. 2018. *The Oxford Handbook of Environmental Criminology*. Oxford, England: Oxford Univerisity Press.

Brunsdon, Chris, and Lex Comber. 2015. *R for Spatial Analysis and Mapping*. 1st ed. Thousand Oaks, CA: SAGE.

Buil-Gil, David, Juan Medina Ariza, and Natalie Shlomo. 2021. "Measuring the Dark Figure of Crime in Geographic Areas: Small Area Estimation from the Crime Survey for England and Wales." *British Journal of Criminology* 61 (2): 364–88.

Buil-Gil, David, Angelo Moretti, Natalie Shlomo, and Juan Medina Ariza. 2019. "Worry about Crime in Europe: A Model-Based Small Area Estimation from the European Social Survey." *European Journal of Criminology*, 1477370819845752.

Bycoffe, Aaron, Ella Koeze, David Wasserman, and Julia Wolfe. 2018. "The Atlas of Redistricting." https://projects.fivethirtyeight.com/redistricting-maps/.

Cahill, Meagan, and Gordon Mulligan. 2007. "Using Geographically Weighted Regression to Explore Local Crime Patterns." *Social Science Computer Review* 25 (2): 174–93.

Cairo, Alberto. 2016. *The Truthful Art: Data, Charts, and Maps for Communication*. New Riders.

Cameron, James. 2005. In *Mapping Crime: Understanding Hot Spots*, edited by John Eck, Spencer Chainey, James Cameron, Michael Leitner, and Ronald Wilson, 35–64. Washington, DC: National Institute of Justice.

Camoes, Jorge. 2016. *Data at Work: Best Practices for Creating Effective Charts and Information Graphics in Microsoft Excel*. New Riders.

Caplan, Joel, and Leslie Kennedy. 2016. *Risk Terrain Modeling: Crime Prediction and Risk Reduction*. University of California Press.

Carr, Daniel, and Linda Pickle. 2010. *Visualizing Data Patterns with Micromaps*. Boca Raton, FL: CRC Press.

Ceccato, Vania. 2013. *Moving Safely: Crime and Perceived Safety in Stockholm's Subway Stations*. Lanham: Lexington Books.

Centers for Disease Control and Prevention. 2021. "Shapefiles." https://www.cdc.gov/epiinfo/support/downloads/shapefiles.html.

Chainey, Spencer. 2013a. "Examining the Influence of Cell Size and Bandwidth Size on Kernel Density Estimation Crime Hotspot Maps for Predicting Spatial Patterns of Crime." *Bulletin of the Geographical Society of Liege* 60: 7–19.

————. 2013b. "Methods and Techniques for Understanding Crime Hot Spots." In *Mapping Crime: Understanding Hot Spots*, edited by John Eck, Spencer Chainey, James Cameron, Michael Leitner, and Ronald Wilson, 15–34. Washington DC: National Institute of Justice.

————. 2014. "Examining the Extent to Which Hotspot Analysis Can Support Spatial Predictions of Crime." PhD thesis, UCL (University College London).

————. 2021. *Understanding Crime: Analyzing the Geography of Crime*. Redlands, CA: ESRI Press.

Chainey, Spencer, and Braulio-Figueiredo Alves-da-Silva. 2016. "Examining the Extent of Repeat and Near Repeat Victimisation of Domestic Burglaries in Belo Horizonte, Brazil." *Crime Science* 5 (1): 1–10.

Chainey, Spencer, and Jerry Ratcliffe. 2005. *GIS and Crime Mapping*. Chichester, England: John Wiley & Sons.

Chainey, Spencer, Lisa Tompson, and Sebastian Uhlig. 2013. "The Utility of Hotspot Mapping for Predicting Spatial Patterns of Crime." *Security Journal* 21: 4–28.

Chan, Martin. 2020. "RStudio Projects and Working Directories: A Beginner's Guide." https:

//www.r-bloggers.com/2020/01/rstudio-projects-and-working-directories-a-beginners-guide/.

Chatfield, Chris, and Haipeng Xing. 2019. *The Analysis of Time Series: An Introduction with r*. 7th ed. Boca Raton, FL: CRC Press.

Cheng, Tao, and Monsuru Adepeju. 2014. "Modifiable Temporal Unit Problem (MTUP) and Its Effect on Space-Time Cluster Detection." *Plos One*. https://doi.org/10.1371/journal.pone.0100465.

Chi, Guangqing, and Jun Zhu. 2020. *Spatial Regression Models for the Social Sciences*. Thousand Oaks, CA: Sage.

Clayton, David, and John Kaldor. 1987. "Empirical Bayes Estimates of Age-Standardized Relative Risks for Use in Disease Mapping." *Biometrics*, 671–81.

Cohen, Lawrence, and Marcus Felson. 1979. "Social Change and Crime Rate Trends: A Routine Activity Approach." *American Sociological Review*, 588–608.

Collazos, Daniela, Eduardo García, Daniel Mejía, Daniel Ortega, and Santiago Tobón. 2020. "Hot Spots Policing in a High-Crime Environment: An Experimental Evaluation in Medellín." *Journal of Experimental Criminology*, doi.org/10.1007/s11292-019-09390-1.

Cornish, Derek, and Ronald Clarke. 1986. *The Reasoning Criminal: Rational Choice Perspectives on Offending*. Transaction Publishers.

Darmofal, David. 2015. *Spatial Analysis for the Social Sciences*. New York, NY: Cambridge University Press.

Davies, T, and Kate Bowers. 2018. "Street Networks and Crime." In *The Oxford Handbook of Environmental Criminology*, edited by Gerben Bruinsma and Shane Johnson, 545–78. New York: Oxford University Press.

Davies, T, Jonathan Marshall, and Martin Hazelton. 2017. "Tutorial on Kernel Estimation of Continuous Spatial and Spatiotemporal Relative Risk with Accompanying Instruction in r."

Dent, Borden, J Torguson, and T Hodler. 2008. *Thematic Map Design*. New York, New York, NY: McGraw-Hill.

Dorling, Daniel. 1991. "The Visualisation of Spatial Social Structure." PhD thesis, University of Newcastle upon Tyne PhD thesis.

———. 1996. *Area Cartograms: Their Use and Creation, Concepts and Techniques in Modern Geography*. Institute of British Geographers.

Dougenik, James, Nicholas Chrisman, and Duane Niemeyer. 1985. "An Algorithm to Construct Continuous Area Cartograms." *The Professional Geographer* 37 (1): 75–81.

Eck, John, and William Spelman. 1997. "Problem Solving: Problem-Oriented Policing in Newport News." Research Report. Police Executive Research Forum.

Eck, John, and David Weisburd. 2015. "Crime Places in Crime Theory." *Crime and Place: Crime Prevention Studies* 4.

Elhorst, Paul. 2010. "Applied Spatial Econometrics: Raising the Bar." *Spatial Economic Analysis* 5 (1): 9–28.

———. 2014. *Spatial Econometrics: From Cross-Sectional Data to Spatial Panels*. Heildeberg: Springer.

Ellis, Peter. 2018. *Ggseas: 'Stats' for Seasonal Adjustment on the Fly with 'Ggplot2'*. https://CRAN.R-project.org/package=ggseas.

Emch, Michael, Elisabeth Root, Sophia Giebultowicz, Mohammad Ali, Carolina Perez-Heydrich, and Mohammad Yunus. 2010. "Integration of Spatial and Social Network Analysis in Disease Transmission Studies." *CAnnals of the Association of American Geographers* 102 (5): 1004–15.

Farrell, Graham, and Ken Pease. 2017. "Preventing Repeat and Near Repeat Crime Concentrations." In *Handbook of Crime Prevention and Community Safety*, edited by Nick Tilley and Aiden Sidebottom, 143–56. London: Routledge.

Field, Kenneth. 2007. "Map Evaluation Guidelines." http://downloads.esri.com/Mapping Center2007/arcGISResources/more/MapEvaluationGuidelines.pdf.

———. 2012. "Using a Binning Technique for Point-Based Multiscale Web Maps." https://www.esri.com/arcgis-blog/products/arcgis-online/mapping/using-a-binning-technique-for-point-based-multiscale-web-maps/.

———. 2015. "When Is a Heat Map Not a Heat Map." http://cartonerd.blogspot.com/2015/02/when-is-heat-map-not-heat-map.html.

———. 2018. *Cartography.* Redlands, CA: ESRI Press.

Field, Kenneth, and Damien Demaj. 2012. "Reasserting Design Relevance in Cartography: Some Concepts." *The Cartographic Journal* 49 (1): 70–76.

Field, Kenneth, and Daniel Francis Luke Dorling. 2016. "UK Election Cartography." *International Journal of Cartography* 2: 202–32.

Floch, Jean-Michel, and Ronan LeSaout. 2018. "Spatial Econometrics : Common Models." In *Handbook of Spatial Analysis: Theory and Application with r*, edited by Vincent Loonis and Marie-Pierre de Bellefon, 149–77. Montrouge: Institut national de la statistique et des études économiques.

Foster, Ian, Rayid Ghani, Ron Jarmin, Frauke Kreuter, and Julia Lane. 2016. *Big Data and Social Science: A Practical Guide to Methods and Tools.* CRC Press.

Fotheringham, AS, Chriss Brunsdon, and Martin Charlton. 2003. *Geographically Weighted Regression: The Analysis of Spatially Varying Relationships.* John Wiley & Sons.

Fox, John. 2003. "Effect Displays in r for Generalised Linear Models." *Journal of Statistical Software* 8 (15).

Friedman, Jerome, Trevor Hastie, and Robert Tibshirani. 2001. *The Elements of Statistical Learning.* Vol. 1. 10. Springer series in statistics New York.

Gastwirth, Joseph. 1972. "The Estimation of the Lorenz Curve and Gini Index." *The Review of Economics and Statistics,* 306–16.

Gelb, Jeremy. 2021. *spNetwork: Spatial Analysis on Network.* https://CRAN.R-project.org/package=spNetwork.

Gelman, Andrew. 2008. "Scaling Regression Inputs by Dividing by Two Standard Deviations." *Statistics in Medicine* 27: 2865–73.

Gelman, Andrew, and Jenniffer Hill. 2007. *Data Analysis Using Regression and Multilevel/Hierarchical Models.* Cambridge: Cambridge University Press.

Gelman, Andrew, Jenniffer Hill, and Aki Vehtari. 2020. *Regression and Other Stories.* Cambridge: Cambridge University Press.

Gelman, Andrew, and Phillip Price. 1999. "All Maps of Parameters Estimates Are Misleading." *Statistics in Medicine* 18: 3221–34.

Gerell, Manne. 2017. "Smallest Is Better? The Spatial Distribution of Arson and the Modifiable Areal Unit Problem." *Journal of Quantitative Criminology* 33 (2): 293–318.

Getis, Arthur, and JK Ord. 1992. "The Analysis of Spatial Association by Use of Distance Statistics." *Geographical Analysis* 24 (3): 189–206.

Goldstein, Herman. 1990. *Problem-Oriented Policing.* McGraw Hill.

Gomez-Rubio, Virgilio, Juan Ferrandiz, and Antonio Lopez. 2005. "Detecting Clusters of Disease with r." *Journal of Geographical Systems* 7: 189–206.

Gorr, Wilpen, and Kristen Kurland. 2012. *GIT Tutorial for Crime Analysis.* Redland, CA: ESRI Press.

Gorr, Wilpen, and YongJei Lee. 2015. "Early Warning System for Temporary Crime Hot Spots." *Journal of Quantitative Criminology* 31: 25–47. https://doi.org/10.1371/journal.pone.0100465.

Goudriaan, Heike, Karin Witterbrood, and Paul Nieuwbeerta. 2006. "Neigbourhood Characteristics and Reporting Crime." *British Journal of Criminology* 46: 719–42.

Grossenbacher, Timo. 2019. "Bivariate Maps with Ggplot2 and Sf." Blog. Tamedia. https://timogrossenbacher.ch/2019/04/bivariate-maps-with-ggplot2-and-sf/.

Grove, Louise, Graham Farrell, David Farrington, and Shane Johnson. 2014. "Preventing Repeat Victimisation: A Systematic Review." Swedish National Council for Crime Prevention.

Guerry, Andre-Michelle. 1833. "Essay on the Moral Statistics of France, Trans. HP Whitt and VW Reinking." Lewiston, NY: Edwin Mellen Press.

Haining, Robert. 2003. *Spatial Data Analysis: Theory and Practice*. Cambridge: Cambridge University Press.

Haining, Robert, and Guangquan Li. 2020. *Modelling Spatial and Spatial-Temporal Data: A Bayesian Approach*. Boca Raton, FL: CRC Press.

Harries, Keith. 1999. "Mapping Crime: Principle and Practice." Research Report. National Institute of Justice.

Hart, Timothy, and Paul Zandbergen. 2012. "Effects of Data Quality on Predictive Hotspot Mapping." Final Report. National Institute of Justice.

Healy, Kieran. 2019. *Data Visualization. A Practical Introduction*. Princeto, NJ: Princeton University Press.

Her Majesty Inspectorate of Constabulary. 2014. "Crime Recording: Making the Victim Count." Final Report. HMIC.

Herrmann, Christopher R. 2013. "Street-Level Spatiotemporal Crime Analysis: Examples from Bronx County, NY (2006–2010)." In *Crime Modeling and Mapping Using Geospatial Technologies*, 73–104. Springer.

Hijmans, Robert, Edwin Rojas, Mariana Cruz, Rachel OBrien, and Israel Barrantes. 2021. "DIVA-GIS." http://www.diva-gis.org/gData.

Hipp, John, and Adam Boessen. 2013. "Egohoods as Waves Washing Across the City: A New Measure of "Neighborhoods"." *Criminology* 51 (2): 287–327.

Hipp, John, Robert Faris, and Adam Boessen. 2012. "Measuring 'Neighborhood': Constructing Network Neighborhoods." *Social Networks* 34 (1): 128–40.

Hyndman, Rob, and George Athanasopoulos. 2021. *Forecasting: Principles and Practice*. 3rd ed. Melbourne, Australia: OTexts.

INEGI. 2021. "National Geostatistical Framework." https://en.www.inegi.org.mx/temas/mg/#Downloads.

James, Gareth, Daniela Witten, Trevor Hastie, and Robert Tibshirani. 2013. *An Introduction to Statistical Learning with Applications in r*. New York: Springer.

Jiang, Bin. 2013. "Head/Tail Breaks: A New Classification Scheme for Data with a Heavy-Tailed Distribution." *The Professional Geographer* 65 (3): 482–94.

Johnson, Shane D, Wim Bernasco, Kate J Bowers, Henk Elffers, Jerry Ratcliffe, George Rengert, and Michael Townsley. 2007. "Space–Time Patterns of Risk: A Cross National Assessment of Residential Burglary Victimization." *Journal of Quantitative Criminology* 23 (3): 201–19.

Johnson, Zachary-Forest. 2011. "Hexbins!" http://indiemaps.com/blog/2011/10/hexbins/.

Kelejian, Harry, and Gianfranco Piras. 2017. *Spatial Econometrics*. London, UK: Academic Press.

Khalid, Shoaib, Fariha Shoaib, Tianlu Qian, Yikang Rui, Arezu Imran Bari, Muhammad Sajjad, Muhammad Shakeel, and Jiechen Wang. 2018. "Network Constrained Spatio-Temporal Hotspot Mapping of Crimes in Faisalabad." *Applied Spatial Analysis and Policy* 11: 599–622.

Kim, Young-A, and John Hipp. 2019. "Pathways: Examining Street Network Configurations, Structural Characteristics and Spatial Crime Patterns in Street Segments." *Journal of Quantitative Criminology* 36: 725–52.

Kirk, Andy. 2016. *Data Visualisation: A Handbook for Data Driven Design*. London, UK: Sage.

Knox, Ernest G, and Maurice S Bartlett. 1964. "The Detection of Space-Time Interactions." *Journal of the Royal Statistical Society. Series C (Applied Statistics)* 13 (1): 25–30.

Kopczewska, Katarzyna. 2021. *Applied Spatial Statistics and Econometrics: Data Analysis in r*. Milton Park: Routledge.

Kounadi, Ourania. 2009. "Assessing the Quality of OpenStreetMap Data." *Msc Geographical Information Science, University College of London Department of Civil, Environmental And Geomatic Engineering*, 19.

Kulldorff, Martin. 1997. "A Spatial Scan Statistic." *Communications in Statistics, Theory, and Methods* 26 (6): 1481–96.

Kulldorff, Martin, Richard Heffernan, Jessica Hartman, Renato Assunção, and Farzad Mostashari. 2005. "A Space–Time Permutation Scan Statistic for Disease Outbreak Detection." *Plos One*, DOI: 10.1371/journal. pmed.0020059.

Kulldorff, Martin, and Ulf Hjalmars. 1999. "The Knox Method and Other Tests for Space-Time Interaction." *Biometrics* 55: 544–52.

Kulldorff, Martin, and Neville Nagarwalla. 1995. "Spatial Disease Clusters: Detection and Inference." *Statistics in Medicine* 14: 799–810.

Lamigueiro, Oscar Perpiñan. 2014. *Displaying Time Series, Spatial, and Space-Time Data with r*. Boca Raton, FL: CRC Press.

Langton, Samuel, and Reka Solymosi. 2018. "Open Data for Crime and Place Research: A Practical Guide in r." *In "The Study of Crime and Place: A Methods Handbook" Eds: Elisabeth Groff and Corey Haberman. Temple University Press.*

———. 2021. "Cartograms, Hexograms and Regular Grids: Minimising Misrepresentation in Spatial Data Visualisations." *Environment and Planning B: Urban Analytics and City Science* 48 (2): 348–57.

Lawson, Andrew. 2021a. *Bayesian Disease Mapping*. 3rd ed. Boca Raton, FL: CRC Press.

———. 2021b. *Statistical Methods in Spatial Epidemiology*. 2nd ed. Chichester, UK: John Wiley & Sons.

———. 2021c. *Using r for Bayesian Spatial and Spatio-Temporal Health Modeling*. Boca Raton, FL: CRC Press.

LeSage, James, and Kelley Pace. 2009. *Introduction to Spatial Econometrics*. Boca Raton, FL: CRC Press.

Levine, Ned. 2013. *Crime Stat IV: A Spatial Statistics Program for the Analysis of Crime Incident Locations*. Washington, DC: National Institute of Justice.

Leyser, Ottoline, Danny Kingsley, and Jim Grange. 2015. "The Science 'Reproducibility Crisis' – and What Can Be Done about It." https://theconversation.com/the-science-reproducibility-crisis-and-what-can-be-done-about-it-74198.

Li, Huiyang, and Nadine Moacdieh. 2014. "Is 'Chart Junk' Useful? An Extended Examination of Visual Embellishment." In *Proceedings of the Human Factors and Ergonomics Society Annual Meeting*, 58:1516–20. 1. SAGE Publications Sage CA: Los Angeles, CA.

Lorenz, Max. 1905. "Methods of Measuring the Concentration of Wealth." *Publications of the American Statistical Association* 9 (70): 209–19.

Lovelace, Robin, Jakub Nowosad, and Jannes Muenchow. 2019. *Geocomputation with r*. Boca Raton, FL: Chapman & Hall/CRC Press.

Lu, Yongmei, and Xuwei Chen. 2007. "On the False Alarm of Planar k-Function When Analyzing Urban Crime Distributed Along Streets." *Social Science Research* 36 (2): 611–32.

Mantel, N. 1967. "The Detection of Disease Clustering and a Generalized Regression Approach." *Cancer Research* 27 (2): 209–20.

Marchione, Elio, and Shane D Johnson. 2013. "Spatial, Temporal and Spatio-Temporal

Patterns of Maritime Piracy." *Journal of Research in Crime and Delinquency* 50 (4): 504–24.

Marshall, Roger. 1991a. "A Review of Methods for the Statistical Analysis of Spatial Patterns of Disease." *Journal of the Royal Statistical Society* 154 (3): 421–41.

———. 1991b. "Mapping Disease and Mortality Rates Using Empirical Bayes Estimators." *Applied Statistics* 40 (2): 283–94.

Mayhew, Henry. 1861. *London Labour and the London Poor: A Cyclopaedia of the Condition and Earnings of Those That Will Work, Those That Cannot Work, and Those That Will Not Work. Those That Will Not Work: Comprising Prostitutes, Thieves, Swindlers, Beggars/by Several Contributors; with Introductory Essay on the Agencies at Present in Operation in the Metropolis for the Suppression of Vice and Crime by William Tuckniss; with Illustrations.* Charles Griffin.

McElreath, Richard. 2018. *Statistical Rethinking: A Bayesian Course with Examples in r and Stan.* Chapman & Hall/CRC.

McSwiggan, Greg, Adrian Baddeley, and Gopalan Nair. 2016. "Kernel Density Estimation on a Linear Network." *Scandinavian Journal of Statistics* 44 (2): 324–45.

Meyer, Sebastian, Leonhard Held, and Michael Höhle. 2017. "Spatio-Temporal Analysis of Epidemic Phenomena Using the R Package surveillance." *Journal of Statistical Software* 77 (11): 1–55. https://doi.org/10.18637/jss.v077.i11.

MIT. 2021. "Geographic Information Systems (GIS): Data for the World." https://libguides.mit.edu/gis/world.

Moehler, GO, MB Short, PJ Brantingham, FP Schoenberg, and GE Tita. 2011. *Journal of the American Statistical Association* 106 (493): 100–108.

Monmonier, Mark. 1996. *How to Lie with Maps.* 2nd ed. Chicago: The University of Chicago Press.

Mooney, Peter, and Marco Minghini. 2017. *A Review of OpenStreetMap Data.* Ubiquity Press.

Moraga, Paula, Dilinie Seimon, Varsha Ujjinni VIijay Kumar, and Andre Ribiero-Amaral. 2021. "Rspatialdata: A Collection of Data Sources and Tutorials on Visualising Spatial Data Using r." https://rspatialdata.github.io/index.html.

Moss, Stephen. 2013. "Which Is the Most Dangerous Day of the Week?" https://www.theguardian.com/lifeandstyle/2013/may/29/most-dangerous-day-of-week.

Neville, Ned. 2013. "CrimeStat IV. A Spatial Statistical Program for the Analysis of Incident Locations." Software Documentation. National Institute of Justice.

Newton, Andrew. 2008. "A Study of Bus Route Crime Risk in Urban Areas: The Changing Environs of a Bus Journey." *Built Environment* 34 (1): 88–103.

———. 2018. "Macro-Level Generators of Crime, Including Parks, Stadiums, and Transit Stations." In *The Oxford Handbook of Environmental Criminology*, edited by Gerben Bruinsma and Shane Johnson, 497–517. Oxford: Oxford University Press.

Nusrat, Sabrina, and Stephen Kobourov. 2016. "The State of the Art in Cartograms." In *Computer Graphics Forum*, 35:619–42. 3. Wiley Online Library.

Nussbaumer, Cole. 2013. "Strategies for Avoiding the Spaghetti Graph." https://www.storytellingwithdata.com/blog/2013/03/avoiding-spaghetti-graph.

O'Sullivan, David. 2014. "Spatial Network Analysis." In *Handbook of Regional Science*, edited by Manfred Fische and Peter Nijkamp, 1253–73. Heidelberg: Springer.

O'Sullivan, David, and David Unwin. 2010. *Geographic Information Analysis.* 2nd ed. Hoboken, NJ: John Wiley & Sons.

Okabe, Atsuyuki, and Kokichi Sugihara. 2012. *Spatial Analysis Along Networks.* Chichester: John Wiley & Sons.

Okabe, Atsuyuki, and Ikuho Yamada. 2001. "The k-Function Method on a Network and Its Computational Implementation." *Geographical Analysis* 33 (3): 271–90.

Oliveira, Mark. 2021. "More Crime in Cities? On the Scaling Laws of Crime and the Inadequacy of Per Capita Rankings—a Cross-Country Study." *Crime Science*, no. 10. https://doi.org/10.1186/s40163-021-00155-8.

Olmedilla, M, M Martínez-Torres, and SL Toral. 2016. "Harvesting Big Data in Social Science: A Methodological Approach for Collecting Online User-Generated Content." *Computer Standards & Interfaces* 46: 79–87.

Open Street Map. 2021a. "Overpass API/Overpass QL." https://wiki.openstreetmap.org/wiki/Overpass_API/Overpass_QL.

———. 2021b. "Tag:boundary=administrative." https://wiki.openstreetmap.org/wiki/Tag:boundary%3Dadministrative#10_admin_level_values_for_specific_countries.

Openshaw, Stan. 1981. "The Modifiable Areal Unit Problem." *Quantitative Geography: A British View*, 60–69.

Openshaw, Stan, Martin Charlton, Colin Wymer, and Alan Craft. 1987. "A Mark 1 Geographical Analysis Machine for the Automated Analysis of Point Data Sets." *International Journal of Geographical Information Systems* 1 (4): 335–58.

Ord, JK, and Arthur Getis. 1995. "Local Spatial Autocorrelation Statistics: Distributional Issues and Applications." *Geographical Analysis* 27 (4): 286–306.

Padgham, Mark, Bob Rudis, Robin Lovelace, and Maëlle Salmon. 2017. "Osmdata." *The Journal of Open Source Software* 2 (14). https://doi.org/10.21105/joss.00305.

Park, Robert, and Ernest Burgess. 1925. *The City*. University of Chicago Press.

Pease, Ken. 1998. "Repeat Victimisation: Taking Stock." Home Office; https://www.ojp.gov/ncjrs/virtual-library/abstracts/repeat-victimisation-taking-stock.

Pease, Ken, and Graham Farrell. 2017. "Repeat Victimisation." In *Enviromental Criminology and Crime Analysis*, edited by Richard Wortley and Mike Townsley. London: Routledge.

Pease, Ken, Dainis Ignatans, and Lauren Batty. 2018. "Whatever Happened to Repeat Victimisation?" *Crime Prevention and Community Safety* 20: 256–67.

Pebesma, Edzer. 2012. "Spacetime: Spatio-Temporal Data in r." *Journal of Statistical Software* 51 (7): 1–30.

———. 2018. "Simple Features for r: Standardized Support for Spatial Vector Data." *The R Journal* 10 (1): 439–46.

Pebesma, Edzer, and Roger Bivand. 2020. "R Spatial Follows GDAL and PROJ Development." https://r-spatial.org/r/2020/03/17/wkt.html.

———. 2021. "Spatial Data Science with Applications in r." https://keen-swartz-3146c4.netlify.app/.

Police.hu. 2020. "Közrendvédelem." http://www.police.hu/hu/a-rendorsegrol/statisztikak/kozrendvedelem.

Quetelet, Adolphe. 1842. *Instructions Pour l'observation Des Phénoménes périodiques*. Vol. 1. 22.

Radil, Steven. 2016. "Spatial Analysis of Crime." In *The Handbook of Measurement Issues in Criminology and Criminal Justice*, edited by Beth Huebner and Timothy Bynum, 535–54. John Wiley & Sons.

Radil, Steven, Colin Flint, and George Tita. 2010. "Spatializing Social Networks: Using Social Network Analysis to Investigate Geographies of Gang Rivalry, Territoriality, and Violence in Los Angeles." *Annals of the Association of American Geographers* 100 (2): 307–26.

Rakshit, Suman, Tilman Davies, M Mehdi-Moradi, Greg McSwiggan, Gopalan Nair, Jorge Mateu, and Adrian Baddeley. 2019. "Fast Kernel Smoothing of Point Patterns on a

Large Network Using Two-Dimensional Convolution." *International Statistical Review* 87 (3): 531–56.

Ramos, Rafael, Bráulio Silva, Keith Clarke, and Marcos Prates. 2021. "Too Fine to Be Good? Issues of Granularity, Uniformity and Error in Spatial Crime Analysis." *Journal of Quantitative Criminology* 37: 419–43.

Ratcliffe, Jerry. 2004. "Geocoding Crime and a First Estimate of a Minimum Acceptable Hit Rate." *International Journal of Geographical Information Science* 18 (1): 61–72.

———. 2010. "Crime Mapping: Spatial and Temporal Challenges." In *Handbook of Quantitative Criminology*, edited by Alex Piquero and David Weisburd, 5–24. New York, NY: Springer.

———. 2020. "Near Repeat Analysis." https://www.jratcliffe.net/near-repeat-analysis.

Ratcliffe, Jerry, and Michael McCullagh. 1998. "Aoristic Crime Analysis." *International Journal of Geographical Information Science* 12 (7): 751–64.

Rengert, George, and Brian Lockwood. 2009. "Geographical Units of Analysis and the Analysis of Crime." In *Putting Crime in Its Place*, edited by David Weisburd, Wim Bernasco, and Gerben J. N. Bruinsma, 109–22. New York, NY: Springer.

Ritchie, Stuart. 2020. *Science Fictions: Exposing Fraud, Bias, Negligence and Hype in Science*. Random House.

Roser, Gabriel, Toby Davies, Kate Bowers, Shane Johnson, and Tao Cheng. 2017. "Predictive Crime Mapping: Arbitrary Grids or Street Networks?" *Journal of Quantitative Criminology* 33: 569–94.

Roth, Robert, Kevin Ross, Benjamin Finch, Wei Luo, and Alan MacEachren. 2013. "Spatiotemporal Crime Analysis in US Law Enforcement Agencies: Current Practices and Unmet Needs." *Government Information Quarterly* 30 (3): 226–40.

Salmon, Maëlle, Dirk Schumacher, and Michael Höhle. 2016. "Monitoring Count Time Series in R: Aberration Detection in Public Health Surveillance." *Journal of Statistical Software* 70 (10): 1–35. https://doi.org/10.18637/jss.v070.i10.

Sampson, Robert, Stephen Raudenbush, and Felton Earls. 1997. "Neighborhoods and Violent Crime: A Multilevel Study of Collective Efficacy." *Science* 277 (5328): 918–24.

Schwabish, Jonathan. 2021. *Better Data Visualizations. A Guide for Scholars, Researchers, and Wonks*. New York: Columbia University Press.

Shaw, Clifford, and Henry McKay. 1942. "Juvenile Delinquency and Urban Areas."

Sherman, Lawrence, Patrick Gartin, and Michael Buerger. 1989. "Hot Spots of Predatory Crime: Routine Activities and the Criminology of Place." *Criminology* 27 (1): 27–56.

Shiode, Shino, and Narushige Shiode. 2020. "Crime Geosurveillance in Microscale Urban Environments: NetSurveillance." *Annals of the American Association of Geographers* 110 (5): 1386–406.

Shiode, Shino, Narushige Shiode, Richard Block, and Carolyn Block. 2015. "Space-Time Characteristics of Micro-Scale Crime Occurrences: An Application of a Network-Based Space-Time Search Window Technique for Crime Incidents in Chicago." *International Journal of Geographical Information Science* 29 (5): 697–719.

Silverman, BW. 1986. "3 the Kernel Method for Univariate Data." In *Density Estimation for Statistics and Data Analysis*, edited by BW Silverman, 34–74. London: Chapman & Hall/CRC.

Singleton, Alex, and Chris Brunsdon. 2014. "Escaping the Pushpin Paradigm in Geographic Information Science: (Re)presenting National Crime Data." *Area* 46 (3): 294–304.

Smith, Michael De, Michael Goodchild, and Paul Longley. 2007. *Geospatial Analysis: A Comprehensive Guide to Principles, Techniques and Software Tools*. Troubador publishing ltd.

Smith, Nate. 2014. "Binning: An Alternative to Point Maps." https://blog.mapbox.com/binning-an-alternative-to-point-maps-2cfc7b01d2ed.

Smith, Susan, and Christopher Bruce. 2008. *CrimeStat III. User Workbook.* Washington, DC: National Institute of Justice.

Smith, Tony. 2009. "Estimation Bias in Spatial Models with Strongly Connected Weight Matrices." *Geographical Analysis* 41 (3): 307–32.

Solymosi, Reka. 2021. "Minimizing Misrepresentation in Spatial Data Visualizations." https://www.doi.org/10.4135/9781529774047.

Solymosi, Reka, Kate Bowers, and Taku Fujiyama. 2015. "Mapping Fear of Crime as a Context-Dependent Everyday Experience That Varies in Space and Time." *Legal and Criminological Psychology* 20 (2): 193–211.

Solymosi, Reka, David Buil-Gil, Laura Vozmediano, and Inês Sousa Guedes. 2020. "Towards a Place-Based Measure of Fear of Crime: A Systematic Review of App-Based and Crowdsourcing Approaches." *Environment and Behavior*, 0013916520947114.

South, Andy. 2017. *Rnaturalearth: World Map Data from Natural Earth.* https://CRAN.R-project.org/package=rnaturalearth.

Spicer, Valerie, Justin Song, Patricia Brantingham, Andrew Park, and Martin Andresen. 2016. "Street Profile Analysis: A New Method for Mapping Crime on Major Roadways." *Applied Geography* 69: 65–74.

Statistics Canada. 2021. "Census Geography." https://www12.statcan.gc.ca/census-recensement/2016/geo/index-eng.cfm.

Steenbeek, Wouter. 2021. "NearRepeat: Near Repeat Calculation Using the Knox Test (Monte Carlo Permutation)." https://github.com/wsteenbeek/NearRepeat.

Steenbeek, Wouter, and David Weisburd. 2016. "Where the Action Is in Crime? An Examination of Variability of Crime Across Different Spatial Units in the Hague, 2001–2009." *Journal of Quantitative Criminology* 32: 442–69.

Stevens, Joshua. 2015. "Bivariate Choropleth Maps: A How-to Guide." Blog. NASA's Earth Observatory. https://www.joshuastevens.net/cartography/make-a-bivariate-choropleth-map/.

Tanimura, Susumu, Chusi Kuroiwa, and Tsutomu Mizota. 2006. "Propotional Symbol Mapping in r." *Journal of Statistical Software* 15 (5): 1–7.

Tennekes, Martijn. 2018. "tmap: Thematic Maps in R." *Journal of Statistical Software* 84 (6): 1–39. https://doi.org/10.18637/jss.v084.i06.

Thioulouse, Jean, Stéphane Dray, Anne-Béatrice Dufour, Aurélie Siberchicot, Thibaut Jombart, and Sandrine Pavoine. 2018. "Multivariate Analysis of Ecological Data with Ade4."

Tita, George, and Steven Radil. 2010. "Spatial Regression Models in Criminology: Modeling Social Processes in the Spatial Weights Matrix." In *Handbook of Quantitative Criminology*, edited by Alex Piquero and David Weisburd, 101–21. New York: Springer.

———. 2011. "Spatializing the Social Networks of Gangs to Explore Patterns of Violence." *Journal of Quantitative Criminology* 27: 521–45.

Tobler, Waldo. 2004. "Thirty Five Years of Computer Cartograms." *ANNALS of the Association of American Geographers* 94 (1): 58–73.

Tompson, Lisa, and Kate Bowers. 2013. "A Stab in the Dark? A Research Note on Temporal Patterns of Street Robbery." *Journal of Research in Crime and Delinquency* 50 (4): 616–31.

Tompson, Lisa, Shane Johnson, Matthew Ashby, Chloe Perkins, and Philip Edwards. 2015. "UK Open Source Crime Data: Accuracy and Possibilities for Research." *Cartography and Geographic Information Science* 42 (2): 97–111.

Tompson, Lisa, Henry Partridge, and Naomi Shepherd. 2009. "Hot Routes: Developing a New Technique for the Spatial Analysis of Crime." *Crime Mapping: A Journal of Research and Practice* 1 (1): 77–96.

Townsley, Michael, Ross Homel, and Janet Chaseling. 2003. "Infectious Burglaries: A Test of the Near Repeat Hypothesis." *British Journal of Criminology* 43: 615–33.

Tufte, Edward. 2001. *The Visual Display of Quantitative Information*. 2nd ed. Cheshire, Conn: Graphics Press.

Tukey, John. 1979. "Statistical Mapping: What Should Not Be Plotted." In *Proceedings of the 1976 Workshop on Automated Cartography and Epidemiology*, edited by DHEW, 18–21. Washington, DC: US Department of Health, Education,; Welfare.

Uittenbogaard, Adriann, and Vania Ceccato. 2012. "Space-Time Clusters of Crime in Stockholm, Sweden." *Review of European Studies* 4 (5): 148–56.

UK Data Service. 2013. "Generalisation in Census Support, UK Data Service Census Support Help." https://borders.ukdataservice.ac.uk/webhelp/index.htm#t=Data_informa tion%2FGeneralisation_in_Census_Support.htm.

United States Census Bureau. 2021. "Census Mapping Files." https://www.census.gov/geo graphies/mapping-files.html.

van der Meer, Lucas, Lorena Abad, Andrea Gilardi, and Robin Lovelace. 2021. *Sfnetworks: Tidy Geospatial Networks*. https://CRAN.R-project.org/package=sfnetworks.

Vaughan, Davis. 2021. "Comprehensive Date-Time Handling for r." https://www.tidyverse. org/blog/2021/03/clock-0-1-0/.

Veaux, RD De, PF Velleman, and DE Bock. 2012. *Stats: Data and Models with Statistical Methods for the Social Sciences*. Boston, MA: Pearson.

Walker, Kyle. 2021. *Crsuggest: Obtain Suggested Coordinate Reference System Information for Spatial Data*. https://CRAN.R-project.org/package=crsuggest.

Wall, Melanie. 2004. "A Close Look at the Spatial Structure Implied by the CAR and SAR Model." *Journal of Statistical Planning and Inference* 121: 311–24.

Waller, HLance, and Carol Gotway. 2004. *Applied Spatial Statistics for Public Health Data*. Chichester, UK: John Wiley & Sons.

Ward, Michael, and Kristian Gleditsch. 2008. *Spatial Regression Models*. Thousand Oakds: Sage.

Weisburd, David. 2015. "The Law of Crime Concentration and the Criminology of Place." *Criminology* 53 (2): 133–57.

Weisburd, David, Wim Bernasco, and Gerben Bruinsma. 2008. *Putting Crime in Its Place*. Springer.

Weisburd, David, and Chester Britt. 2014. *Statistics in Criminal Justice*. Springer.

Weisburd, David, Gerben Bruinsma, and Wim Bernasco. 2009. "Units of Analysis in Geographic Criminology: Historical Development, Critical Issues, and Open Questions." In *Putting Crime in Its Place*, 3–31. Springer.

Wheeler, Andrew. 2016. "Tables and Graphs for Monitoring Temporal Crime Trends: Translating Theory into Practical Crime Analysis Advice." *International Journal of Police Science & Management* 18 (3): 159–72.

Wheeler, Andrew, and Wouter Steenbeek. 2021. "Mapping the Risk Terrain for Crime Using Machine Learning." *Journal of Quantitative Criminology* 37: 445–80.

Wickham, Hadley. 2010. "A Layered Grammar of Graphics." *Journal of Computational and Graphical Statistics* 19 (1): 3–28.

———. 2019. *Advanced r*. CRC Press.

———. 2021. *The Tidyverse Style Guide*. https://style.tidyverse.org/index.html.

Wickham, Hadley, Mara Averick, Jennifer Bryan, Winston Chang, Lucy McGowan, Roman François, Garrett Grolemund, Alez Hayes, Lionel Henry, and Jim Hester. 2019. "Welcome to the Tidyverse." *Journal of Open Source Software* 4 (43): 1686.

Wickham, Hadley, Romain François, Lionel Henry, and Kirill Müller. 2021. *Dplyr: A Grammar of Data Manipulation*. https://CRAN.R-project.org/package=dplyr.

Wickham, Hadley, and Garrett Grolemund. 2017. *R for Data Science: Import, Tidy, Transform, Visualize, and Model Data*. Boston, MA: O'Reilly.

Wikle, Christopher, Andrew Zammit-Mangion, and Noel Cressie. 2019. *Spatio-Temporal Statistics with r*. Boca Raton, FL: CRC Press.

Wikström, Per-Olof, Andromachi Tseloni, and Dimitris Karlis. 2011. "Do People Comply with the Law Because They Fear Getting Caught?" *European Journal of Criminology* 8 (5): 401–20.

Wilson, James, and George Kelling. 1982. "Broken Windows." *Atlantic Monthly* 249 (3): 29–38.

Woodworth, JT, GO Moehler, AL Bertozzi, and PJ Brantingham. 2014. "Non-Local Crime Density Estimation Incorporating Housing Information." *Philosophical Transactions of the Royal Society* 372 (2028).

Yutani, Hiroaki. 2018. "Geom_sf_text() and Geom_sf_label() Are Coming." https://yutani.rbind.io/post/geom-sf-text-and-geom-sf-label-are-coming/.

Zeileis, Achim, Christian Kleiber, and Achim Zeileis. 2012. "Package 'Ineq.'" *Vienna: Comprehensive R Archive Network*.

Index